Applied Plant Virology

Applied Plant Virology

Editor

Jaidev Sharma

scitus
academics

Applied Plant Virology

Edited by **Jaidev Sharma**

ISBN: 978-1-68117-044-2

Library of Congress Control Number: 2015935439

www.scitusacademics.com
Box No. 4766, 616 Corporate Way,
Suite 2, Valley Cottage,
NY 10989

Notice

Printed in United States of America on Acid Free Paper ∞

Contents

Preface vii

Chapter 1 **Ecology and Management of Grapevine Leafroll Disease** 1
Rodrigo P. P. Almeida, Kent M. Daane, Vaughn A. Bell, G. Kai Blaisdell, Monica L. Cooper, Etienne Herrbach, and Gerhard Pietersen

Chapter 2 **Galling Aphids (Hemiptera: Aphidoidea) in China: Diversity and Host Specificity** 39
Jing Chen, and Ge-Xia Qiao

Chapter 3 **Crop Immunity Against Viruses: Outcomes and Future Challenges** 73
Valérie Nicaise

Chapter 4 **Activity of the Novel Fungicide SYP-Z048 against Plant Pathogens** 129
Fengping Chen, Ping Han, Pengfei Liu, Naiguo Si, Junli Liu, and Xili Liu

Chapter 5 **Production of Platinum atom Nanoclusters at one end of Helical Plant Viruses** 149
Yuri Drygin, Olga Kondakova, and Joseph Atabekov

Chapter 6 **The Induction of Stromule Formation by a Plant DNA-Virus in Epidermal Leaf Tissues Suggests a Novel Intra- and Intercellular Macromolecular Trafficking Route** 169
Björn Krenz, Holger Jeske, and Tatjana Kleinow

Chapter 7 **Real-Time PCR in Virology** 205
Ian M. Mackay, Katherine E. Arden, and Andreas Nitsche

Chapter 8 **Sequence Exploration Reveals Information bias Among Molecular Markers Used in Phylogenetic Reconstruction for Colletotrichum Species** 251
Sephra N Rampersad, Fazeeda N Hosein, and Christine VF Carrington

Chapter 9 **Effect of ATG Initiation Codon Context Motifs on the Efficiency of Translation of mRNA Derived from Exogenous Genes in the Transgenic Silkworm, Bombyx mori 287**

Ken-ichiro Tatematsu, Keiro Uchino, Hideki Sezutsu, and Toshiki Tamura

Citations 313

Index 317

Preface

In Pasteur's day, and for many years after his death, the word "virus" was used to describe any cause of infectious disease. Many bacteriologists soon discovered the cause of numerous infections. However, some infections remained, many of them horrendous, for which no bacterial cause could be found. These agents were invisible and could only be grown in living animals. The discovery of viruses was the key that unlocked the door that withheld the secrets of the cause of these mysterious infections. And, although Koch's postulates could not be fulfilled for many of these infections, this did not stop the pioneer virologists from looking for viruses in infections for which no other cause could be found.

Editor

Chapter 1

Ecology and Management of Grapevine Leafroll Disease

Rodrigo P. P. Almeida[1], Kent M. Daane[1], Vaughn A. Bell[2], G. Kai Blaisdell[1], Monica L. Cooper[3], Etienne Herrbach[4,5], and Gerhard Pietersen[6]

[1]Department of Environmental Science, Policy and Management, University of California at Berkeley, Berkeley, CA, USA

[2]The New Zealand Institute for Plant and Food Research Limited, Havelock North, New Zealand

[3]Division of Agriculture and Natural Resources, University of California at Napa, Napa, CA, USA

[4]UMR1131 Santé de la Vigne et Qualité du Vin, Institut National de la Recherche Agronomique, Colmar, France

[5]UMR1131, Université de Strasbourg, Strasbourg, France

[6]Agricultural Research Council-Plant Protection Research Institute, c/o Department of Microbiology and Plant Pathology, University of Pretoria, Pretoria, South Africa

ABSTRACT

Grapevine leafroll disease (GLD) is caused by a complex of vector-borne virus species in the family Closteroviridae. GLD is present in all grape-growing regions of the world, primarily affecting wine grape varieties. The disease has emerged in the last two decades as one of the major factors affecting grape fruit quality, leading to research efforts aimed at reducing its economic impact. Most research has focused on the pathogens themselves, such as improved detection protocols, with limited work directed toward disease ecology and the development of management practices. Here we discuss the ecology and management of GLD, focusing primarily on *Grapevine leafroll-associated virus 3*, the most important virus species within the complex. We contextualize research done on this system within an ecological framework that forms the backbone of the discussion regarding current and potential GLD management strategies. To reach this goal, we introduce various aspects of GLD biology and ecology, followed by disease management case studies from four different countries and continents (South Africa, New Zealand, California-USA, and France). We review ongoing regional efforts that serve as models for improved strategies to control this economically important and worldwide disease, highlighting scientific gaps that must be filled for the development of knowledge-based sustainable GLD management practices.

INTRODUCTION

Emerging plant diseases are a global threat to the food supply, environmental sustainability, and economic stability of regions and nations. In this paper, we discuss the ecology and management of grapevine leafroll disease (GLD), a worldwide disease that is caused by a complex of virus species in the family Closteroviridae, which contains emerging and re-emerging plant pathogens of economic importance. GLD is present in virtually all commercial grape (*Vitis vinifera*) growing regions; its distribution is thought to be due to regional, continental, and intercontinental transport of virus-infected plant material. While GLD has long been present in the major grape-growing regions, it has only recently been recognized as a disease of economic importance. Various hypotheses have been proposed to explain this (e.g., Golino et al., 2008), but none have been well supported. For example, there is no evidence of the emergence of a new virus species or strain (Wang et al., 2011), or introduction of a rapidly moving or efficient insect vector species associated with the increased incidence of GLD. The only common factors are the observation of vector-mediated pathogen spread in vineyards and an increased GLD awareness by academics, farmers, and other stakeholders. Regardless of the driving forces, GLD is now considered a disease of importance in viticulture, especially to wine grape growers who aim for a high quality uniform crop.

Herein, we will not focus on factors that have made GLD such a pre-eminent disease, although studies are needed to address this. We propose that the integration of disciplines is necessary to address GLD, and to devise disease management practices that are practical, sustainable, financially viable, and environmentally sound. Within this interdisciplinary context, our goal is to discuss various components of GLD that are relevant to its ecology, epidemiology, and management. Much of this review focuses on the mealybug-transmitted ampeloviruses, more specifically *Grapevine leafroll-associated virus 3* (GLRaV-3), which is the most widespread species in the virus complex causing GLD. We highlight notable gaps in the current body of knowledge that need to be addressed for the development of sustainable disease control practices. Then we discuss management strategies being implemented in each of four countries in four continents by presenting case studies.

GRAPEVINE LEAFROLL DISEASE

Grapevine leafroll disease has been described from different regions in Europe and elsewhere for over a century (Hoefert and Gifford, 1967), and was first shown to be transmissible to vines in 1936 (Scheu, 1936). The demonstration of graft transmissibility opened early avenues of GLD research, including the search for etiological agents and the impact of abiotic factors on symptom development. Even today the etiology and symptomatology of GLD is not completely clear, as multiple virus species cause GLD, and symptoms result from complex biotic and abiotic interactions. Furthermore, there is no infectious clone for any agent associated with GLD.

Grapevine leafroll disease is most obvious and problematic in cool-climate regions, where fruit on infected vines has delayed ripening that results in lowered brix, which in turn affects wine quality (Over de Linden and Chamberlain, 1970; Goheen, 1988). The most obvious GLD symptoms appear in the fall, when red cultivars display leaf reddening with green venation (Figure 1). While symptoms are not as apparent in white cultivars, there is a slight leaf chlorosis. Both red and white cultivars develop downward rolling of leaf margins and phloem disruption. Significant losses result from a combination of factors including yield reductions of up to 40%, increased management costs, shortened vineyard life spans, and adverse impacts on wine quality resulting from decreased fruit quality and delayed maturation (Woodrum et al., 1984; Goheen, 1988; Credi and Babini, 1997;Martelli et al., 2002). The economic impact of GLD is still poorly understood, as are the implications of various control strategies. A recent study by Atallah et al. (2012) estimated the economic impact of GLD to range from US$25,000 to US$40,000 per hectare for vineyards with a 25-year lifespan. The authors analyzed various scenarios, incorporating disease prevalence, yield reduction and fruit quality; at low levels of disease incidence (1–25%), roguing can significantly decrease economic losses, which was identified as an economically important practice together with planting of virus-free plant material. The economic impact of vector management has not been explored.

Figure 1: Leaf symptoms of grapevine leafroll disease include inter-veinal reddening and leafrolling in red-fruited varieties. Symptoms are most pronounced around the harvest period. These photographs were taken in the

fall (September) in Napa, CA, USA. Photographs show symptomatic leaf (A), group of leaves (B), and whole plant (C).

Grapevine leafroll disease has three essential biological components: (1) a complex of viruses in the Closteroviridae, (2) grapevine host plants, and (3) species of mealybugs (Pseudococcidae) and soft scales (Coccidae) that transmit GLRaVs. Much of this review will focus on GLRaV-3, which is the best studied species worldwide and has been implicated in a majority of GLD spread that has been mediated by known insect vectors. While GLRaV-2 is of economic importance, this *Closterovirus*species has no known vectors (Martelli et al., 2002). In addition, GLRaV-7, a member of the proposed genus Velarivirus (Al Rwahnih et al., 2012), does not appear to cause GLD and also has no known vectors (Tsai et al., 2010).

GRAPEVINE LEAFROLL-ASSOCIATED VIRUSES

Virus species causing GLD are sequentially named *Grapevine leafroll-associated virus 1, Grapevine leafroll-associated virus 2,* and so on (GLRaV-1, GLRaV-2, GLRaV-n). All GLRaVs are in the genus*Ampelovirus*, except for GLRaV-2 and GLRaV-7, as previously discussed. GLRaVs in the genus*Ampelovirus* are divided into two phylogenetic groups, one that includes GLRaV-4, -5, -6, -9, and others, and another comprising GLRaV-1 and -3 (Maliogka et al., 2009). The taxonomy of GLRaVs is undergoing significant changes with recent proposals awaiting International Committee on Taxonomy of Viruses (ICTV) approval; the most relevant proposal is a change in sequence similarity thresholds for delineating species that would collapse GLRaV-4, -5, -6, -9, and other proposed species and divergent variants into one species, GLRaV-4 (Martelli et al., 2012; Thompson et al., 2012).

Both groups of GLRaV ampeloviruses, like other species in the Closteroviridae, are filamentous virions with a large (13–18 kb) positive-sense single-stranded RNA genome (Fuchs et al., 2009;Martelli et al., 2012). However, there are important differences in genome structure between the groups. The genomes of GLRaV-4-like species are ~5 kb smaller and lack several open reading frames on their 3′ ends that are present in GLRaV-1 and -3 (Thompson et al., 2012). Despite the large genetic diversity among GLRaV species, little is known about the phenotypic variability in disease symptoms among or within species. One careful study of GLRaV-2 demonstrated that disease symptoms were associated with the phylogenetic clustering of variants (Bertazzon et al., 2010), but similar work has not been performed with other viruses. Despite this gap in knowledge, GLRaV-3 has emerged as the key species causing GLD worldwide. The reasons behind the prominence of GLRaV-3 are poorly understood, especially because other GLD-causing species also co-exist with GLRaV-3, often within one vineyard or plant (Sharma et al., 2011), and some can be transmitted by the same vector species (Le Maguet et al., 2012). Notably, GLRaV-3 has been identified as the main species being transmitted by vectors throughout the world (see case studies).

The importance of GLRaV-3 genetic diversity is not understood from a phenotypic or ecological perspective. However, some important insights into GLRaV-3 ecology have been gained from genetic diversity studies. First, it appears that most variants are present in major grape-growing regions worldwide (Gouveia et al., 2011; Jooste et al., 2011; Sharma et al., 2011). Second, it is likely that much of the diversity within the species has yet to be discovered, given the increasing number of well-supported phylogenetic clades (e.g., Sharma et al., 2011; Seah et al., 2012). Lastly, there is no evidence of positive selection in GLRaV-3 field populations (Wang et al., 2011), suggesting that the virus is not undergoing novel selective pressures.

HOST PLANTS

Although ampeloviruses colonize a wide range of plant taxa, GLRaVs appear to be limited to grapevines (*Vitis*). To our knowledge, GLRaVs have only been isolated from *Vitis* spp. Focus on the commercially widespread *Vitis vinifera* may have limited our knowledge of potential host range, although a recent survey in Napa Valley, California, which included 41 plant species in 12 families in addition to wild grapes (*Vitis californica* and *Vitis californica* × *Vitis vinifera* hybrids), showed that wild *Vitis* can be infected with GLRaV-2 and GLRaV-3 (Klaassen et al., 2011). Because of extensive exchange of easily propagated plant material that has occurred worldwide (Rowhani et al., 2005), transport of virus-infected plant material has been identified as a major factor responsible for the global spread of GLD and its etiological agents. Quarantine regulations and national programs aimed at reducing the import of pathogens have been established in several countries, and are responsible for providing virus-free plant material to farmers. The integration of these practices into management of GLD is discussed below.

INSECT VECTORS

Plant to plant transmission of GLRaV-3 by the mealybug *Planococcus ficus* (Signoret) was the first demonstration of an insect vector of a GLD pathogen (Engelbrecht and Kasdorf, 1990). Since then, several species of mealybugs have been shown to transmit GLRaV species, including *Pseudococcus maritimus* (Ehrhorn), *Pseudococcus viburni* (Signoret), *Pseudococcus longispinus* (Targioni-Tozzetti), *Pseudococcus calceolariae* (Maskell), *Pseudococcus comstocki* (Kuwana), *Planococcus citri* (Risso), *Phenacoccus aceris* (Signoret), and *Heliococcus bohemicus* Sulc (reviewed in Daane et al., 2012; Herrbach et al., 2013). Additionally, the soft scales *Pulvinaria vitis* (L.), *Parthenolecanium corni* (Bouché), *Ceroplastes rusci* (L.), *Neopulvinaria innumerabilis* (Rathvon), *Coccus longulus*(Douglas), *Parasaissetia nigra* (Nietner), and *Saissetia* sp. are also vectors (Belli et al., 1994;Mahfoudhi et al., 2009; Le Maguet, 2012; Herrbach et al., 2013; Krüger and Douglas, 2013). Impressive here is the breadth of vector species, which is essentially inclusive of all common mealybugs and soft scales found worldwide where GLD is of concern.

Recognition of insect vectors is essential for the development of disease management practices, including control of the correct vector species. However, the ecological relevance of different mealybug or soft scale species to GLD spread has yet to be properly addressed. Tsai et al. (2010)found no evidence of strict vector–virus species specificity for transmission and, to date, it appears that all GLRaV species can be transmitted by the different grape-associated mealybug species tested. This hypothesis was further supported with the demonstration that *Ph. aceris* transmits six*Ampelovirus* species (Le Maguet et al., 2012). Therefore, all mealybugs colonizing grapevines should be considered potential GLRaV vectors until proven otherwise, and vector biology rather than species becomes most important.

Vineyard mealybugs generally have four larval instars for the female and five for the male (Ben-Dov, 1995). The small (~0.5 mm), unsettled first instar, or crawler, is considered to be the dispersal stage, and can be easily moved on personnel, equipment, infested nursery stock (Daane et al., 2012), and carried by the wind (Barrass et al., 1994). Whereas all mealybug and soft scale life stages may be capable of transmitting GLRaV-3, the younger nymphs appear to be more efficient (Petersen and Charles, 1997; Tsai et al., 2008). Vector species with more annual generations or higher fecundity would pose a greater threat. Variability in annual number of generations and fecundity exists. For example, in coastal California wine grape vineyards there are approximately one, two, three, and four annual generations of *Pa. corni*, *Ps. maritimus*, *Ps. viburni*, and *Pl. ficus*, respectively (Geiger and Daane, 2001; Gutierrez et al., 2008; Varela et al., 2013).

Mealybugs and soft scales are phloem feeders that use long, slender mouthparts to suck plant fluids (Daane et al., 2012). Most vineyard mealybug species can feed on the trunk, canes, leaves, and berries; however, there is variation in seasonal feeding location and movement on the vine among and within species, as described for *Ps. maritimus* (Geiger and Daane, 2001; Grasswitz and James, 2008), *Pl. citri* (Cid et al., 2010), and *Pl. ficus* (Becerra et al., 2006). Some mealybug species commonly maintain a portion of their population on vine roots, such as *Ps. calceolariae* (Bell et al., 2009) and *Pl. ficus* (Walton and Pringle, 2004). This presents a considerable replant problem as after the vine is removed, remnant roots can remain viable for years, supporting GLRaVs and mealybugs that bridge the old infested vineyard to the new replants (Pietersen, 2006).

Control of different vector species can vary considerably. Monitoring insect populations is an essential component of pest control; however, visually monitoring for mealybugs, especially at low densities, is too labor intensive to be cost effective. Sex pheromones for numerous species have recently been identified, including *Pl. ficus*, *Ps. viburni*, *Ps. maritimus*, *Ps. longispinus*, and *Ps. calceolariae* (reviewed in Daane et al., 2012), and trap counts can be used to predict berry damage (Walton and Pringle, 2004); however, there are no economic injury levels determined for these insects as GLD vectors. To control GLD spread, most vineyard managers have adopted a zero tolerance for vectors, and monitoring manifests as presence/absence scores. Efficient insecticides for mealybugs and soft scales exist, particularly some neonicotinoids and biosynthesis inhibitors (Daane et al., 2012). However, *Pl. ficus* first instar nymphs can both acquire and inoculate GLRaV-3 in less than 1 h (Tsai et al., 2008). Because the more effective insecticides are systemic, and the vector must feed on the plant to be killed, the applications may reduce mealybug densities in the treated vineyard but not necessarily protect it from virus spread by dispersing mealybugs. For some mealybug species, insecticides alone do not provide complete control, and additional control is provided by natural enemies. In New Zealand, for example, *Ps. viburni* was brought under exceptional control by release of the parasitoid *Acerophagus (Pseudaphycus) maculipennis* Signoret (Charles et al., 2010). In contrast, *Anagyrus pseudococci* Signoret is the primary parasitoid of *Pl. citri* and *Pl. ficus* around the world (Daane et al., 2012), but parasitism alone does not deliver control sufficient to reduce the spread of GLD. Mating disruption, which works best at lower pest densities, is being investigated for*Pl. ficus* (Walton et al., 2006) and may become an integral part of future control measures.

Of those countries reported in this review, *Pl. ficus* and *Ps. calceolariae* appear to be of greatest concern in most regions, but all mealybugs and soft scales should be viewed as potential vectors. The role played by different vector species in GLD epidemiology and vector ecology is still poorly understood. Regardless of the species, for GLD management through vector control, it is likely that multiple monitoring and control techniques must be employed to maintain the exceptionally low pest densities needed to suppress and control GLD.

VIRUS TRANSMISSION BIOLOGY

Among insect-borne plant viruses, those transmitted by mealybugs and soft scales are among the least understood. These insects transmit other viruses of economic importance to a range of crops such as cassava, banana, pineapple, and cocoa (Herrbach et al., 2013). The characterization of transmission parameters has rarely been performed, severely limiting our understanding of disease epidemiology. Nevertheless, the importance in understanding the transmission of ampeloviruses infecting grapevines has recently become apparent, and several insect vectors species are now being studied for their efficiency to transmit different GLRaV species. Most work did not go beyond the identification of new insect species as vectors. In many cases virus source plants were infected with multiple virus species, which presents a challenge because multiple infections may lead to cases of virus facilitation or competition. While a better picture of GLRaV transmission by vectors is emerging, much remains undone. Nevertheless, trends can be inferred and used to generate testable hypotheses. The transmission of GLRaV-3 appears to be more efficient than that of other GLRaV species; based on inferences from studies designed to identify new vector–virus combinations rather than compare transmission efficiency. Competing hypotheses may explain these observations. First, viruses that are transmitted less efficiently may reach lower populations within plants than GLRaV-3. Therefore they may be acquired less frequently from the phloem, resulting in lower transmission rates. Alternatively, molecular interactions between virus and vector may affect transmission efficiency. Lastly, GLRaVs may be transmitted with similar efficiency, but those with observed lower transmission may require a higher number of virions inoculated to generate a successful infection.

In the Closteroviridae, all vector-borne viruses studied so far are transmitted in a semi-persistent manner (Karasev, 2000), but in this regard GLRaVs are poorly characterized. Cabaleiro and Segura (1997b) provide important insights into the biology of GLRaV-3 transmission by *Pl. citri*; however, they mentioned that their results were not conclusive to characterize transmission as semi-persistent. Conclusive evidence of semi-persistent transmission of GLRaV-3 was only obtained byTsai et al. (2008). Transmission efficiency of GLRaV-3 by *Pl. ficus* first instars peaked with 24-h acquisition and inoculation access periods, with a leveling-off after 48 h (Tsai et al., 2008). *Pl. ficus*mealybugs lost the ability to transmit GLRaV-3 four days after acquisition (Tsai et al., 2008). It is imperative that similar experiments with more virus and vector species be performed, although given the phylogenetics of the group (Tsai et al., 2010), it is expected that all GLRaV ampeloviruses will be transmitted in a semi-persistent manner.

Reported transmission rates are difficult to compare given the varied experimental methods and generally low number of replicates used. For example, with semi-persistent transmission, a vector can lose the ability to transmit a virus upon molting to the next life stage, and longer experimental acquisition access periods used may have resulted in insects being moved to a new plant immediately after molting (and losing acquired virus). Such a protocol would effectively result in a shorter acquisition access period. Here, we report calculated *Ps* values, following Swallow (1985), which provide an estimate of infection rate or probability of transmission by a single insect derived from experiments that used insect groups (Figure 2), based on existing transmission studies. When any one particular published study included multiple experiments, we combined those experiments to report one *Ps* per published study, only including those experiments relevant to the question (e.g., mealybug life stage).

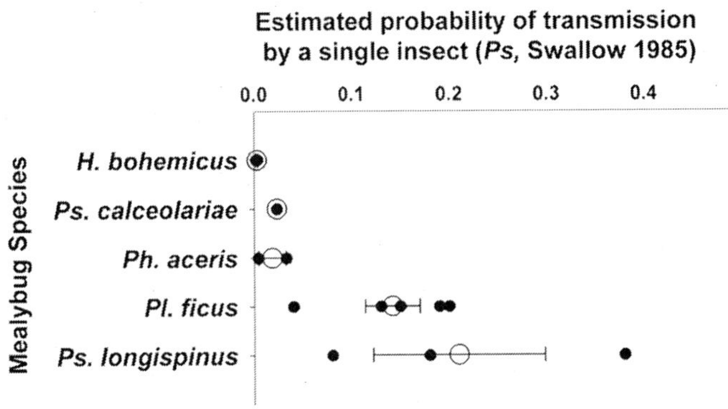

Figure 2: Estimated transmission efficiency of GLRaV-3 per individual per day (*Ps*; per Swallow, 1985) by different mealybug species, including studies that tested transmission by first and/or second instars, nymphs, adults, or mixed life stages, and used access periods of 1 day or longer. When available, only earliest life stages tested are included. From individual publications including multiple experiments, those experiments are combined to produce one estimate of *Ps* for each mealybug species. Results are based on a limited number of peer-reviewed publications per species; *Pl. ficus* – 5, *Ps. longispinus* – 3, *H. bohemicus* and *Ph. aceris* – 2, and *Ps. calceolariae* – 1. Figure shows *Ps* for individual publications (dark circles), mean (open circles), and standard error.

Earlier life stages of mealybugs have higher reported transmission efficiency than more mature life stages. *Pl. ficus* first and second instar nymphs have reported *Ps* = 0.04–0.2 (Tsai et al., 2008;Mahfoudhi et al., 2009), and adults have *Ps* = 0.009–0.02, about 10-fold lower than the nymphs.*Ph. aceris* first and second instars have *Ps* = 0.05 and 0.02, respectively (Le Maguet et al., 2012). *Ps. longispinus* first instar nymphs transmit GLRaV-3 at *Ps* = 0.08, and *Ps. calceolariae* first instar nymphs have *Ps* = 0.02, while no transmission was found by third instars of either mealybug species (Petersen and Charles, 1997). Because nymphs settle and feed more quickly than adults (Sandanayaka et al., 2012), it is possible that transmission by adult mealybugs would increase with longer access periods, although most studies appeared to use sufficiently long periods that this should not have confounded the results.

There also appears to be variation in transmission efficiency of GLRaV-3 among mealybug species (Figure 2). Three different research groups found similar *Ps* values for *Pl. ficus* nymphs, ranging from 0.04 to 0.2 (Douglas and Krüger, 2008; Tsai et al., 2008, 2010, 2011; Mahfoudhi et al., 2009). Estimated *Ps* for *Ph. aceris* was 0.004–0.03, for *Ps. calceolariae* 0.02, and for *H. bohemicus* 0.002–0.003 (Petersen and Charles, 1997; Sforza et al., 2003; Zorloni et al., 2006; Le Maguet et al., 2012); but different life stages were used and the results are probably not directly comparable. Widely variable results were obtained within *Ps. longispinus*, with transmission ranging from *Ps* = 0.08–0.38 (Petersen and Charles, 1997; Kuniyuki et al., 2005; Douglas and Krüger, 2008). The variation found within *Ps. longispinus* and among studies in general could be due to varied experimental techniques, to differences in transmission efficiency among insect populations or species, or to differences in GLRaV-3 variants that were tested.

It is not known whether GLRaV-3 populations within a donor plant affects transmission by mealybugs, but many viruses are transmitted at higher rates when the donor plant has higher viral infection (Froissart et al., 2010). GLRaV-3 populations vary seasonally in magnitude and distribution within a host plant, but the general trends are not well understood; virus population in leaves may increase during the growing season before dropping as leaves senesce (Tsai et al., 2012). Differences in transmission efficiency when mealybugs either acquire from, or inoculate to different plant tissues have not been found, although there is evidence that acquisition from stems may lead to lower transmission than from petioles or leaves (Tsai et al., 2011). Transmission by *Ps. longispinus*and *Ps. calceolariae* nymphs, for example, was tested early and late in the growing season from known infected vines in a vineyard, and no difference was found between the two time points (Petersen and Charles, 1997). While a change in transmission with viral population, plant tissue, or season has not been found, this possibility should not be ignored.

DISEASE ECOLOGY

Evidence of GLD spread in vineyards was first found in South Africa (Engelbrecht and Kasdorf, 1985), and confirmed there using an interplant study with healthy vines among established infected vines (Engelbrecht and Kasdorf, 1990). A similar interplant study in Spain also provided evidence of GLD spread (Cabaleiro and Segura, 1997a; Cabaleiro et al., 2008) following observations that older vineyards tended to have higher GLD incidence. In both cases mealybugs were recorded present at the interplant study sites. Controlled greenhouse tests of GLRaV-3 transmission by *Pl. ficus*(Engelbrecht and Kasdorf, 1990) and *Pl. citri* (Cabaleiro and Segura, 1997b) linked mealybugs to the observed vineyard spread. GLD spread in established vineyards, 8–10 years after the initial planting, has been documented in Australia (Habili et al., 1995; Habili and Nutter, 1997), California-USA (Golino et al., 2008), and France (Le Maguet et al., 2013). The rate of spread was similar in these studies, close to 10% increase per year once GLD infections were identified as being present, and newly infected vines were spatially aggregated, indicating vine-to-vine spread.

Leafroll spread through newly planted blocks adjacent to highly infected blocks has been documented in South Africa (Pietersen, 2006; Figure 3), New Zealand (Charles et al., 2009), and Italy (Gribaudo et al., 2009). In New Zealand, populations of *Ps. longispinus* were monitored in nearby older leafroll-infected blocks, and the number of newly infected vines tended to increase more dramatically one growing season after large mealybug populations were found in neighboring blocks with 100% GLD incidence. Spatial analysis indicated that infected vines were randomly distributed throughout the blocks in early years, but aggregated toward the end of the study, indicating that long distance dispersal, such as wind-borne crawlers, as well as vine-to-vine movement of mealybugs was contributing to leafroll spread. In Italy, 20% virus prevalence was found 10 years after planting, indicating notably less apparent spread than in other regions (Engelbrecht and Kasdorf, 1990; Habili et al., 1995; Pietersen, 2006; Cabaleiro et al., 2008; Charles et al., 2009).

Figure 3: (A) Vineyards with high GLD incidence (dark red) serve as source of inoculum for adjacent blocks, in which disease spatial distribution is patchy, suggesting initial introduction of virus into uninfected blocks followed by within-block spread. (B) Example of secondary spread within rows, where an initial infection spread to neighboring plants. Both photographs were taken from the wine-producing region of Western Cape, South Africa.

Grapevine leafroll disease is caused by a number of virus species, and within those species, there are genetically distinct variants. Within a growing region, for example, the geographical distribution differs among genetically distinct GLRaV-3 variants (Jooste et al., 2011; Sharma et al., 2011), yet little is known about what processes have led to this variation, or its impact on GLD. Furthermore, mixed variant infections within one plant are common and differential transmission of the variants may occur. In this complex system, interactions need to be considered among multiple virus and vector species. Potential virus and vector exchange with neighboring unmanaged communities needs to be evaluated. The effects of abiotic factors such as climate and nutrient availability need to be considered. Finally, a holistic view of the effects of various management practices is needed.

GLRaVs and their variants may vary in severity and may interact with each other during transmission and establishment in the host (Jooste et al., 2011). Some studies have also implicated GLRaV-1, -3, -4, and -9 in facilitating transmission of *Grapevine virus A* (GVA, *Vitivirus*; Zorloni et al., 2006; Hommay et al., 2008; Tsai et al., 2010; Le Maguet et al., 2012; Herrbach et al., 2013) but the evidence is inconclusive. These and other potential interactions could lead to changes in symptomatic disease prevalence and spread in vineyards. Some plant viruses can actually be beneficial to plants (Roossinck, 2011), and environmental conditions can alter the nature of effects a virus has on its host. The responses of GLD severity to varied environmental conditions, some of which can be controlled by changing management practices, remain largely unknown. Specific horticultural practices that are expected to affect the impact of GLD on yield and fruit quality should be studied. For example, partial defoliation of vines, which is expected to improve ripening, has been shown to improve the quality of must (freshly pressed fruit juice) from grapes infected with GLRaV-3 (Pereira-Crespo et al., 2012).

Pathogen-vector specificity can affect regional patterns of disease caused by vector-borne pathogens. Different genetic variants of a pathogen can differ in transmission efficiency by one vector species (Power, 1996; Tsetsarkin et al., 2011). Alternatively, one virus can be transmitted more or less efficiently by different vector species. GLRaV-3 is transmitted by many vector species and can be regarded as a "vector generalist" (Tsai et al., 2010), but GLRaV-3 transmission efficiency can differ among vector species (Douglas and Krüger, 2008). Adaptation to a vector that is already present, or introduction of a new vector into an area, can lead to dramatic changes in the prevalence of a vector-borne pathogen (Purcell and Feil, 2001). Furthermore, introduction of a new vector with a higher transmission efficiency of one pathogen variant than another can lead to changes in the relative prevalence of pathogen variants in a region, which can be as devastating as the introduction of a new pathogen. More knowledge is needed about the interactions of GLRaVs with their vectors.

CURRENT DISEASE MANAGEMENT OPTIONS

Despite the economic impact of GLD on the world's wine industry, efforts to manage this disease are still being developed or have only recently been implemented over large agricultural areas (e.g.,Pietersen et al., 2013). Here we provide a summary of current management strategies being utilized or tested in four countries – South Africa, New Zealand, California-USA, and France. Our goal is to highlight management options that have been used to address both shared and unique challenges associated with this disease, with the expectation that each case study provides novel insights into the complexities of controlling GLD in the field.

A CASE STUDY OF DISEASE MANAGEMENT OPTIONS–SOUTH AFRICA

In South Africa, GLRaV-3 is the most important virus causing GLD (Pietersen and Kasdorf, 1993) and is transmitted predominantly by *Pl. ficus* and to a lesser extent by *Ps. longispinus* and multiple soft scale insect species (Walton and Pringle, 2004; Douglas and Krüger, 2008; Krüger and Douglas, 2013). Management of GLD is primarily through the provision of healthy planting material via the South African Vine Improvement Association (VIA). The VIA supplies the majority of planted vines utilized in the industry, and all VIA wine grape cultivars or clones are subjected to virus elimination via heat therapy and *in vitro* meristem tip propagation (Engelbrecht and Schwerdtfeger, 1979). Hardened off plantlets are established and maintained in insect-free greenhouses as *nuclear plants*(i.e., plant material of the highest level of sanitation in the certification scheme). On establishment, and every 5 years thereafter, these plants are subjected to compulsory tests for GLRaV-1, -2, and -3 (Goszczynski et al., 1995), *Grapevine fanleaf virus*, GVA and GVB by ELISA and by immunoelectron microscopy (Pietersen and Kasdorf, 1993) for GLRaV-4 and -5 in addition to the previously listed viruses. Furthermore, these plants are subjected to hardwood indexing on seven *Vitis* indicators (for 2 or 3 years depending on the disease). For plants to be certified as nuclear material, they must be negative for all viruses tested as well as GLD, grapevine stem grooving disease, grapevine corky bark disease, Shiraz disease, grapevine fleck disease, grapevine vein necrosis, and grapevine vein mosaic disease.

Planting material from nuclear blocks is propagated to establish *foundation blocks*, either in greenhouses or open field plantings. Open field foundation block vineyards must be on virgin soil (i.e., not previously planted in grapevines) that must test free of *Xiphinema index* (California dagger nematode), and must be at least 25 m from other vineyards. The vines from these blocks are tested every year by

ELISA for GLRaV-1, -2, and -3 if they are located in high risk areas (less than 25 m from other grapevines if mealybugs are recorded in the vicinity) or every 3 years if in low risk areas (no mealybugs trapped or observed, and the block is at least 25 m from vineyards of lower phytosanitary status). Plants testing negative for GLD in foundation blocks may be used to establish*mother-blocks*. Mother-blocks are typically commercial grape-growing vineyards and only need to be 3 m away from other vineyards. They can be planted in untested virgin soil or on soil that has previously been planted to *Vitis* but tests free of *X. index*. Visual inspection for GLD symptoms is conducted annually in autumn on red cultivars, Chardonnay, Cape Riesling, and Semillon.

The use of certified planting material and the above plan, however, do not rid South African vineyards of GLD. Mother-blocks in traditional grape production areas become infected rapidly with GLD. For example, during a 2001–2006 spatio-temporal study of 55 red cultivar mother-block vineyards in which no specific GLD control was applied, once GLD infections were initially found there was an average annual GLD increase of 1.94 times (Pietersen, 2006). Because of this, South African mother-blocks are only utilized for planting material if GLD infection levels of less than 5% exist in the vineyard. At infection levels below 5%, the producer may permit the removal of infected vines, or canes from infected and single adjacent vines within the row may be cut and dropped annually before planting material is collected. In spite of these measures, GLD-infected planting material can still be found within the certified material, with randomly occurring GLD-infected vines in newly established vineyards observed in 3% of all the mother-blocks (Pietersen, 2006). Based on the average rate of infection amongst the 55 mother-blocks monitored, it was estimated that the initial GLD incidence in the planting material was less than 1%. Since the mid-2000s many local plant improvement organizations have been propagating mother-block material in areas in which grapevines have not been grown previously. Certified material is therefore now differentiated as mother-blocks in low risk areas (three-star rated material) and in areas at risk to GLD re-infection (one-star material). At this time, three-star material is still relatively scarce; therefore responsible producers apply systemic insecticides at planting, and rogue GLD-infected vines in the newly established vineyards.

Secondary spread from a GLD-infected vine to adjacent vines in a row is the major cause of new GLD infections in the industry and occurred in all mother-blocks monitored (Pietersen, 2006; Figure 3). Roguing of infected vines is feasible and effective on an experimental scale (Pietersen et al., 2003). Removal of infected vines, combined with mealybug control, is extremely effective at controlling GLD in commercial vineyards, and this practice is becoming more widely applied (Pietersen and Walsh, 2012). Pietersen (2006) also presented circumstantial evidence of GLD spread in a replanted vineyard from a preceding vineyard, either through the presence of viruliferous mealybugs on remnant root material, or on volunteer hosts. The persistence of GLRaV-3 in remnant roots and potential of transmission by mealybugs from these has subsequently been demonstrated (Bell et al., 2009). Fallow periods of up to two seasons, during which remnant roots are removed, have been utilized in a number of commercial vineyards locally (Pietersen and Walsh, 2012). A clear demonstration of the efficacy of this strategy on its own must still be shown.

Gradients of GLD infection from the edges of a vineyard are commonly observed. Pietersen (2006) recorded gradients of various slopes from 70% of the 55 mother-blocks analyzed. These gradients reflect initial introduction of the virus from a source external to the vineyard, and in 32% of the blocks monitored the gradient could clearly be ascribed to an adjacent GLD-infected vineyard. These gradients are likely due to immigrating first instar mealybugs, either by their own motility over short distances, or on farm workers' clothing, on implements, by wind, ants, or possibly even by birds. A number of strategies have been employed to reduce the introduction of the disease from external sources (Pietersen and Walsh, 2012), including stringent control of mealybugs in all vineyards within the region, planting new vineyards far from heavily infected vineyards, avoiding traffic (implements and workers) from infected to healthy vineyards, or if unavoidable, washing implements with soapy water when moving between vineyards, and conducting work in healthy vineyards before moving into an infected vineyard. Following such a program, the near-eradication of GLD has been achieved at a commercial wine estate in the Somerset West district, from 100% infection on 41.26 ha in 2002 to 0.027% infection on 77.84 ha in 2012 (Pietersen et al., 2013). This result provided strong evidence that by using the full suite of GLD and mealybug control strategies available, disease incidence and its progression can be reversed. Further studies are required to determine the relative efficiency of individual components of the integrated control strategy.

A CASE STUDY OF DISEASE MANAGEMENT OPTIONS – NEW ZEALAND

Grapevine leafroll disease was first described in New Zealand in the early twentieth century (Bragato, 1902), but it was not until the 1960s that research to quantify its impact on vine performance and fruit quality started (Chamberlain et al., 1970; Over de Linden and Chamberlain, 1970). Today, GLRaV-3 is the most widespread and economically damaging disease affecting grapevines (Bonfiglioli and Hoskins, 2006). Concerned with the long-term impact of GLRaV-3 on wine quality, the national sector body, New Zealand Winegrowers (NZW), developed the grafted grapevine standard, with one of its aims being to minimize the probability of plant material with diseases such as GLRaV-3 being released to the industry.

A grower survey in 2005 revealed few respondents were well informed about the threats posed by GLRaV-3 or the options available for limiting its spread. Furthermore, a review of local and international literature, aimed to identify GLRaV-3 research priorities and knowledge gaps, was prepared (Charles et al., 2006). Of the numerous recommendations generated by these NZW initiatives, a plan for grower education and communication was prioritized. A collaborative program was established in which viticulturists, winemakers, and vine nursery groups collaborated with plant virologists, vine physiologists, and entomologists in a multi-disciplinary integrated approach to establish a GLRaV-3 control program.

A GLRaV-3 control pilot project began in 2009 in two North Island winegrowing regions: the Gimblett Gravels, a winegrowing sub-region in Hawke's Bay, and Martinborough. The project focused on controlling GLRaV-3 in red grape varieties because symptomatic vines are relatively easily identified visually by the dark red downward curling leaves with green veins. The project had three aims: (1) to visually identify and map the presence of GLRaV-3 in vines in both regions; (2) to control

GLRaV-3 through a combination of vine removal, hygiene practices, and improved vector management; and (3) to enable eventual vine replacement, whilst incorporating the new knowledge into "best practice" guidelines for nationwide dissemination (Hoskins et al., 2011). Here, we summarize the process and some of the achievements of the first 3 years of a 6-year project.

The control of GLRaV-3 in the field has focused on two strategies. The first was the removal (or roguing) of individual symptomatic vines (or small clusters of symptomatic vines), with most vineyard owners roguing symptomatic vines only. The second strategy, whole block removal, was adopted in blocks where roguing individual vines was considered unlikely to contain or control the disease. In New Zealand, the economic threshold of GLD incidence beyond which roguing was thought to be practical was ~20% of vines (Hoskins et al., 2011).

In the Gimblett Gravels and Martinborough regions, participating vineyards supplying ~40 individual wineries encompassed an area of ~1,100 ha. Training of vineyard personnel to identify GLRaV-3 symptomatic vines accurately was initiated. Once trained, vineyard personnel systematically moved through every red grape variety block late in the season identifying symptomatic vines, plotting their position with GPS and marking vines to guide the roguing done in winter. While the regional mapping of symptomatic vines is ongoing and the data have yet to be fully interpreted, individual vineyard owners are provided annually with preliminary block-specific results. The provision of this information has substantially aided the profile of the project and its educational goals, particularly in measuring the incidence and changes to the spread of GLRaV-3 (Hoskins et al., 2011).

Augmenting the regional mapping of GLRaV-3 were block-specific studies focused on GLRaV-3 identification together with monitoring the disease vectors, mealybugs. Data were analyzed from nine blocks in the Gimblett Gravels planted in various red grape varieties (~21,000 vines). The objective was to determine if a combined approach of GLRaV-3 visual identification and roguing, supported by good vector control could reduce disease incidence to a point where less than 1.0% of vines per block were rogued annually. While this study is still underway, preliminary results are presented here (V. A. Bell, unpublished results). In the nine study blocks, the percentage of symptomatic vines identified and rogued per year steadily declined from an average of

11.8% in 2009 to 2.7% in 2012. Over this period, a total of 4,902 symptomatic vines were rogued across the nine study blocks (23.8% of the original plantings). After 3 years, the evidence suggests roguing can successfully control GLRaV-3, although as discussed, good vector management was integral to a successful outcome.

In 2011 and 2012, mapping the positions of symptomatic vines in each block revealed 82.4 and 88.6%, respectively, were in close proximity to a vine rogued since 2009, supporting similar findings in earlier studies (Habili and Nutter, 1997; Cabaleiro and Segura, 2006; Pietersen, 2006). Of these neighboring vines, most at risk of acquiring GLRaV-3 were the "first" vines, the within-row immediate neighbors of a vine rogued at least 12 months earlier. This pattern of GLRaV-3 spread suggested the infection pathway, mediated by vector dispersal, was from the vine rogued at least 12 months earlier. In 2010, an average of 78% of all "first" vines had no visual symptoms of GLRaV-3, indicating they were either healthy or if infected, the visual symptoms were yet to express. By 2012, "first" vines relative to other "nearest neighbors" remained most at risk of GLRaV-3, although on average, 92% of "first" vines were symptomless. Based on the results of this study, the risk of a "first" vine acquiring GLRaV-3 was low, especially as the benefits of roguing and effective vector management accumulated over time. Consequently, good control of GLRaV-3 was achieved under almost all circumstances by roguing symptomatic vines only.

A further important aspect of the project was to determine the extent to which vector populations influenced GLRaV-3 control outcomes. Throughout this study, the vector most commonly encountered was the mealybug *Ps. calceolariae*, which colonizes all aerial parts and the roots of grapevines. Monitoring indicated mealybug numbers declined in most blocks over time as vineyard managers heeded warnings to improve vector control and to adopt better hygiene practices, such as removing the remnant roots of rogued vines. Being long-term reservoirs of GLRaV-3 (Bell et al., 2009), remnant roots colonized by *Ps. calceolariae* provide a likely pathway for the disease to infect young replacement vines.

In 2012, GLRaV-3 incidence in three of the nine study blocks was reduced to less than 1.0%, and in these blocks since 2010, mealybug counts from the third and final generation in late summer (March) were consistently low, ranging from two to eight mealybugs per 100 vine leaves inspected. Significantly, in two of these blocks (identified as A and B), disease incidence in 2009 was relatively high at 10.1 and 16.0%, respectively, so to have effectively controlled GLRaV-3 in just 3 years was an encouraging result. Given the known economic impacts of GLRaV-3, it was not possible to include an "unmanaged control" component in any of the study blocks. Despite this position, the finding of significant mealybug populations (78–175 mealybugs per 100 vine leaves inspected) in another two study blocks (C and D) provided useful comparisons with blocks A and B. In 2009, GLRaV-3 incidence in blocks C and D was 9.9 and 15.1%, respectively, but by 2012 cumulative vine losses due to GLRaV-3 were ca. 40%, culminating in the removal of all residual vines in both blocks.

With symptomatic vines identified and rogued each year in all nine study blocks, what most distinguished blocks C and D from the other seven was the high number of mealybugs. In this instance, poor mealybug control was probably due to non-adherence to insecticide (i.e., buprofezin) best practice with water volumes about one-third the label recommendations, thus compromising coverage and vine wetting. These contrasts in vector abundance demonstrated that roguing symptomatic vines alone provided relatively unsuccessful control of GLRaV-3 when it was not supported by effective mealybug management. In other words, while total eradication of *Ps. calceolariae* was not a prerequisite for controlling GLRaV-3, containing this disease was only achieved in those blocks where mealybug numbers were consistently low.

A CASE STUDY OF DISEASE MANAGEMENT OPTIONS – CALIFORNIA-USA

California accounts for 89.5% of domestic U.S. wine grape production – a total of 3.6 million tons in 2010 – with a farm gate value of US$2.06 billion. In a survey conducted by the American Vineyard Foundation in 2009, grape growers considered mealybug control and GLD one of their top priorities, solidifying this as a high priority research issue that threatens the sustainability of the industry. California grape growers have begun implementing multiple tactics in an effort to minimize current and future losses attributed to GLD. Although various GLRaV species are present in California, GLRaV-3 has been identified as the most important in the premiere wine-producing region of Napa Valley (Sharma et al., 2011).

California growers aim to minimize incidence of GLD and other grapevine diseases by planting material certified through the California Grapevine Certification and Registration (CGC&R) Program. Established in 1956, the CGC&R Program is administered by the California Department of Food and Agriculture (CDFA; Alley and Golino, 2000). It targets the elimination of grapevine diseases that spread from vine-to-vine by grafting and/or vegetative propagation. Under the auspices of the CGC&R Program, correctly named grape materials that pass specific disease tests are identified and/or created, and maintained as Foundation materials by Foundation Plant Services (FPS) at the University of California, Davis, for use by California commercial nurseries.

The CGC&R Program includes provisions for three levels of planting stock: California Foundation stock, California Registered stock, and California Certified grapevines. FPS at University of California, Davis maintains vines in the FPS Foundation block; materials derived from FPS Foundation vines are California Foundation stock. Vineyards planted by participants in the CGC&R Program using California Foundation stock material are known as California Registered increase blocks. They are inspected annually and tested for pathogens as needed by inspectors from CDFA. Material derived from the California Registered increase blocks is California Registered stock. When California Registered cuttings are rooted, or California Registered scion cuttings are grafted to California Registered rootstock cuttings, the resulting vines are classified as California Certified grapevines and are sold to growers for commercial planting. Nursery participation in the CGC&R Program is strongly encouraged but not mandatory. Other limitations to the CGC&R Program include the use of traditional screening methods (ELISA, RT-PCR, qPCR), which require prior knowledge of pathogens and are incapable of detecting unknown variants or agents. The variable population of GLRaV species in plant tissue, including rootstock and scion, is also a limitation to the production of reliable laboratory test results, and therefore material that is free of known viruses. To manage GLD spread, California wine grape growers identify symptomatic vines, document annual changes in disease incidence in vineyard blocks, and remove diseased vines. Vine removal occurs only in blocks where disease incidence is below a threshold determined by each grower. Thresholds are typically generated by an economic analysis based on vineyard age, cost of replanting versus redevelopment, grape purchasing contracts, the wine program for which the grapes are destined, and other considerations. Generally, growers identify vineyards with greater than 20–30% disease incidence for redevelopment of the entire block, whereas vine removal occurs in vineyards with less than 20% disease incidence. However, the threshold for roguing versus redevelopment varies considerably among growers, especially when grapes are destined for a high price point wine, or when redevelopment is particularly costly or challenging, such as in hillside blocks. Timely removal of diseased vines is limited by the cost associated with routine and reliable identification of these vines. It is not common practice in California to regularly identify and rogue symptomatic vines, although some growers have made it a regular practice in recent years. Dedicating resources to this effort can be complicated because peak symptom development overlaps with harvest period. There has also been a general lack of awareness of the importance of this practice. Both concerns are being addressed through research and educational programs directed by researchers at the University of California, with the goal of increasing awareness of the importance of this practice while identifying faster and easier ways to do it. In particular, infected vines may be identified using hyperspectral imaging technology that measures differences in leaf spectral reflectance between GLRaV-3 infected and uninfected grapevines (Naidu et al., 2009).

Mealybug management is a major component of GLD control programs in California. Currently five mealybug species cause direct damage and are potential vectors of GLRaV in vineyards: *Pl. ficus*, *Ps. maritimus*, *Ps. viburni*, *Ps. longispinus*, and *Ferrisia gilli* Gullan (Daane et al., 2012). Recently, a multiplex PCR procedure was developed to identify seven species of mealybug typically found in California vineyards (Daane et al., 2011). The ability to identify young mealybug nymphs to species using rapid and sensitive detection techniques helps growers make informed decisions about mealybug management. Trapping programs using pheromone-loaded lures also provide important information on mealybug species presence.

Growers rely on a combination of tactics including insecticides, mating disruption, biological control, and management of some ant species to minimize populations of *Pl. ficus* (Daane et al., 2008). Unfortunately, the Argentine ant (*Linepithema humile*) in particular, which "farms" mealybugs, is very aggressive in California vineyards and growers therefore struggle to maintain the extremely low mealybug populations required to minimize virus transmission. Results of recent investigations suggest that regional management programs for *Pl. ficus*, utilizing a combination of these tools, may provide better long-term control than individual efforts by isolated growers. Efforts are therefore underway to develop and implement similar regional management programs for other vineyard mealybug species. Populations of *Ps. maritimus* are of particular interest in coastal northern California vineyards, where they are commonly associated with spread of GLRaV-3, the most prevalent virus species that is spreading in the area.

A CASE STUDY OF DISEASE MANAGEMENT OPTIONS – FRANCE

In France, GLD is believed to be present since at least the early 1900s, as "rougeau" or "rougeot," and later as "enroulement foliaire," suspected then to be the same as the "*Rollkrankheit*" and "Leafroll" already described in Germany and the United States, respectively (Goheen et al., 1958; Vuittenez, 1958). However, GLD has long been seen as an unimportant problem for French viticulture, at least less crucial than fungal diseases and even *Grapevine fanleaf virus*. One reason may be that GLD symptoms were, and still are, often confused with other diseases or deficiencies, especially on white-berried cultivars. However, management of GLD was soon seen as a matter of sanitary selection (Vuittenez, 1958). At present, three species, GLRaV-1, -2, and -3, are detected in French commercial vineyards. GLRaV-1 and -3 are more frequent in north-eastern (Alsace, Beaujolais, Bourgogne, Champagne) and in southern (Mediterranean regions and Bordelais) vineyards, respectively, whereas GLRaV-2 is more common in the south-west. Over the last decades, infections by GLRaV-1 were recorded from many areas in Burgundy, Beaujolais, and Champagne. At the same time, wider infestations of mealybugs and soft scales were reported from these regions, probably related to the decreasing use of insecticides against the European grapevine moth (*Lobesia botrana*).

In France, sanitary selection was set up in the 1940s with the aim of producing healthy plant material to initially combat the spread of *Grapevine fanleaf virus* (Valat, 1972; Walter and Martelli, 1997). This process was greatly improved since then, due to progress in virological knowledge and detection methods, and is still seen today as the primary way to control GLD, which was incorporated into the system at a later date, among other viral and phytoplasmal diseases. According to French regulations (see www.legifrance.com), which follow a European Commission Directive, all planting material is classified in one of four categories: *initial, base, certified,* or *standard*(FranceAgriMer, 2013). The first three are produced only by specific institutions (initial, base) or nurseries (certified) and are subjected to detection tests to demonstrate the absence of viruses. Indexing is performed for any new clone prior to registration. During the pre-multiplication and multiplication processes, ELISA tests are used for all certified material. So far, only GLRaV-1 and -3 are taken into account among the GLRaVs, and all vines found infected at these steps have to be removed. The production of initial, base, and certified material is under the control of FranceAgriMer, a government agency. Growers can choose between certified and standard planting material, the first being more expensive but tested free of certain viruses (Walter and Martelli, 1997). Standard material is produced either in nurseries, where only visual inspections of symptoms are performed, or by growers who practice mass selection. Therefore, the use of standard material increases the risk of spreading GLD.

The use of insecticides against GLRaV vectors is allowed in France. However, few active ingredients, mainly organophosphates, are specifically registered for controlling scale and mealybug insects on grapevine. Moreover, insecticide implementation is not regulated nationally and will vary according to regional practices and viticultural advisers. Deeper biological and epidemiological knowledge is urgently required in order to adjust the use of insecticides to specific disease risk levels, depending on disease incidence and vector density. While natural enemies of vectors are known and experiments (e.g., using lacewings) are underway, there is no biological control program established in France. The development of vector monitoring by lure traps and mating disruption will first require the identification of the sex pheromones, still unknown, of common species like *Ph. aceris,H. bohemicus*, and *Pa. corni*.

In the course of certification schemes, thermotherapy- or meristem culture-based sanitation methods are sometimes used, especially for high value clones or cultivars. GLD management in France currently relies mainly on the sanitary selection of planting material, so in the long-term, healthy planting material seems the key to controlling GLD. More effort is to be devoted to improve sanitary selection, requiring deeper knowledge of the diversity of viruses and their effect on grapevine. Virological knowledge will also improve both specificity and sensitivity of detection methods used. Moreover, it should be desirable in the future to coordinate the certification schemes among countries producing planting material. In France, growers need to be better informed about GLD symptomatology and the detrimental effects of GLD, especially in case of co-infection of vines with two or more distinct viruses, a common phenomenon for grapevine. In addition, better information should assist nurseries and others producing plant material, particularly those using mass selection, to adopt procedures aimed at producing virus-clean grapevines. Moreover, a recent French study showed the risk to neighboring vineyards posed by leafroll-infected and scale insect-infested plots (Le Maguet et al., 2013). Therefore, a new plantation should take into account the sanitary status of neighbors and the possible spread of vectors from older to younger plots. Better guidelines (such as planting of hedges, vine-free strips between plots, insecticide treatments) should be tested for their efficacy, particularly for isolating mother plant blocks.

Finally, the wide variety of GLD epidemiological scenarios in France (e.g., Le Maguet et al., 2013) and the difficulty to define any damage thresholds hamper determining recommendations on how best to manage virus-infected and/or vector-infested vineyards. There is a crucial need for deeper knowledge of epidemiology, including determinant factors such as infection intensity, sensitivity of cultivars, virus and vector species present, and landscape structure. GLD management is more than a single grower's concern, instead requiring a collective approach across whole communities of growers, advisors, and scientists.

MANAGEMENT STRATEGIES – SUMMARY

The preceding case studies from grape-growing regions worldwide share remarkable similarities, and illustrate a combination of approaches required for GLD control. Importantly, they illustrate that GLD must be managed at a large scale and that a long-term management strategy is needed. Especially in the case of GLRaV-3, infected vines or blocks will continually act as sources of inoculum, perpetuating disease spread. For this reason, a coordinated area-wide approach is required, with education of growers as the first step. A second component of control is access to uninfected propagation material. A centralized service that includes a stringent certification program is needed to provide disease-free planting material for a growing region, as is the case in the countries reported here. In some regions where vectors may live on infected roots from previous crops, extra care is needed to assure that the planting area does not contain a GLRaV-3 source in the remaining live roots of infected vines that were removed. In both newly planted blocks and those with mature vines, roguing of symptomatic vines, and possibly vines immediately adjacent to those symptomatic vines, appears to be effective in preventing future disease spread. A third aspect of GLD management is control of insect vectors. Because mealybug nymphs are the most infective life stage and could travel long distances in air currents, insect control is often needed before large mealybug populations are detected. Therefore, knowledge of the life cycle of vectors can inform decisions regarding implementation of insect control. Finally, effective control of an existing GLD problem cannot be achieved within one growing season. Instead, favorable results are found after multiple years of regional management practices that incorporate eliminating infected plant material and controlling vector populations. Still lacking in GLD management is fundamental knowledge about disease spread. While a majority of management has focused on red varieties, largely because identification of symptoms is easier, GLD also affects white varieties. As long as white varieties continue to be overlooked, they may continue to be a source of virus thus hampering management efforts.

ACKNOWLEDGMENTS

Based in New Zealand but of the world, Dr Rod Bonfiglioli was a talented and widely respected plant virologist. His passing in May 2009 in New Zealand sadly coincided with the start of his long anticipated GLRaV-3 control project. It will remain an enduring legacy to a man unlikely to be forgotten by his many friends and colleagues. Rodrigo P. P. Almeida, Kent M. Daane, G. K. Blaisdell, and Monica L. Cooper acknowledge funding from the American Vineyard Foundation, Viticulture Consortium West, Oregon Wine Board, California Fruit Tree, Nut Tree and Grapevine IAB, California Department of Food and Agriculture, and United States Department of Agriculture (SCRI). Monica L. Cooper gratefully acknowledges the support of Napa wine grape growers and their staff. Vaughn A. Bell gratefully acknowledges funding provided by New Zealand Winegrowers, the Ministry for Primary Industries Sustainable Farming Fund, the Gimblett Gravel Winegrowers' Association, and the support of the vineyard owners and their staff. Etienne Herrbach acknowledges funding from INRA and France AgriMer. Gerhard Pietersen acknowledges funding from Winetech, South Africa.

REFERENCES

1. Alley, L., and Golino, D. A. (2000). "The origins of the grape program at Foundation Plant Materials Service," in*Proceedings of the ASEV 50th Anniversary Meeting,* Seattle, Washington, 222–230.
2. Al Rwahnih, M., Dolja, V. V., Daubert, S., Koonin, E. V., and Rowhani, A. (2012). Genomic and biological analysis of Grapevine leafroll-associated virus 7 reveals a possible new genus within the family Closteroviridae. *Virus Res.* 163, 302–309.
3. Atallah, S. S., Gomez, M. I., Fuchs, M. F., and Martinson, T. E. (2012). Economic impact of grapevine leafroll disease on*Vitis vinifera* cv. Cabernet franc in Finger Lakes vineyards of New York. *Am. J. Enol. Vitic.* 63, 73–79.
4. Barrass, I. C., Jerie, P., and Ward, S. A. (1994). Aerial dispersal of first- and second-instar longtailed mealybug,*Pseudococcus*

longispinus (Targioni Tozzetti) (Pseudococcidae: Hemiptera). *Aust. J. Exp. Agric.* 34, 1205–1208.

5. Becerra, V., Gonzalez, M., Herrera, M. E., and Miano, J. L. (2006). Population dynamics of vine mealybug *Planococcus ficus* Sign. (Hemiptera: Pseudococcidae) in vineyards. *Rev. Fac. Cienc. Agrar. Univ. Nac. Cuyo* 38, 1–6.
6. Bell, V. A., Bonfiglioli, R. G. E., Walker, J. T. S., Lo, P. L., Mackay, J. F., and McGregor, S. E. (2009). Grapevine leafroll-associated virus 3 persistence in *Vitis vinifera* remnant roots. *J. Plant Pathol.* 91, 527–533.
7. Belli, G., Fortusini, A., Casati, P., Belli, L., Bianco, P. A., and Prati, S. (1994). Transmission of a grapevine leafroll associated closterovirus by the scale insect *Pulvinaria vitis* L. *Riv. Patol. Veg.* 4, 105–108.
8. Ben-Dov, Y. (1995). *A Systematic Catalogue of the Mealybugs of the World (Insecta: Homoptera: Coccoidea: Pseudococcidae, and Putoidae) with Data on Geographical Distribution, Host plants, Biology, and Economic Importance*. Hampshire: Intercept Ltd.
9. Bertazzon, N., Borgo, M., Vanin, S., and Angelini, E. (2010). Genetic variability and pathological properties of grapevine leafroll-associated virus 2 isolates. *Eur. J. Plant Pathol.* 127, 185–197.
10. Bonfiglioli, R., and Hoskins, N. (2006). Managing virus in New Zealand vineyards. *Aust. N. Z. Grapegrow. Winemak.* 515, 43–46.
11. Bragato, R. (1902). *Report to the Department of Agriculture New Zealand.* Wellington: New Zealand Government, 451–455.
12. Cabaleiro, C., and Segura, A. (1997a). Field transmission of grapevine leafroll associated virus 3 (GLRaV-3) by the mealybug*Planococcus citri*. *Plant Dis.* 81, 283–287.
13. Cabaleiro, C., and Segura, A. (1997b). Some characteristics of the transmission of grapevine leafroll associated virus 3 by*Planococcus citri* Risso. *Eur. J. Plant Pathol.* 103, 373–378.
14. Cabaleiro, C., and Segura, A. (2006). Temporal analysis of grapevine leafroll associated virus 3 epidemics. *Eur. J. Plant Pathol.* 114, 441–446.

15. Cabaleiro, C., Couceiro, C., Pereira, S., Cid, M., Barrasa, M., and Segura, A. (2008). Spatial analysis of epidemics of Grapevine leafroll associated virus-3. *Eur. J. Plant Pathol.* 121, 121–130.

16. Chamberlain, E. E., Over de Linden, A. J., and Berrysmith, F. (1970). Virus diseases of grapevines in New Zealand. *N. Z. J. Agric. Res.* 13, 338–358.

17. Charles, J. G., Bell, V. A., Lo, P. L., Cole, L. M., and Chhagan, A. (2010). Mealybugs (Hemiptera: Pseudococcidae) and their natural enemies in New Zealand vineyards from 1993–2009. *N. Z. Entomol.* 33, 84–91.

18. Charles, J. G., Cohen, D., Walker, J. T. S., Forgie, S. A., Bell, V. A., and Breen, K. C. (2006). A review of the ecology of Grapevine leafroll associated virus type 3 (GLRaV-3). *N. Z. Plant Prot.* 59, 330–337.

19. Charles, J. G., Froud, K. J., Brink, R. V. D., and Allan, D. J. (2009). Mealybugs and the spread of grapevine leafroll-associated virus 3 (GLRaV-3) in a New Zealand vineyard. *Australas. Plant Pathol.* 38, 576–583.

20. Cid, M., Pereira, S. Cabaleiro, C., and Segura, A. (2010). Citrus mealybug (Hemiptera: Pseudococcidae) movement and population dynamics in an arbor-trained vineyard. *J. Econ. Entomol.* 103, 619–630.

21. Credi, R., and Babini, A. R. (1997). Effects of virus and virus-like infections on growth, yield and fruit quality of Albana and Trebbiano Romagnolo grapevines. *Am. J. Enol. Vitic.* 48, 7–12.

22. Daane, K. M., Almeida, R. P. P., Bell, V. A., Botton, M., Fallahzadeh, M., Mani, M., et al. (2012). "Biology and management of mealybugs in vineyards," in *Arthropod Management in Vineyards*, eds N. J. Bostanian, R. Isaacs, and C. Vincent (Dordrecht: Springer), 271–308.

23. Daane, K. M., Cooper, M. L., Triapitsyn, S. V., Walton, V. M., Yokota, G. Y., Haviland, D. R., et al. (2008). Vineyard managers and researchers seek sustainable solutions for mealybugs, a changing pest complex. *Calif. Agric.* 62, 167–176.

24. Daane, K. M., Middleton, M. C., Sforza, R., Cooper, M. L., Walton, V. M., Walsh, D. B., et al. (2011). Development of a multiplex PCR for identification of vineyard mealybugs. *Environ. Entomol.* 40, 1595–1603.

25. Douglas, N., and Krüger, K. (2008). Transmission efficiency of Grapevine leafroll-associated virus 3 (GLRaV-3) by the mealybugs *Planococcus ficus* and *Pseudococcus longispinus* (Hemiptera: Pseudococcidae). *Eur. J. Plant Pathol.* 122, 207–212.

26. Engelbrecht, D. J., and Kasdorf, G. G. F. (1985). Association of a closterovirus with grapevines indexing positive for grapevine leafroll disease and evidence for its natural spread in grapevine. *Phytopathol. Mediterr.* 24, 101–105.

27. Engelbrecht, D. J., and Kasdorf, G. G. F. (1990). Transmission of grapevine leafroll disease and associated closteroviruses by the vine mealybug *Planococcus ficus*. *Phytophylactica* 22, 341–346.

28. Engelbrecht, D. J., and Schwerdtfeger, U. (1979). In vitro grafting of grapevine shoot apices as an aid to the recovery of virus-free clones. *Phytophylactica* 11, 183–185.

29. FranceAgriMer. (2013). *Certification des bois et plants de vigne*. Available at: www.franceagrimer.fr/filiere-vin-et-cidriculture/Vin/Appui-a-la-filiere/Innovation-et-qualite/Bois-et-plants-de-vigne/Certification-des-bois-et-plants-de-vigne(accessed January 22, 2013).

30. Froissart, R., Doumayrou, J., Vuillaume, F., Alizon, S., and Michalakis, Y. (2010). The virulence–transmission trade-off in vector-borne plant viruses: a review of (non-)existing studies. *Philos. Trans. R. Soc. Lond. B Biol. Sci.* 365, 1907–1918.

31. Fuchs, M., Marsella-Herrick, P., Loeb, G. M., Martinson, T. E., and Hoch, H. C. (2009). Diversity of ampeloviruses in mealybug and soft scale vectors and in grapevine hosts from leafroll-affected vineyards. *Phytopathology* 99, 1177–1184.

32. Geiger, C. A., and Daane, K. M. (2001). Seasonal movement and distribution of the grape mealybug (Homoptera: Pseudococcidae): developing a sampling program for San Joaquin Valley vineyards. *J. Econ. Entomol.* 94, 291–301.

33. Goheen, A. C. (1988). "Leafroll," in *Compendium of Grape Diseases*, eds R. C. Pearson and A. C. Goheen (Saint Paul, MN: APS Press), 93 p.

34. Goheen, A. C., Harmon, F. N., and Weinberger, J. H. (1958). Leafroll (White Emperor disease) of grapes in California. *Phytopathology* 48, 51.

35. Golino, D. A., Weber, E. A., Sim, S. T., and Rowhani, A. (2008). Leafroll disease is spreading rapidly in a Napa Valley vineyard. *Calif. Agric.* 62, 156–160.

36. Goszczynski, D. E., Kasdorf, G. G. F., and Pietersen, G. (1995). Production and use of antisera specific to grapevine leafroll-associated viruses following electrophoretic separation of their proteins and transfer to nitrocellulose. *Afr. Plant Prot.* 1, 1–8.

37. Gouveia, P., Santos, M. T., Eiras-Dias, J. E., and Nolasco, G. (2011). Five phylogenetic groups identified in the coat protein gene of grapevine leafroll-associated virus 3 obtained from Portuguese grapevine varieties. *Arch. Virol.* 156, 413–420.

38. Grasswitz, T. R., and James, D. G. (2008). Movement of grape mealybug, *Pseudococcus maritimus*, on and between host plants. *Entomol. Exp. Appl.* 129, 268–275.

39. Gribaudo, I., Gambino, G., Bertin, S., Bosco, D., Cotroneo, A., and Mannini, F. (2009). Monitoring the spread of viruses after vineyard replanting with heat-treated clones of *Vitis vinifera* "Nebbiolo". *J. Plant Pathol.* 91, 741–744.

40. Gutierrez, A. P., Daane, K. M., Ponti, L., Walton, V. M., and Ellis, C. K. (2008). Prospective evaluation of the biological control of vine mealybug: refuge effects and climate. *J. Appl. Ecol.* 45, 524–536.

41. Habili, N., Fazeli, C. F., Ewart, A., Hamilton, R., Cirami, R., Saldarelli, P., et al. (1995). Natural spread and molecular analysis of grapevine leafroll-associated virus 3 in Australia. *Phytopathology* 85, 1418–1422.

42. Habili, N., and Nutter, F. W. Jr. (1997). Temporal and spatial analysis of grapevine leafroll-associated virus 3 in Pinot Noir grapevines in Australia. *Plant Dis.* 81, 625–628.

43. Herrbach, E., Le Maguet, J., and Hommay, G. (2013). "Virus transmission by mealybugs and soft scales (Hemiptera, Coccoidea)," in *Vector-mediated Transmission of Plant Pathogens*, ed. J. K. Brown (St. Paul, MN: American Phytopathological Society Press) (in press).

44. Hoefert, L. L., and Gifford, E. M. Jr. (1967). Grapevine leafroll virus – history and anatomic effects. *Hilgardia* 38, 403–426.

45. Hommay, G., Komar, V., Lemaire, O., and Herrbach, E. (2008). Grapevine virus A transmission by larvae of*Parthenolecanium corni*. *Eur. J. Plant Pathol.* 121, 185–188.

46. Hoskins, N., Thompson, C., and Andrew, R. (2011). New Zealand Winegrowers working hard to eliminate leafroll. *Aust. N. Z. Grapegrow. Winemak.* 569, 23–25.

47. Jooste, A. E. C., Pietersen, G., and Burger, J. T. (2011). Distribution of grapevine leafroll associated virus-3 variants in South African vineyards. *Eur. J. Plant Pathol.* 131, 371–381.

48. Karasev, A. V. (2000). Genetic diversity and evolution of Closteroviruses. *Annu. Rev. Phytopathol.* 38, 293–324.

49. Klaassen, V. A., Sim, S. T., Dangl, G. S., Osman, F., Al Rwahnih, M., Rowhani, A., et al. (2011). Vitis californica and *Vitiscalifornica* × *Vitis vinifera* hybrids are hosts for Grapevine leafroll-associated virus-2 and-3 and Grapevine virus A and B.*Plant Dis.* 95, 657–665.

50. Krüger, K., and Douglas, N. (2013). Grapevine leafroll-associated virus 3 (GLRaV-3) transmission by three soft scale insect species (Hemiptera: Coccidae) with notes on their biology. *Afr. Entomol.* 21, 1–8.

51. Kuniyuki, H., Rezende, J. A. M., De Willink, C. G., Novo, J. P. S., and Yuki, V. A. (2005). Transmissão do Grapevine leafroll-associated virus 3 pela cochonilha *Pseudococcus longispinus* Targioni-Tozzetti (Hemiptera: Pseudococcidae).*Summa Phytopathol.* 31, 65–68.

52. Le Maguet, J. (2012). *Epidémiologie de l'enroulement viral de la vigne dans les vignobles français septentrionaux et transmission par cochenilles vectrices*. PhD thesis, Université de Strasbourg, Strasbourg, 204 p.

53. Le Maguet, J., Beuve, M., Herrbach, E., and Lemaire, O. (2012). Transmission of six ampeloviruses and two vitiviruses to grapevine by *Phenacoccus aceris*. *Phytopathology* 102, 717–723.

54. Le Maguet, J., Fuchs, J. J., Beuve, M., Chadoeuf, J., Herrbach, E., and Lemaire, O. (2013). The role of the mealybug*Phenacoccus aceris* in the epidemic of Grapevine leafroll-associated virus-1 (GLRaV-1) in two French vineyards. *Eur. J. Plant Pathol.* 135, 415–427.

55. Mahfoudhi, N., Digiaro, M., and Dhouibi, M. H. (2009). Transmission of grapevine leafroll viruses by *Planococcus ficus*(Hemiptera: Pseudococcidae) and *Ceroplastes rusci* (Hemiptera: Coccidae). *Plant Dis.* 93, 999–1002.

56. Maliogka, V. I., Dovas, C. I., Lotos, L., Efthimiou, K., and Katis, N. I. (2009). Complete genome analysis and immunodetection of a member of a novel virus species belonging to the genus *Ampelovirus*. *Arch. Virol.* 154, 209–218.

57. Martelli, G. P., Abou Ghanem-Sabanadzovic, N., Agranovsky, A. A., Al Rwahnih, M., Dolja, V. V., Dovas, C. I., et al. (2012). Taxonomic revision of the family Closteroviridae with special reference to the grapevine leafroll-associated members of the genus *Ampelovirus* and the putative species unassigned to the family. *J. Plant Pathol.* 94, 7–19.

58. Martelli, G. P., Agranovsky, A. A., Bar-Joseph, M., Boscia, D., Candresse, T., Coutts, R. H., et al. (2002). The family Closteroviridae revised. *Arch. Virol.* 147, 2039–2044.

59. Naidu, R. A., Perry, E. M., Pierce, F. J., and Mekuria, T. (2009). The potential of spectral reflectance technique for the detection of Grapevine leafroll-associated virus-3 in two red-berried wine grape cultivars. *Comput. Electron. Agric.* 66, 38–45.

60. Over de Linden, A. J., and Chamberlain, E. E. (1970). Effect of grapevine leafroll virus on vine growth and fruit yield and quality. *N. Z. J. Agric. Res.* 13, 689–698.

61. Pereira-Crespo, S., Segura, A., Garcia-Berrios, J., and Cabaleiro, C. (2012). Partial defoliation improves must quality of cv. Albariño infected by grapevine leafroll associated virus 3. *Phytopathol. Mediterr.* 54, 383–389.

62. Petersen, C. L., and Charles, J. G. (1997). Transmission of grapevine leafroll-associated closteroviruses by *Pseudococcus longispinus* and *P. calceolariae*. *Plant Pathol.* 46, 509–515.

63. Pietersen, G. (2006). "Spatio-temporal distribution dynamics of grapevine leafroll disease in Western Cape vineyards," in*Extended abstracts of the 15th Meeting of the International Council for the Study of Virus and Virus-like diseases of the Grapevine (ICVG)*, April 3–7, Stellenbosch, South Africa, 126–127.

64. Pietersen, G., and Kasdorf, G. G. F. (1993). "Use of IEM for the detection of the viruses of the grapevine leafroll complex in South Africa," in *Extended Abstracts of the 11th Meeting of the International Council for the Study of Viruses and Virus Diseases of Grapevine*, Montreux, Switzerland, September 6–9.
65. Pietersen, G., Kellerman, U., Oosthuizen, W. T., and Spreeth, N. (2003). "Effect of rouging on natural spread of grapevine leafroll disease in two vineyards in South Africa," in *8th International Congress of Plant Pathology*, February 2–7, Christchurch, New Zealand.
66. Pietersen, G., Spreeth, N., Oosthuizen, T., van Rensburg, A., van Rensburg, M., Lottering, D., et al. (2013). Control of grapevine leafroll disease spread at a commercial wine estate in South Africa: a case study. *Am. J. Enol. Vitic.* doi: 10.5344/ajev.2013.12089
67. Pietersen, G., and Walsh, H. A. (2012). "Development of a LAMP technique for the control of grapevine leafroll associated virus type 3 (GLRaV-3) in infected white cultivar vines by roguing," in *Proceedings of the 17th Congress of the International Council for the Study of Virus and Virus-like diseases of the Grapevine (ICVG)*, October 8–11, 2012, Davis, California, 50–51.
68. Power, A. G. (1996). Competition between viruses in a complex plant–pathogen system. *Ecology* 77, 1004–1010.
69. Purcell, A., and Feil, H. (2001). Glassy-winged Sharpshooter. *Pestic. Outlook* 12, 199–203.
70. Roossinck, M. J. (2011). The good viruses: viral mutualistic symbioses. *Nat. Rev. Microbiol.* 9, 99–108Rowhani, A., Uyemoto, J. K., Golino, D. A., and Martelli, G. P. (2005). Pathogen testing and certification of *Vitis* and*Prunus* species. *Annu. Rev. Phytopathol.* 43, 261–278.
71. Sandanayaka, W. R. M., Blouin, A. G., Prado, E., and Cohen, D. (2012). Stylet penetration behaviour of *Pseudococcus longispinus* in relation to acquisition of Grapevine leafroll virus 3. *Arthropod Plant Interact.* 7, 137–146.
72. Scheu, G. (1936). *Mein Winzerbuch*. Berlin: Reichsnährstand Verlags-Ges, 274 p.
73. Seah, Y. M., Sharma, A. M., Zhang, S., Almeida, R. P. P., and Duffy, S. (2012). A divergent variant of Grapevine leafroll-associated virus 3 is present in California. *Virol. J.* 9, 235.

74. Sforza, R., Boudon-Padieu, E., and Greif, C. (2003). New mealybug species vectoring grapevine leafroll-associated viruses-1 and -3 (GLRaV-1 and -3). *Eur. J. Plant Pathol.* 109, 975–981.

75. Sharma, A. M., Wang, J., Duffy, S., Zhang, S., Wong, M. K., Rashed, A., et al. (2011). Occurrence of grapevine leafroll-associated virus complex in Napa Valley. *PLoS ONE* 6:e26227. doi: 10.1371/journal.pone.0026227

76. Swallow, W. H. (1985). Group testing for estimating infection rates and probabilities of disease transmission.*Phytopathology* 75, 882–889.

77. Thompson, J. R., Fuchs, M., and Perry, K. L. (2012). Genomic analysis of grapevine leafroll associated virus-5 and related viruses. *Virus Res.* 163, 19–27.

78. Tsai, C. W., Bosco, D., Daane, K. M., and Almeida, R. P. P. (2011). Effect of host plant tissue on the vector transmission of grapevine leafroll-associated virus 3. *J. Econ. Entomol.* 104, 1480–1485.

79. Tsai, C. W., Chau, J., Fernandez, L., Bosco, D., Daane, K. M., and Almeida, R. P. P. (2008). Transmission of Grapevine leafroll-associated virus 3 by the vine mealybug (*Planococcus ficus*). *Phytopathology* 98, 1093–1098.

80. Tsai, C. W., Daugherty, M. P., and Almeida, R. P. P. (2012). Seasonal dynamics and virus translocation of Grapevine leafroll-associated virus 3 in grapevine cultivars. *Plant Pathol.* 61, 977–985.

81. Tsai, C.W., Rowhani, A., Golino, D.A., Daane, K.M., and Almeida, R. P. P. (2010). Mealybug transmission of grapevine leafroll viruses: an analysis of virus–vector specificity. *Phytopathology* 100, 830–834.

82. setsarkin, K. A., Chen, R., Leal, G., Forrester, N., Higgs, S., Huang, J., et al. (2011). Chikungunya virus emergence is constrained in Asia by lineage-specific adaptive landscapes. *Proc. Natl. Acad. Sci. U.S.A.* 108, 7872–7877.

83. Valat, C. (1972). La sélection clonale de la vigne, facteur d'amélioration de la productivité du vignoble français. *Options Mediterr.* 12, 89–91.

84. Varela, L., Daane, K., Phillips, P., and Bettiga, L. (2013). "European fruit lecanium scale," in *UC IPM Grape Pest Management Manual,*

3rd Edn, ed. L. Bettiga (Davis, CA: University of California, Agriculture and Natural Resources), 3343.

85. Vuittenez, A. (1958). Transmission par greffage d'une virose du type "enroulement foliaire" commune dans les vignobles de l'est et du centre est de la France. *C. R. Acad. Agric. France* 44, 313–316.
86. Walter, B., and Martelli, G. P. (1997). "Clonal and sanitary selection of the grapevine," in *Sanitary Selection of the Grapevine*, ed. B. Walter (Paris: INRA Editions), 43–95.
87. Walton, V. M., Daane, K. M., Bentley, W. J., Millar, J. G., Larsen, T. E., and Malakar-Kuenen, R. (2006). Pheromone-based mating disruption of *Planococcus ficus* (Hemiptera: Pseudococcidae) in California vineyards. *J. Econ. Entomol.* 99, 1280–1290.
88. Walton, V. M., and Pringle, K. L. (2004). Vine mealybug, *Planococcus ficus* (Signoret) (Hemiptera: Pseudococcidae), a key pest in South African vineyards. A review. *S. Afr. J. Enol. Vitic.* 25, 54–62.
89. Wang, J., Sharma, A. M., Duffy, S., and Almeida, R. P. P. (2011). Genetic diversity in the 3′ terminal 4.7-kb region of Grapevine leafroll-associated virus 3. *Phytopathology* 101, 445–450.
90. Woodrum, R. C., Antcliff, A. J., Krake, L. R., and Taylor, R. H. (1984). Yield differences between Sultana clones related to virus status and genetic factors. *Vitis* 23, 73–83.
91. Zorloni, A., Prati, S., Bianco, P. A., and Belli, G. (2006). Transmission of grapevine virus A and grapevine leafroll-associated virus-3 by *Heliococcus bohemicus*. *J. Plant Pathol.* 88, 325–328.

Chapter 2

Galling Aphids (Hemiptera: Aphidoidea) in China: Diversity and Host Specificity

Jing Chen[1,2], and Ge-Xia Qiao

[1]Key Laboratory of Zoological Systematics and Evolution, Institute of Zoology, Chinese Academy of Sciences, No. 1 Beichen West Road, Chaoyang District, Beijing 100101, China
[2]College of Life Sciences, Graduate University of Chinese Academy of Sciences, No. 19 Yuquan Road, Shijingshan District, Beijing 100049, China

ABSTRACT

Gall formation is an interesting plant response to aphid feeding. This paper presents a review of galling aphids in China. Altogether, 157 species and subspecies in ten families and subfamilies are found to induce galls on their host plants. As many as 39% species are endemic to China. The Eriosomatinae include the highest percentage of gall-inducing species. The great diversity of gall morphology may be described in terms of five characteristics: type, site, size, shape, and structure. The host association and host specificity of galling aphids are also discussed.

INTRODUCTION

Aphids (Hemiptera: Aphidoidea) are an important group of phloem-feeding insects that may limit plant productivity [1] and transmit plant viruses [2]. They feed from the phloem sieve elements by penetrating their slender stylets intercellularly during which specific plant responses are triggered [3, 4]. A remarkable and interesting plant response is the formation of galls. These atypical plant tissue growths, the result of interactions between the inducer and the plant [5], reflect the complex and intimate relationship between plants and insects. Galls provide abundant nutrition [6–11], a favorable microenvironment [6, 7, 12, 13], and protection against natural enemies [6, 7, 14] to the inducer and its offspring. They also mitigate clonal mixing and maintain the genetic integrity [15, 16]. Therefore, galling aphids have been viewed as a useful model system for studying herbivore-plant interactions. Approximately 10–20% of the 4,700 aphid species known worldwide can induce galls on their host plants [17]. Gall traits are important biological characteristics of galling aphids. The shape, structure, and site of galls are highly species specific [18]. Thus, galls are commonly regarded as the extended phenotype of aphids [19–21] and hence helpful for aphid identification and phylogenetic study.

A total of approximately 1,100 aphid species are now known from China, constituting at least 23% of the world's aphid fauna [22]. Studies of galling aphids in China focus primarily on the subfamilies Eriosomatinae and Hormaphidinae. Zhang et al. [23] and Chen and Qiao [24] systematically studied the diversity of the galls of these two subfamilies, respectively. The evolution of galls in the tribes Fordini and Pemphigini of Eriosomatinae was also discussed [10, 11]. In this paper, galling aphids in China are reviewed with special emphasis on the diversity of gall-inducing species and gall morphology. Host association and host specificity are also considered.

The aphid species information in this paper was obtained from the species records of specimens deposited in the National Zoological Museum of China, Institute of Zoology, Chinese Academy of Sciences, Beijing, China (NZMCAS) and identified by aphid taxonomists of our research group. The information about galls and host plants was taken from field collection records. The host plant species were identified by plant taxonomists of the Institute of Botany, Chinese Academy of Sciences and Beijing Forestry University. The aphid classification system used in this paper follows G. Remaudière and M. Remaudière [25].

GALLING APHID DIVERSITY

At present, approximately 157 species and subspecies (14% of the total number of Chinese aphid species) are known to induce galls on their specific host plants in China. They are restricted to ten families and subfamilies, that is, Adelgidae, Phylloxeridae, Eriosomatinae, Hormaphidinae, Aphidinae, Chaitophorinae, Mindarinae, Myzocallidinae, Phyllaphidinae, andThelaxinae. Of these gall-inducing aphid species, 83 species belong to the subfamily Eriosomatinae, 37 to Aphidinae, and 19 to Hormaphidinae. The principal groups of galling aphids and the host plants bearing galls in China are listed in Table 1.

Table 1: Taxonomy of galling aphids and host plants bearing galls in China

Family	Subfamily/tribe	Genera	Host plants
Adelgidae		*Adelges, Pineus*	*Picea*
Phylloxeridae		*Acanthochermes*	*Quercus*
		Viteus	*Vitis*
Aphididae	Eriosomatinae		
	Eriosomatini	*Aphidounguis, Colopha, Eriosoma, Kaltenbachiella, Tetraneura*	*Ulmus*
	Fordini	*Aploneura, Baizongia, Chaetogeoica, Forda, Namaforda, Smynthurodes*	*Pistacia*
		Floraphis, Kaburagia, Meitanaphis, Nurudea, Schlechtendalia	*Rhus*
	Pemphigini	*Epipemphigus, Pachypappa, Pachypappella, Pemphigus, Thecabius*	*Populus*
		Prociphilus	Caprifoliaceae, Oleaceae, Rosaceae
	Hormaphidinae		
	Cerataphidini	*Aleurodaphis*	*Sinojackia*
		Astegopteryx, Cerataphis, Ceratoglyphina, Ceratovacuna, Ktenopteryx,Pseudoregma	*Styrax*
		Tuberaphis	*Styrax, Viscum*
	Hormaphidini	*Hamamelistes*	*Betula*

	Nipponaphidini	*Asiphonipponaphis, Metanipponaphis, Neothoracaphis, Nipponaphis*	*Distylium*
	Aphidinae		
	Aphidini	*Aphis*	Malvaceae, *Prunus*
		Cryptosiphum	*Artemisia*
		Hyalopterus	Rosaceae
	Macrosiphini	*Cryptaphis*	*Geranium*
		Cryptomyzus, Hyperomyzus	*Ribes*
		Diuraphis	Pooideae
		Dysaphis, Myzus, Ovatus, Sappaphis, Sorbaphis, Tuberocephalus	Rosaceae
		Hayhurstia	*Chenopodium*
		Lipaphis	Cruciferae
		Neorhopalomyzus	*Lonicera*
		Rhopalosiphoninus	*Deutzia, Syringa*
		Semiaphis	*Lonicera*, Umbelliferae
	Chaitophorinae	*Chaitophorus*	*Populus*
		Periphyllus	*Acer, Aesculus, Koelreuteria*
	Mindarinae	*Mindarus*	*Abies*
	Myzocallidinae	*Tinocallis*	*Sapindus, Lagerstroemia*
	Phyllaphidinae	*Phyllaphis*	*Fagus*
	Thelaxinae	*Kurisakia*	*Quercus*

Out of these galling aphids, 61 species and subspecies are endemic to China, accounting for 39% of the total galling species. Endemic species, localities (province) where galls were collected, host plants bearing galls, and gall morphology are listed in Table 2.

Table 2: Endemic galling aphid species in China

Species	Distribution	Host plants bearing galls	Gall morphology
Adelgidae			
Adelges (Gilletteella) glandulae (Zhang)	Sichuan, Yunnan	*Picea brachytyla, P. likiangensis, P. likiangensis*var. *balfouriana, P. purpurea*	Leaf gall, cone-like, multiple-cavity, without associated needles [33].
Adelges (Sacchiphantes) roseigallis (Li & Tsai)	Gansu	*Picea asperata, P. crassifolia*	Leaf gall, cone-like [34].
Pineus (Pineus) armandicola Zhang, Zhong & Zhang	Yunnan	*Picea likiangensis*	Leaf gall, cone-like or torch-like, multiple-cavity [35].
Pineus (Pineus) sichunanus Zhang	Sichuan, Yunnan	*Picea brachytyla, P. likiangensis, P. likiangensis*var. *balfouriana, P. purpurea*	Leaf gall, broom-like or torch-like, multiple-cavity [33].
Phylloxeridae			
Acanthochermes similiquercus Jiang, Huang & Qiao	Sichuan	*Quercus* sp.	Small spherical gall on the leaf with an opening on the lower surface [36].
Eriosomatinae Eriosomatini			

*Eriosoma cerum*Zhang	Beijing, Hebei, Inner Mongolia	*Ulmus pumila*	Leaf edge curls downward [37].
*Eriosoma fukangense*Zhang	Xinjiang	*Ulmus pumila*	Leaf rolls, screw-like [37].
Eriosoma multilocularis Zhang & Zhang	Beijing, Hebei	*Ulmus pumila*	Multiple-cavity gall on the underside of leaves, with longitudinal ridges on lateral sides [38].
*Eriosoma spirifolium*Zhang	Xinjiang	*Ulmus pumila*	Leaf curls, spiral-shaped [37].
*Eriosoma togrogum*Zhang	Inner Mongolia, Shaanxi	*Ulmus pumila*	Leaf pseudogall [39].
Eriosoma ulmipumilicola Zhang & Zhang	Hebei	*Ulmus pumila*	Leaf rolls and swellings [37].
*Eriosoma usuense*Zhang & Qiao	Xinjiang	*Ulmus pumila*	Leaf spirally twisted [37].
Kaltenbachiella glabra Akimoto	Taiwan	*Ulmus uyematsui*	Almost globular gall, without hairs, arising upward from the midrib of leaves [40].
Tetraneura (Indotetraneura) asymmachia Zhang & Zhang	Liaoning	*Ulmus parvifolia, U. pumila*	Leaf gall, irregular pouch-like, stalked, smooth, with secondary exit laterally [37].
Tetraneura (Tetraneura) aequiunguis Zhang & Zhang	Inner Mongolia, Shandong	*Ulmus* sp.	Leaf gall [41].

*Tetraneura (Tetraneura) persicina*Zhang & Zhang	Liaoning, Shandong	*Ulmus pumila*	Leaf gall, peach-shaped, stalked [37].
*Tetraneura (Tetraneura) triangula*Zhang & Zhang	Hebei, Shandong, Tianjin	*Ulmus pumila*	Irregular-shaped gall on the upper side of leaves [37].
Tetraneura (Tetraneura) ulmicema Zhang	Ningxia	*Ulmus glabra*	Leaf gall, smooth, without hairs [37].
Eriosomatinae Fordini			
Chaetogeoica foliodentata (Tao)	Jiangsu, Shaanxi, Shandong, Sichuan, Yunnan, Zhejiang	*Pistacia chinensis, P. weinmannifolia*	Cockscomb-like gall, arising from the midrib on the upper surface of leaves [37].
Chaetogeoica ovagalla (Zhang)	Shandong	*Pistacia chinensis*	Elongate oval gall, arising from the midrib of leaves [42].
Chaetogeoica sensucopia (Zhang)	Shandong	*Pistacia chinensis*	Small elongate oval gall, arising from the midrib of leaves [42].
Chaetogeoica ulmidrupa Zhang	Shandong	*Pistacia aculeata*	Dimidiate gall, with a longitudinal ridge [43].
Chaetogeoica yunlongensis (Zhang & Zhong)	Shaanxi, Yunnan	*Pistacia chinensis*	Cucumber-shaped gall, arising from the midrib of leaves [42].

Floraphis choui Xiang	Shaanxi	*Rhus potaninii*	Gall produced on leaflets, branched with every branch obconical [44].
*Floraphis meitanensis*Tsai & Tang	Guizhou, Hunan, Sichuan	*Rhus punjabensis* var.*sinica*	Gall produced on the main axis of a compound leaf, flattened, branched from base [37, 45].
Kaburagia rhusicola ensigallis (Tsai & Tang)	Guizhou, Hubei, Hunan, Sichuan	*Rhus punjabensis* var.*sinica*	Elongate jujube-like gall, with a pointed tip, arising from the main vein or secondary vein on the base of leaves [37].
Kaburagia rhusicola ovatirhusicola Xiang	Hubei, Shaanxi, Yunnan	*Rhus potaninii*	Gall obovate, outer surface with prominent net-like veins, produced on leaflet [44].
Kaburagia rhusicola ovogallis (Tsai & Tang)	Guizhou, Hubei, Hunan, Sichuan	*Rhus chinensis, R. punjabensis* var. *sinica*	Gall ovate, tip rounded, with short hairs, arising from the secondary vein on the base of leaves [37].

*Meitanaphis elongallis*Tsai & Tang	Guizhou, Hubei, Hunan	*Rhus punjabensis* var.*sinica*	Gall jujube-like, outer surface with many fine longitudinal ridges, arising from the midrib on the lower surface of leaves [37, 45].
Schlechtendalia peitan(Tsai & Tang)	Guizhou, Hunan, Sichuan	*Rhus chinensis*, *R. chinensis* var. *roxburghii*,*R. punjabensis* var. *sinica*	Gall ovate, somewhat flattened, with short soft hairs, arising from the secondary vein on the base of leaves [37].
Eriosomatinae Pemphigini			
Epipemphigus chomoensis (Zhang)	Tibet	*Populus* sp.	Gall cockscomb-like, outer surface uneven with 3 small protuberances, situated along the midrib on the upper surface of leaves, with a primary exit on the lower surface [46].

Epipemphigus yunnanensis (Zhang)	Guizhou, Yunnan	*Populus bonatii, P. cathayana, P. yunnanensis*	Gall cockscomb-like, outer surface uneven with 3–5 acute angled protuberances, situated along the midrib on the upper surface of leaves, with a primary exit on the lower surface [46].
Pachypappella aliquipila Zhang	Hebei	*Populus* sp.	Leaf folds [37].
*Pemphigus circellatus*Zhang & Zhong	Yunnan	*Populus tremula* var.*davidiana, P. yunnanensis*	Big opened gall on the branch [47].
*Pemphigus cylindricus*Zhang	Tibet	*Populus* sp.	Cylinder-shaped leaf gall [48].
Pemphigus mangkamensis Zhang	Tibet	*Populus* sp.	Cockscomb-like gall on the upper side of leaves [48].
Pemphigus sinobursarius Zhang	Heilongjiang, Inner Mongolia, Liaoning, Ningxia, Yunnan	*Populus cathayana, P. euphratica, P. simonii*	Round pouch-like gall, arising from the base of midrib on the under surface of leaves [46].
*Pemphigus tibetensis*Zhang	Beijing, Gansu, Hebei, Tibet, Xinjiang	*Populus cathayana*	Pomegranate-like gall, with thick wall and a primary exit, produced on the twig [46].

*Pemphigus turritus*Zhang	Gansu	*Populus purdomii*	Sharp chilli-like or sharp horn-like gall, arising from the midrib on the upper surface of leaves, with a primary exit on the lower surface [37, 49].
*Pemphigus wuduensis*Zhang	Gansu	*Populus purdomii*	Flattened globular gall, arising from the base of midrib on the upper surface of leaves, with a secondary exit [37, 49].
*Pemphigus yangcola*Zhang	Yunnan	*Populus yunnanensis*	Semispherical gall on the twig, smooth, with primary exit [46].
Prociphilus (Prociphilus) gambosae Zhang & Zhang	Hebei	*Syringa oblata*	Leaf curls [50].
Prociphilus (Prociphilus) ligustrifoliae (Tseng & Tao)	Guizhou, Shaanxi, Sichuan, Yunnan	*Ligustrum japonicum, L. lucidum*	Leaf curls [51].
Thecabius (Oothecabius) sequelus Zhang	Hebei, Xinjiang	*Populus simonii*	Leaf folds and swellings along the midrib, dumpling-shaped [52].
Thecabius (Parathecabius) zhongi Zhang	Gansu	*Populus cathayana*	Dumpling-shaped pseudogall, formed by folding of the two halves of a leaf together [52].

Thecabius (Thecabius) beijingensis Zhang	Beijing, Hebei, Heilongjiang, Liaoning, Shanxi	*Populus beijingensis, P. cathayana, P. koreana*	Leaf folds and rolls up along the midrib, forming a swollen sausage-shaped pseudogall [52].
Hormaphidinae			
Aleurodaphis sinojackiae Qiao & Jiang	Jiangsu, Zhejiang	*Sinojackia xylocarpa*	Leaf curls, boat-shaped [53].
Asiphonipponaphis vasigalla Chen, Sorin & Qiao	Hunan	*Distylium chinense*	Vase-shaped gall, arising from or near the midrib on the upper surface of leaves, with a flower-shaped opening at the tip when mature [54].
Cerataphis jamuritsu(Takahashi)	Hong Kong, Taiwan	*Styrax suberifolia*	Large bell-shaped gall on the branch, covered with much white wax [55].
Hamamelistes similibetulae (Qiao & Zhang)	Tibet	*Betula albosinensis*	Small conical gall on the upper side of leaves [56].
*Ktenopteryx eosocallis*Qiao & Zhang	Fujian, Guangxi	*Styrax odoratissima*	Leaf deformation [57].
Tuberaphis cymigalla(Qiao & Zhang)	Fujian	*Distylium racemosum*(unlikely, requiring further confirmation)	Branched gall on the twig [58].
Tuberaphis viscisucta(Zhang)	Yunnan	*Viscum album*	Leaf curls [59].

Aphidinae			
*Cryptosiphum artemisiae linanense*Zhang	Gansu, Hebei, Heilongjiang, Jilin, Liaoning, Zhejiang	*Artemisia argyi, A. atrovirens, A. mongolica,A. selengensis*	Leaf rolls, fist-like [60].
Cryptosiphum atriplicivorum Zhang	Gansu	*Artemisia* sp.	Leaf rolls and swellings [39].
Neorhopalomyzus lonicerisuctus Zhang, Zhong & Zhang	Sichuan, Yunnan	*Lonicera* sp.	Leaf curls [61].
*Sappaphis dipirivora*Zhang	Beijing	*Pyrus* sp.	Leaf rolls and swellings [62].
*Sappaphis sinipiricola*Zhang	Hebei, Henan	*Pyrus betulaefolia*	Leaf rolls [62].
*Tuberocephalus (Trichosiphoniella) tianmushanensis*Zhang	Beijing, Zhejiang	*Prunus pauciflora*	Leaf curls, cylinder-shaped [60].
Tuberocephalus (Trichosiphoniella) tuberculus Su, Jiang & Qiao	Sichuan	An unidentified species of Rosaceae	Pseudogall along the lateral vein of young leaves [63].
Chaitophorinae			
Periphyllus acerihabitans Zhang	Jiangsu, Zhejiang	*Acer buergerianum*	Leaf curls [64].

GALL MORPHOLOGICAL DIVERSITY

Type

Aphids can induce both pseudogalls and true galls on their host plants. Pseudogalls are commonly found on leaves and appear as leaf folds, leaf rolls, leaf curls, or leaf blisters. In the subfamilies Aphidinae, Chaitophorinae, Mindarinae, Myzocallidinae, Phyllaphidinae, and Thelaxinae, only pseudogalls are produced. Hyalopterus pruni (Geoffroy) produces noticeable leaf-roll pseudogalls on Prunus armeniaca in the spring (Figure 1(n)). Cryptosiphum artemisiaeBuckton causes leaves to swell and roll on Artemisia (Figure 1(o)). Different from pseudogalls, true galls are more diverse in shape, complex in structure, and not only located on the leaves. The Adelgidae and Phylloxeridae induce true galls on their host plants. In the Eriosomatinae and Hormaphidinae, both pseudogalls and true galls are formed. A species of Hormaphidinae,Tuberaphis takenouchii (Takahashi), forms broccoli-head-like galls on the twigs of Styrax formosana, found in Taiwan. However, its congeneric species T. viscisucta (Zhang), which occurs in Yunnan, feeds on the leaves of Viscum album and causes the leaf edges to curl downward.

(a)

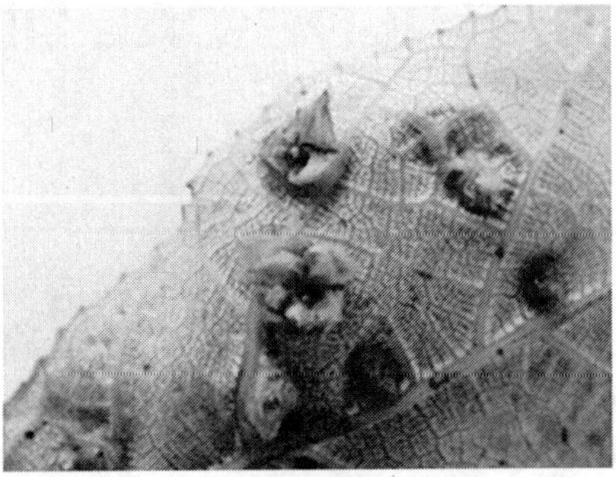

(b)

(c)

(d)

(e)

(f)

(g)

(h)

(i)

(j)

(k)

(l)

(m)

(n)

(o)

Figure 1: (a) A gall of Adelges sp. on Picea crassifolia (Huangzhong, Xining, Qinghai, China; 8 July 2009): pineapple-like, multiple-cavity, closed. (b) Galls of Acanthochermes similiquercus Jiang, Huang & Qiao on a leaf of Quercus sp. (Baoding, Panzhihua, Sichuan, China; 15 April 2005): spherical, single-cavity, some matured ones are open. (c) Galls ofTetraneura (Tetraneurella) nigriabdominalis (Sasaki) on a leaf of Ulmussp. (Hulun Buir, Inner Mongolia, China; 1 August 2004): spindle-shaped, single-cavity, closed. (d) A pseudogall of Epipemphigus imaicus(Cholodkovsky) on Populus cathayana (Shangri-La, Yunnan, China; 27 July 2010): silkworm-like, single-cavity, open on the undersurface of the leaf. (e) A gall of Pemphigus bursarius (Linnaeus) on a petiole of Populus simonii (Miyun, Beijing, China; 25 July 2009): bag-like, single-cavity, with a natural lateral opening. (f) A gall of Pemphigus immunis Buckton on a twig of Populus sp. (Miyun, Beijing, China; 25 July 2009): long saccate, single-cavity, closed. (g) A dumpling-shaped leaf-fold pseudogall ofThecabius (Oothecabius) populi (Tao) on Populus ussuriensis (Miyun, Beijing, China; 25 July 2009). (h) A gall of Chaetogeoica foliodentata (Tao) on the leaf midrib of Pistacia weinmannifolia (Qishan, Baoji, Shaanxi, China; 12 July 2004): cockscomb-like, single-cavity, closed. (i) A gall of-Tuberaphis owadai Kurosu & Aoki on a twig of Styrax japonica(Yuanyang, Kunming, Yunnan, China; 10 June 2009): coral-like, single-cavity, with many small openings on the projections. (j) Galls ofAsiphonipponaphis vasigalla Chen, Sorin & Qiao on the leaf midribs ofDistylium chinense (Jishou, Hunan, China; 15 April 2010): vase-shaped, single-cavity, open on the bottom. (k) A gall of Metanipponaphis sp. on a petiole of Distylium chinense (Jishou, Hunan, China; 6 September 2009): global, single-cavity, closed. (l) Galls of Neothoracaphis yanonis(Matsumura) on the leaves of Distylium chinense (Jishou, Hunan, China; 21 May 2009): spherical with a pointed bottom, protruding from both sides of the leaves, single-cavity, closed. (m) A gall of Nipponaphis sp. on a twig of Distylium chinense (Jishou, Hunan, China; 5 September 2009): bottle-shaped, single-cavity, closed. (n) Leaf rolls caused by Hyalopterus pruni (Geoffroy) on Prunus armeniaca (Huangzhong, Xining, Qinghai, China; 8 July 2009). (o) Leaf rolls and swellings caused by Cryptosiphum artemisiae Buckton on Artemisia sp. (Miyun, Beijing, China; 25 July 2009).

Site

The fundatrix (the first spring generation that hatches from overwintering eggs) of each galling aphid species induces its gall at a specific site on a specific host [26]. Galls may occur on the leaf blade, the leaf vein, the petiole, the main axis of a compound leaf, twigs, branches, and roots. Viteus vitifoliae (Fitch) causes galls on grape leaves and gall-like swellings on grape roots. Pemphigus bursarius (Linnaeus) produces galls on the petioles of Populus simonii(Figure 1(e)). The galls of Floraphis meitanensis Tsai & Tang are located on the main axis of the compound leaf of Rhus punjabensis var. sinica. Among these galls, the leaf gall is the most common type.

The galling site is closely related to the sink strength of galls and therefore has important effects on the reproductive success of the fundatrix [18, 27–29]. Fundatrices compete for galling sites. Intraspecific fights have been reported in different aphid species [28, 30, 31]. Furthermore, one plant species can host several galling aphid species. These species coexist by attacking different organs of a single host plant. In China, Distylium chinense harbors at least four species of Hormaphidinae, that is, Asiphonipponaphis vasigalla Chen, Sorin &

Qiao,Neothoracaphis yanonis (Matsumura), Metanipponaphis sp., and Nipponaphis sp.. They form specific galls on the leaf midrib (Figure 1(j)), the leaf blade (Figure 1(l)), the petiole (Figure1(k)), and the twig (Figure 1(m)), respectively. This observation indicates an adaptive radiation of galling aphids that exploit different ecological niches within a host plant.

Size

Gall size is also highly variable. Some aphids make very small galls. For example, the galls ofAcanthochermes similiquercus Jiang, Huang & Qiao (Figure 1(b)) are small bean sized. Each gall contains only one aphid. In contrast, some galls are extremely large. The galls ofCeratoglyphina styracicola (Takahashi) in Taiwan can reach a diameter of 9 cm and contain 100,000 aphids [32]. It is obvious that gall size is directly linked to aphid colony size. Like the galling site, size is also an important factor influencing the gall sink strength [18, 29]. The species of Hormaphidinae on Distylium chinense serve to illustrate this principle. Large galls of Nipponaphis sp. (Figure 1(m)) divert nutrients from neighboring shoots. The smaller galls of Metanipponaphis sp. (Figure 1(k)) divert nutrients only from neighboring leaves on the same shoot. The smallest galls of Neothoracaphis yanonis (Figure 1(l)) use only the leaf on which they are located.

Shape

The shapes of aphid galls vary and have been described as chilli-like, cockscomb-like, bag-like, spherical, cucumber-like, flower-like, boat-like, broccoli-head-like, banana-bundle-shaped, and so on. Gall shape is highly species specific. Galls produced by individuals of the same species are distinctively similar in shape [18]. Chaetogeoica foliodentata (Tao) produces cockscomb-like galls on Pistacia weinmannifolia (Figure 1(h)). Epipemphigus imaicus(Cholodkovsky) forms silkworm-like galls on Populus cathayana (Figure 1(d)). Different species induces different shaped galls even on the same organ of the same host plant. On the leaves of Distylium chinense, for example, Asiphonipponaphis vasigalla forms vase-shaped galls (Figure 1(j)), whereas the galls of Neothoracaphis yanonis are spherical with a pointed bottom and protrude from both sides of the leaves (Figure 1(l)). This observation suggests that gall shape is determined by the aphids rather than by the plants and gall is an extended phenotype of the aphids [20]. Fundatrix behavior during gall initiation is assumed to play a decisive role in regulating gall shape, especially complex and peculiar shapes [18, 29].

Structure

Based on the number of cavities, galls can be divided into single-cavity and multiple-cavity types. The single-cavity gall is the simplest and most common type of aphid gall. The multiple-cavity gall is composed of several subgalls with more complex ontogeny and structure [24, 65]. The Adelgidae and many galling species in the tribe Cerataphidini of Hormaphidinae produce multiple-cavity galls (see [65]). The pineapple-like galls of Adelges sp. (Figure 1(a)) are formed from spruce buds in which the developing leaves enlarge laterally and the margins merge into each other, forming multiple chambers within which the aphids feed [66]. Galls can also be categorized into two types of structure: closed and open. Some galls are initially closed and do not open until maturity. However, other galls remain open during their entire period of development. True galls include both closed and open types. In Shandong, the galls of Tetraneura (Tetraneurella) nigriabdominalis (Sasaki) (Figure 1(c)) are closed from early May through early June and then split laterally. From these slits, the alatae leave the gall to found colonies on the roots of Gramineae. However, the galls of Pemphigus bursarius have a natural lateral opening over an extended period from late May through September (Figure 1(e)). In contrast, pseudogalls are generally open and relatively simple in structure. On Populus ussuriensis, Thecabius (Oothecabius) populi (Tao) causes the leaves to fold downward, forming open dumpling-shaped pseudogalls (Figure 1(g)).

Gall shape and structure are closely related to gall fitness, the number of offspring produced within a gall. Together with gall size, gall shape and structure determine the inner surface area of the gall, the most accurate and practical measure of gall fitness [67]. The inner surface area determines the feeding area available to the nymphs within the gall and is thus correlated with the aphid colony size. It would be interesting to investigate the possibility that aphid galls tend to be more complex in shape and structure (e.g., multiple-cavity galls) to support larger colonies. Aphid galls show great diversity of type, site, size, shape, and structure. However, these gall traits are also highly species specific. Therefore, they help to identify species, especially species that are difficult to distinguish morphologically [68]. Knowledge of the evolutionary driving forces behind the divergence of gall morphology is limited. Natural enemies [6, 7, 14, 21], competition for galling sites [26], and increasing sink strength [8, 10, 11] might have influenced the divergence of gall traits. The great variety of galls is hypothesized to be related to aphid adaptive radiation [5, 69]. Several case studies of selected aphid groups have been conducted to investigate how gall diversification has influenced aphid speciation. The gall-inducing aphid group Fordini is supposed to have radiated primarily by using different sites on the same plant organ and by enlarging the gall size [8, 10]. Its sister group, the gall-inducing Pemphigini, is assumed to have diversified by attacking different plant organs [11].

HOST ASSOCIATION AND HOST SPECIFICITY

Gall-bearing host plants are richly abundant in China. These gall-bearing plants belong to at least 64 genera and 27 families. The common gall-bearing plants include Picea, Ulmus,Populus, Styrax, and Distylium (Table 1). Gall-inducing aphids are generally heteroecious, alternate between primary host plants (woody), where galls are produced and the sexual phase of life cycle is completed, and secondary host plants (herbaceous) where only parthenogenetic generations occur. Some aphids, however, are capable of inducing galls on secondary host plants, for example, Hamamelistes similibetulae on Betula albosinensis and Cryptosiphum artemisiae on Artemisia.

Galling aphids are strictly associated with their primary hosts. The host specificity is well defined and represented particularly in the Adelgidae, Eriosomatinae, and Hormaphidinae (Table 1). The pattern of host association differs among aphid lineages. The Adelgidae are restricted to Picea (Pinaceae). The subfamily Eriosomatinae includes three host-specific tribes: Eriosomatini on Ulmus (Ulmaceae), Fordini on Pistacia and Rhus (Anacardiaceae) (Fordina on Pistacia, Melaphidina on Rhus, resp.), and Pemphigini on Populus (Salicaceae).

The Hormaphidinae include three tribes: Cerataphidini on Styrax (Styracaceae), Hormaphidini onHamamelis (Hamamelidaceae), and Nipponaphidini on Distylium (Hamamelidaceae). But the galling species of Hormaphidini in China are restricted to their secondary host Betula(Betulaceae) owing to the absence of primary host in their distribution areas. In Aphidinae, the association is not so rigid. Most gall-inducing species are primarily associated with Rosaceae. Some other remotely related plant species, such as Lonicera (Caprifoliaceae) and Ribes(Saxifragaceae), are also occupied.

The association of galling aphids with their specific host plants is supposed to be ancient [70]. One hypothesis about what plays an important role in shaping aphid-host plant association is "fundatrix specialization" [71–74]. This hypothesis suggests that the fundatrix is highly specialized to the ancestral host and is less able to acquire new hosts than are other morphs over long evolutionary periods. It should be noted that, with few exceptions, the fundatrix, the morph that hatches from the overwintering egg on the primary host in the spring, is the only morph that can induce a gall [18, 29]. Thus, the phylogenetic constraints on the fundatrix would in some degree explain the high host specificity of galling aphids to their primary hosts.

Aphids have very intimate associations with their host plants. The aphids obtain food resources and habitat from their hosts. It is commonly assumed that host plants have a great influence on aphid diversification [75]. Two of the major hypothesized pathways of diversification in phytophagous insects are cospeciation with host plants [76, 77] and speciation through host shift [78]. Both hypotheses rely on some degree of host specialization and suggest that host plant diversity plays a major role in insect diversification [75, 79]. In particular, galling aphids are closely linked with their host plants and, as a whole, occupy a wide range of plant species. Primary host specificity is strong. Different aphid families, subfamilies, or tribes are strictly associated with different plant families or genera. We think that the high host plant diversity should have considerable influence on the diversification of galling aphids and the adaptation to different host plants in different galling lineages might have driven the divergence of galling aphids at higher taxonomic levels.

CONCLUSIONS

This paper has presented a preliminary review of gall-inducing aphids in China. The major groups of galling aphids were surveyed. The great diversity of gall morphology was analyzed systematically according to type, site, size, shape, and structure. These traits are highly species specific, and most are closely related to gall fitness. The host plants bearing galls are rich in species, and strong primary host specificity was found. We suggest that host association and varied gall morphology might jointly have driven the diversification and adaptive radiation of galling aphids.

ACKNOWLEDGMENTS

The work was supported by the National Natural Sciences Foundation of China (Grant no. 30830017), National Science Funds for Distinguished Young Scientists (no. 31025024), National Science Fund for Fostering Talents in Basic Research (no. J0930004), and a grant (no. O529YX5105) from the Key Laboratory of the Zoological Systematics and Evolution of the Chinese Academy of Sciences.

REFERENCES

1. A. F. G. Dixon, Aphid Ecology: An Optimization Approach, Chapman & Hall, London, UK, 2nd edition, 1998.
2. R. E. F. Matthews, "Relationships between plant viruses and invertebrates," in Plant Virology, R. E. F. Matthews, Ed., pp. 520–561, Academic Press, New York, NY, USA, 3rd edition, 1991.
3. D. G. Pollard, "Plant penetration by feeding aphids (Hemiptera, Aphidoidea): a review,"Bulletin of Entomological Research, vol. 62, pp. 631–714, 1973.
4. L. L. Walling, "The myriad plant responses to herbivores," Journal of Plant Growth Regulation, vol. 19, no. 2, pp. 195–216, 2000.
5. J. D. Shorthouse, D. Wool, and A. Raman, "Gall-inducing insects—Nature's most sophisticated herbivores," Basic and Applied Ecology, vol. 6, no. 5, pp. 407–411, 2005.

6. P. W. Price, G. L. Waring, and G. W. Fernandes, "Hypotheses on the adaptive nature of galls," Proceedings of the Entomological Society of Washington, vol. 88, pp. 361–363, 1986.
7. P. W. Price, G. W. Fernandes, and G. L. Waring, "Adaptive nature of insect galls,"Environmental Entomology, vol. 16, pp. 15–24, 1987.
8. M. Inbar, M. Wink, and D. Wool, "The evolution of host plant manipulation by insects: molecular and ecological evidence from gall-forming aphids on Pistacia," Molecular Phylogenetics and Evolution, vol. 32, no. 2, pp. 504–511, 2004.
9. Y. Koyama, I. Yao, and S. I. Akimoto, "Aphid galls accumulate high concentrations of amino acids: a support for the nutrition hypothesis for gall formation," Entomologia Experimentalis et Applicata, vol. 113, no. 1, pp. 35–44, 2004.
10. H. C. Zhang and G. X. Qiao, "Molecular phylogeny of Fordini (Hemiptera: Aphididae: Pemphiginae) inferred from nuclear gene EF-1α and mitochondrial gene COI," Bulletin of Entomological Research, vol. 97, no. 4, pp. 379–386, 2007.
11. H. C. Zhang and G. X. Qiao, "Systematic status of the genus Formosaphis Takahashi and the evolution of galls based on the molecular phylogeny of Pemphigini (Hemiptera: Aphididae: Eriosomatinae)," Systematic Entomology, vol. 32, no. 4, pp. 690–699, 2007.
12. E. P. Felt, Plant Galls and Gall Makers, Comstock Publishing Company, Ithaca, NY, USA, 1940.
13. D. G. Miller, C. T. Ivey, and J. D. Shedd, "Support for the microenvironment hypothesis for adaptive value of gall induction in the California gall wasp, Andricus quercuscalifornicus," Entomologia Experimentalis et Applicata, vol. 132, no. 2, pp. 126–133, 2009.
14. H.V. Cornell, "The secondary chemistry and complex morphology of galls formed by the Cynipinae (Hymenoptera): why and how?" American Midland Naturalist, vol. 110, no. 2, pp. 225–234, 1983.
15. W. A. Foster and P. A. Northcott, "Galls and the evolution of social behaviour in aphids," in Plant Galls: Organisms, Interactions, Populations, M. A. J. Williams, Ed., pp. 161–182, Clarendon Press, Oxford, UK, 1994.

16. D. L. Stern and W. A. Foster, "The evolution of soldiers in aphids," Biological Reviews of the Cambridge Philosophical Society, vol. 71, no. 1, pp. 27–79, 1996.
17. S. Chakrabarti, "Diversity and biosystematics of gall-inducing aphids (Hemiptera: Aphididae) and their galls in the Himalaya," Oriental Insects, vol. 41, pp. 35–54, 2007.
18. D. Wool, "Galling aphids: specialization, biological complexity, and variation," Annual Review of Entomology, vol. 49, pp. 175–192, 2004.
19. R. Dawkins, The Extended Phenotype: The Gene as the Unit of Selection, Oxford University Press, Oxford, UK, 1982.
20. D. L. Stern, "Phylogenetic evidence that aphids, rather than plants, determine gall morphology," Proceedings of the Royal Society B, vol. 260, no. 1357, pp. 85–89, 1995.
21. G. N. Stone and K. Schönrogge, "The adaptive significance of insect gall morphology,"Trends in Ecology and Evolution, vol. 18, no. 10, pp. 512–522, 2003.
22. Z. Liu, X. L. Huang, L. Y. Jiang, and G. X. Qiao, "The species diversity and geographical distribution of aphids in China (Hemiptera, Aphidoidea)," Acta Zootaxonomica Sinica, vol. 34, no. 2, pp. 277–291, 2009.
23. H. C. Zhang, G. X. Qiao, and G. X. Zhang, "A study on diversity of aphid's galls in Pemphigidae," Acta Zootaxonomica Sinica, vol. 31, no. 1, pp. 48–54, 2006.
24. J. Chen and G. X. Qiao, "A study on diversity of aphid's galls of Hormaphidinae," Acta Zootaxonomica Sinica, vol. 34, no. 2, pp. 269–276, 2009.
25. G. Remaudière and M. Remaudière, Catalogue of the World's Aphididae, Institut National de la Recherche Agronomique, Paris, France, 1997.
26. M. Inbar and D. Wool, "Phloem-feeding specialists sharing a host tree: resource partitioning minimizes interference competition among galling aphid species," Oikos, vol. 73, no. 1, pp. 109–119, 1995.
27. T. G. Whitham, "Territorial behaviour of Pemphigus gall aphids," Nature, vol. 279, no. 5711, pp. 324–325, 1979.

28. T. G. Whitham, "Costs and benefits of territoriality: behavioral and reproductive release by competing aphids," Ecology, vol. 67, no. 1, pp. 139–147, 1986.
29. D. Wool, "Gall-inducing aphids: biology, ecology, and evolution," in Biology, Ecology, and Evolution of Gall-Inducing Arthropods, A. Raman, C. W. Schaefer, and T. M. Withers, Eds., pp. 73–132, Science Publishers, New Hampshire, USA, 2005.
30. S. Aoki and S. Makino, "Gall usurpation and lethal fighting among fundatrices of the aphid Epipemphigus niisimae (Homotera, Pemphigidae)," Kontyû, vol. 50, pp. 365–376, 1982.
31. M. Inbar, "Competition, territoriality and maternal defense in a gall-forming aphid,"Ethology Ecology and Evolution, vol. 10, no. 2, pp. 159–170, 1998.
32. S. Aoki, "Further observations on Astegopteryx styracicola (Homoptera: Pemphigidae), an aphid species with soldiers biting man," Kontyû, vol. 47, no. 2, pp. 99–104, 1979.
33. G. X. Zhang, T. S. Zhong, and Z. J. Tian, "Two new species and a new subspecies of Adelgidae from Sichuan, China (Homoptera: Adelgidae)," Zoological Research, vol. 1, no. 3, pp. 381–388, 1980.
34. P. Du, F. Ren, and Y. Zhang, "Preliminary research on biological characteristics and control of Sacchiphates roreigallis Li et Tsai," Journal of Gansu Agricultural University, vol. 33, no. 3, pp. 294–301, 1998.
35. Y. H. Li, K. L. Xie, K. G. Cao, Y. H. Gan, and C. W. Luo, "A study on the biology of thePineus armandicola (Homoptera: Adelgidae)," Scientia Silvae Sinicae, vol. 38, no. 1, pp. 82–87, 2002.
36. L. Y. Jiang, X. L. Huang, and G. X. Qiao, "Review of the Chinese new record genusAcanthochermes Kollar (Hemiptera: Phylloxeridae), with a description of one new species," Pan-Pacific Entomologist, vol. 85, no. 2, pp. 43–50, 2009.
37. G. X. Zhang, G. X. Qiao, T. S. Zhong, and W. Y. Zhang, Fauna Sinica, Insecta Vol. 14, Homoptera: Mindaridae and Pemphigidae, Science Press, Beijing, China, 1999.
38. G. X. Zhang and W. Y. Zhang, "Studies on Chinese species of Eriosoma Leach with descriptions of new species (Homoptera: Pemphigidae)," Sinozoologia, vol. 10, pp. 143–152, 1993.

39. G. X. Zhang, X. L. Chen, G. X. Qiao, T. S. Zhong, and J. H. Li, Fauna of Agricultural and Forestry Aphids of Northwest China (Insecta: Homoptera: Aphidinea), China Environmental Science Press, Beijing, China, 1999.
40. S. Akimoto, "Taxonomic study on gall aphids, Colopha, Paracolopha andKaltenbachiella (Aphidoidea: Pemphigidae) in East Asia, with special reference to their origins and distributional patterns," Insecta Matsumurana, vol. 31, pp. 1–79, 1985.
41. G. X. Zhang, W. Y. Zhang, and T. S. Zhong, "Studies on the genus Tetraneura Hartig, 1841 from China (Homoptera: Pemphigidae) with descriptions of new species and subspecies," Sinozoologia, vol. 8, pp. 205–236, 1991.
42. G. X. Zhang, "A study on Baizongia Rondani, 1848 with descriptions of a new subgenus and two new species (Homoptera: Pemphigidae)," Entomologia Sinica, vol. 2, no. 4, pp. 291–299, 1995.
43. G. X. Zhang and G. X. Qiao, "Two new genera, five new species and one new subspecies of Fordinae (Homoptera: Pemphigidae) from China," Entomologia Sinica, vol. 5, no. 1, pp. 1–9, 1998.
44. H. Xiang, "Studies of Chinese gall-nut aphids on Rhus potaninii Maxim,"Entomotaxonomia, vol. 2, no. 4, pp. 303–313, 1980.
45. P. H. Tsai and C. Tang, "The classification of the Chinese gall aphids with description of three genera and six new species from Meitan, Kweichow," Transactions of the Royal Entomological Society of London, vol. 97, no. 16, pp. 405–418, 1946.
46. G. X. Zhang and T. S. Zhong, "Five new species of Pemphigus Hartig and a new species of Sanpupemphigus gen. nov. from China (Aphidoidea, Pemphigidae)," Acta Entomologica Sinica, vol. 22, no. 3, pp. 324–332, 1979.
47. G. X. Zhang and T. S. Zhong, "New species and new records of Pemphigidae from Yunnan, China (Homoptera: Aphidoidea)," Acta Entomologica Sinica, vol. 28, no. 1, pp. 94–96, 1985.
48. G. X. Zhang and T. S. Zhong, "Homoptera: Aphidoidea," Insects of Xizang, vol. 1, pp. 233–282, 1981.
49. G. X. Zhang and G. X. Qiao, "Nine new species of Pemphiginae (Homoptera: Pemphigidae) from China," Entomologia Sinica, vol. 4, no. 4, pp. 283–294, 1997.

50. G. X. Zhang, W. Y. Zhang, T. S. Zhong, and S. B. Tian, "On Chinese Prociphilus Koch and description of new species (Homoptera: Pemphigidae)," Acta Zoologica Sinica, vol. 39, no. 4, pp. 368–373, 1993.
51. S. Tseng and C. C. Tao, "New and unrecorded aphids of China," Journal of the West China Border Research Society, vol. 10, pp. 195–224, 1938.
52. G. X. Zhang, T. S. Zhong, and G. X. Qiao, "A study on Chinese Thecabius Koch 1857, with description of new subgenus, new species and new subspecies," Entomologia Sinica, vol. 2, no. 3, pp. 206–224, 1995.
53. L. Y. Jiang and G. X. Qiao, "A review of Aleurodaphis (Hemiptera, Aphididae, Hormaphidinae) with the description of one new species and keys to species," ZooKeys, vol. 135, pp. 41–56, 2011.
54. J. Chen, M. Sorin, and G. X. Qiao, "Asiphonipponaphis, a new genus of Hormaphidinae (Hemiptera, Aphididae) causing galls on Distylium chinense from China, with description of a new species," ZooKeys, vol. 111, pp. 1–10, 2011.
55. S. Aoki, U. Kurosu, T. Fukatsu, and H. Ishikawa, "Cerataphis jamuritsu, a subtropical aphid producing soldiers in large, hard galls (Homoptera)," Entomological Science, vol. 1, no. 3, pp. 327–333, 1998.
56. G. X. Qiao and G. X. Zhang, "Record of Hormaphis Osten-Sacken (Aphididae: Hormaphidinae) from China, with description of a new species," Oriental Insects, vol. 38, pp. 277–282, 2004.
57. G. X. Qiao and G. X. Zhang, "Ctenopteryx (Hemiptera: Hormaphidinae), a new genus from China, and description of a new species," Pan-Pacific Entomologist, vol. 79, no. 2, pp. 145–150, 2003.
58. G. X. Qiao and G. X. Zhang, "Study on new record genus Rappardiella Noordam with description of new species (Homoptera: Aphidoidea: Hormaphididae)," Acta Zootaxonomica Sinica, vol. 23, no. 3, pp. 289–292, 1998.
59. K. Guo, Systematic taxonomy of aphids of Hormaphididae from China (Homoptera: Aphidoidea), M.S. thesis, Shaanxi Normal University, 2005.

60. G. X. Zhang and T. S. Zhong, "New species and new subspecies of Chinese Macrosiphinae (I) (Homoptera: Aphididae)," Entomotaxonomia, vol. 2, no. 1, pp. 53–63, 1980.
61. X. M. Su, Systematics of Myzini from China (Hemiptera: Aphididae: Macrosiphinae), Ph.D. thesis, Institute of Zoology, Chinese Academy of Sciences, 2011.
62. L. Y. Jiang, G. X. Qiao, G. X. Zhang, and T. S. Zhong, Fauna of Agricultural and Forestry Aphids in Northeast China (Aphidinea, Hemiptera, Insecta), Science Press, Beijing, China, 2011.
63. X. M. Su, L. Y. Jiang, and G. X. Qiao, "Tuberocephalus (Hemiptera: Aphididae) from China with description of a new species," Oriental Insects, vol. 44, no. 1, pp. 157–191, 2010.
64. G. X. Qiao, Studies on the phylogeny of Hormaphididae, Thelaxidae, Chaitophoridae and Greenideidae etc. 7 families, vol. II, Ph.D. thesis, Institute of Zoology, Chinese Academy of Sciences, 1996.
65. S. Aoki and U. Kurosu, "A review of the biology of Cerataphidini (Hemiptera, Aphididae, Hormaphidinae), focusing mainly on their life cycles, gall formation, and soldiers," Psyche, vol. 2010, Article ID 380351, 34 pages, 2010.
66. O. Rohfritsch and M. Anthony, "Strategies in gall induction by two groups of homopterans," in Biology of Insect-Induced Galls, J. D. Shorthouse and O. Rohfritsch, Eds., pp. 102–117, Oxford University Press, Oxford, UK, 1992.
67. Y. Tosaka and T. Nishida, "Gall surface area is a simple and accurate measure of fitness in Nipponaphidini galling aphids (Homoptera: Aphididae)," Applied Entomology and Zoology, vol. 42, no. 2, pp. 217–221, 2007.
68. M. Sorin, "Aphids on the planted trees (4)," Shokubutsu Boeki, vol. 57, no. 10, pp. 480–483, 2003.
69. P. W. Price, "Adaptive radiation of gall-inducing insects," Basic and Applied Ecology, vol. 6, no. 5, pp. 413–421, 2005.
70. N. A. Moran, "A 48-million-year-old aphid-host plant association and complex life cycle: biogeographic evidence," Science, vol. 245, no. 4914, pp. 173–175, 1989.

71. A. K. Mordvilko, "The evolution of cycles and the evolution of heteroecy (migrations) in plant-lice," Annals and Magazine of Natural History, vol. 2, pp. 570–582, 1928.
72. G. C. Shaposhnikov, "The main features of the evolution of aphids," in Evolution and Biosystematics of Aphids. Proceedings of the International Aphidological Symposium at Jablonna, 1981, H. Szelegiewicz, Ed., pp. 19–99, Polska Academia Nauk, Warsaw, Poland, 1985.
73. N. A. Moran, "The evolution of host-plant alternation in aphids: evidence for specialization as a dead end," American Naturalist, vol. 132, no. 5, pp. 681–706, 1988.
74. N. A. Moran, "The evolution of aphid life cycles," Annual Review of Entomology, vol. 37, no. 1, pp. 321–348, 1992.
75. J. Peccoud, J. C. Simon, C. von Dohlen et al., "Evolutionary history of aphid-plant associations and their role in aphid diversification," Comptes Rendus Biologies, vol. 333, no. 6-7, pp. 474–487, 2010
76. C. Mitter, B. Farrell, and D. J. Futuyma, "Phylogenetic studies of insect-plant interactions: insights into the genesis of diversity," Trends in Ecology and Evolution, vol. 6, no. 9, pp. 290–293, 1991.
77. B. D. Farrell, C. Mitter, and D. J. Futuyma, "Diversification at the insect-plant interface: insights from phylogenetics," BioScience, vol. 42, no. 1, pp. 34–42, 1992.
78. T. Jermy, "Evolution of insect/host plant relationships," American Naturalist, vol. 124, no. 5, pp. 609–630, 1984.
79. I. S. Winkler and C. Mitter, "The phylogenetic dimension of insect-plant interactions: a review of recent evidence," in Specialization, Speciation, and Radiation: The Evolutionary Biology of Herbivorous Insects, K. J. Tilmon, Ed., pp. 240–263, University of California Press, Berkeley, Calif, USA, 2008.

Chapter 3

Crop Immunity against Viruses: Outcomes and Future Challenges

Valérie Nicaise

Fruit Biology and Pathology, Virology Laboratory, Institut National de la Recherche Agronomique, University of Bordeaux, UMR 1332, Villenave d'Ornon, France

ABSTRACT

Viruses cause epidemics on all major cultures of agronomic importance, representing a serious threat to global food security. As strict intracellular pathogens, they cannot be controlled chemically and prophylactic measures consist mainly in the destruction of infected plants and excessive pesticide applications to limit the population of vector organisms. A powerful alternative frequently employed in agriculture relies on the use of crop genetic resistances, approach that depends on mechanisms governing plant–virus interactions. Hence, knowledge related to the molecular bases of viral infections and crop resistances is key to face viral attacks in fields. Over the past 80 years, great advances have been made on our understanding of plant immunity against viruses. Although most of the known natural resistance genes have long been dominant R genes (encoding NBS-LRR proteins), a vast number of crop recessive resistance genes were cloned in the last decade, emphasizing another evolutive strategy to block viruses. In addition, the discovery of RNA interference pathways highlighted a very efficient antiviral system targeting the infectious agent at the nucleic acid level. Insidiously, plant viruses evolve and often acquire the ability to overcome the resistances employed by breeders. The development of efficient and durable resistances able to withstand the extreme genetic plasticity of viruses therefore represents a major challenge for the coming years. This review aims at describing some of the most devastating diseases caused by viruses on crops and summarizes current knowledge about plant–virus interactions, focusing on resistance mechanisms that prevent or limit viral infection in plants. In addition, I will discuss the current outcomes of the actions employed to control viral diseases in fields and the future investigations that need to be undertaken to develop sustainable broad-spectrum crop resistances against viruses.

INTRODUCTION

As obligatory intracellular parasites, plant viruses depend on the host machinery to multiply and invade their hosts. In their simplest form, viruses consist of a DNA or RNA genomic segment encoding only few genes and encapsided into a protein shell, called the capsid. Because of natural physical barriers (cuticle, cell wall), viruses are delivered into plant cells through wounds or through the action of vectors (insects, nematodes, fungi) that feed on or infect the plants. Following entry into a host cell and genome decapitation, the infectious cycle includes translation and replication of the viral genome, assembly of progeny virus particles, generalized invasion of the host through cell-to-cell and long-distance movements of viral particles or rib nucleoprotein complexes and finally, transmission to new hosts by vectors. In some cases, transmission to the following generation of host plant is also observed as a result of seed infection. In 2012, the International Committee on Taxonomy of Viruses reported 92 genera of plant viruses of which 82 were assigned in 21 different families (King et al., 2012).

Plant infection by viruses causes physiological disorders responsible for plant diseases of economic and agronomic significance in many crops. Widely employed in fields, the use of plant varieties carrying genetic resistances constitutes the most effective, economic and ecological measure to control viral infections. The last decades have seen substantial advances on the molecular dialog between viral pathogens and their plant hosts, bringing new strategies directly exploitable in crop improvement programs. Nevertheless, the spread of crop viral pests has increased dramatically in recent years. Globalization, trade and climate change, as well as reduced resilience in production systems due to decades of agricultural intensification have all played a part.

This review intends: (i) to provide a brief overview of severe virus-associated plant diseases and their impact on crop production, (ii) to summarize prophylactic strategies employed to control viral epidemics in fields, (iii) to bring an update on current knowledge about plant resistances against viruses, (iv) to present the technical approaches currently employed in crop improvement programs, and (v) to discuss how antiviral mechanisms based on PAMP-triggered immunity (PTI) could be the source of novel plant resistances in fields.

WHAT IS THE REAL IMPACT OF VIRAL DISEASES ON CROPS?

Management of plant virus diseases is a matter of vital importance and concern to farmers, horticulturists, foresters, manufacturers, as well as consumers. It is well-established that virus diseases in different crops cause enormous losses all over the world in terms of quantity and/or quality of products. Although it's very difficult to put a clear figure on the financial impact of plant viruses in agriculture, the yield losses that can be ascribed to plant viruses are estimated to cost worldwide more than $\$_{30}$ billion annually (Sastry and Zitter, 2014).

Amongst the most damaging virus-associated threats, strains of Cassava mosaic begomovirusescause more than 25 million tons of losses per year in Africa, India, and Sri Lanka (Legg and Thresh, 2000; Calvert and Tresh, 2002; Thresh and Cooter, 2005). Because the Cassava crop represents the daily staple for more than 500 million people all over the world, epidemics are often associated to famine events (Legg, 1999; Legg and Thresh, 2000; Calvert and Tresh, 2002). Potato leafroll polerovirus is responsible for an annual potato loss of $\$_{100}$ million in the US and $£_{30-50}$ millions in UK (Wale et al., 2008; Sastry and Zitter, 2014). The losses in citrus tree cultures attributed to Citrus Tristeza closterovirus (CTV) were estimated to over 100 million trees worldwide (Moreno et al., 2008; Harper, 2013). Barley yellow dwarf is the most widely distributed viral disease of cereals, affecting oats, rice, barley, maize, and wheat. It is caused by Barley yellow dwarf luteovirus that costs producers in UK alone about $£_{10}$ million a year in lost production (Ordon et al., 2009; Sastry and Zitter, 2014). It is estimated that the international costs of managing the Sharka disease [caused by Plum pox poty virus (PPV)] since the 1970s have exceeded 10 billion euros (Cambra et al., 2006). Viruses affecting rice cultures result in yield losses estimated at more than $1.5 billion in South-East Asia alone (Abo and Sy, 1998; Hull, 2013; Sasaya et al., 2013). Discovered in Ghana in the middle of the 20th century, Cacao swollen shoot badnavirus is currently endemic in Togo, Ghana, and Nigeria. Over 200 million trees have already been eradicated, representing the most costly effort of any country in the world against a viral plant disease (Dzahini-Obiatey et al., 2010).

It's important to mention that diseases of perennial and fruit crops not only lead to the loss of the crop but the loss of time and cost in bringing the trees to bearing, the losses of other crops that could have been grown on the land during that time, and the differences in the value of the land with and without a productive orchard. Moreover, virus-associated losses in fields are highly under-estimated as some viral infections are asymptomatic alone but contribute in a synergic manner to damages due to attacks from other pathogens (Hull, 2013). More importantly, viruses have been described causing half of the reported emerging infectious diseases from plants (Anderson et al., 2004). Hence, viruses represent serious crop threats, responsible for considerable agronomical losses, global food problem, and human life overall all over the world.

PROPHYLACTIC MEASURES IN FIELDS

Because viral agents are obligate intracellular parasites, curative treatments of virus infections are impossible, making viral diseases very difficult to control in fields. Prophylactic control measures are therefore crucial in combating epidemics on crops. They consist mainly on combining cultural practices, biosecurity measures and organism-vector management (Figure 1).

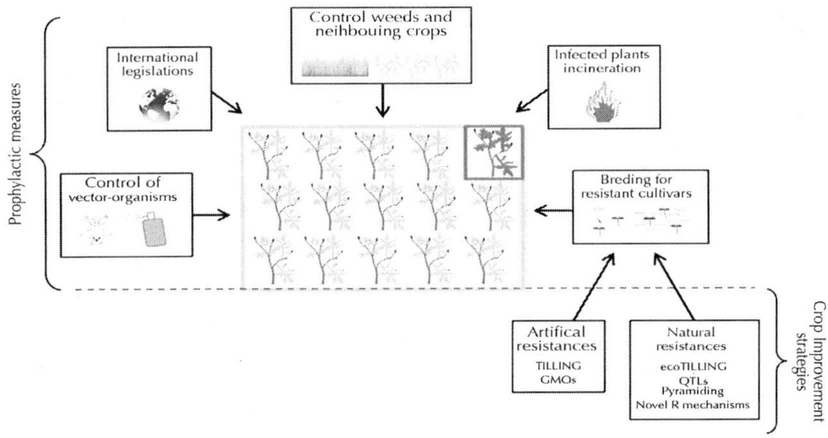

Figure 1: Prophylactic measures and main crop improvement strategies employed to control plant viral diseases.

Perform Regular Inspection for the Presence of Viral Pathogens

In this domain, the rising up of molecular biology techniques combined to continuous characterization of new etiological agents improved significantly the sensitivity, the specificity and the rapidity required to an accurate diagnosis of plant pathogenic viruses (Boonham et al., 2014). The reliability of the available diagnosis tests is a key point in viral disease management in fields, as infected plants need to be eradicated as fast as possible to minimize the virus spread.

Monitor Organism-Vector Populations

Plant viruses need to be transmitted by an organism-vector (insects, nematodes, zoosporic endoparasites) for their plant-to-plant spread. Hence, viral diseases can be efficiently controlled by limiting the populations of their vectors with the applications of appropriate pesticides. The use of non-host "trap plants" may be also considered to attract vectors to reduce the number of individuals feeding on the crop of interest and thus, the transmission of the disease (Bragard et al., 2013).

Set Up a Rigorous Control Program on Weeds and Other Host Plants in the Vicinity of the Field

Epidemics often arise from new viruses or new variants of classic viruses that spilled over from reservoir species to crops. Although this phenomenon results from a complex evolutionary process in which the main players are ecological factors, virus genetic plasticity and host factors, viral diseases can be controlled by managing the spatial structure and composition of field parcels, which impacts resistance durability (Elena et al., 2011; Fabre et al., 2012).

Respect the Phytosanitary Measures Decreed by Various International Commissions

Minimizing viral epidemics involves the respect of international legislations concerning worldwide trade of virus-free plant material, which applies to any development stage of a plant that can be carrier of viruses (seeds or fruit stones, grafts, rootstocks, seedlings, flowers,...), as well as manipulation of decontaminated horticultural tools.

Use Crop Cultivars That are Resistant to Viruses

The use of genetically resistant plants is one of the most efficient, sustainable and frequently employed strategies to control virus infections in fields. For centuries, it has involved plants selected by breeders for their agronomic proprieties combined to the absence of disease symptoms. However, from the middle of the 20th century, plant improvement programs capitalize strongly on the knowledge associated to plant–virus interactions to develop resistant varieties exploitable in agriculture.

WHAT DO WE KNOW ABOUT PLANT IMMUNITY AGAINST VIRUSES?

Faced with viral attacks, plants defend themselves through several resistance layers that are complementary in terms of defense timing (at early or late infection steps), location (in the first infected leaf or in systemic tissues) and targeting the virus-derived molecules (the viral genome or the viral proteins; Figure 2).

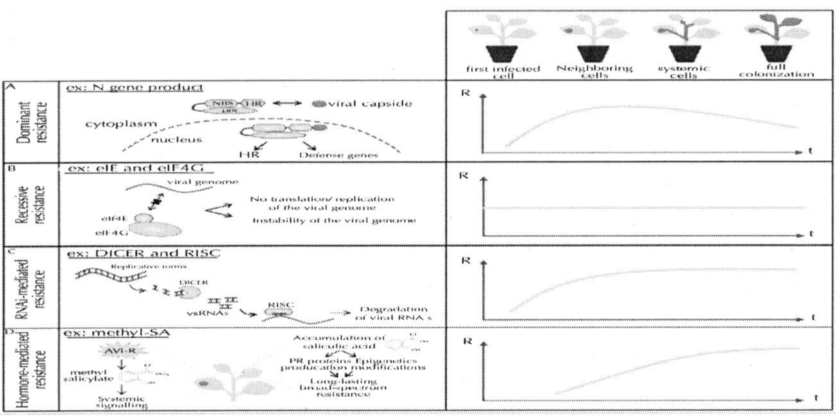

Figure 2: Known antiviral immune mechanisms in plants. Plant resistance mechanisms against viruses are complementary in terms of plant defense timing, location (from the first infected cell to the generalized colonization) and

targeting the virus-derived molecules (genome or proteins from viruses). **(A)** NBS-LRR dominant resistance relies on the interaction between an avirulence factor and a specific R gene product, and is effective several days after the virus entry into the plant. The HR-associated phenomenon confines the viral pathogen in the infected and neighboring cells. **(B)** Recessive resistance that corresponds to the absence of appropriate host factors that are required for the virus cycle is a non-inducible resistance, passive and effective throughout plant colonization. It confers resistance at the infection step that requires the cellular factor of interest. **(C)** RNA interference (RNAi) targets viral nucleic acids. Once set up after few days, the effectiveness of this defense mechanism increases and spreads to the whole plant through a relay-amplification process. **(D)**Hormone-mediated resistance against viral pathogens is represented here by the role of salicylic acid (SA) and methyl-salicylate (Me-SA) in systemic acquired resistance (SAR). On graphs, R, resistance level; t, infection timing.

Dominant Resistances

The majority of dominant resistance genes (R genes) identified in plant–virus interactions belong to the nucleotide binding site-leucine-rich repeat (NBS-LRR) class, that specifically recognize the viral avirulence (avr) gene products, through the establishment of the so-called "gene-for-gene" interaction. Although a direct physical interaction between the avr and R gene products had originally been suggested, the current understanding favors the more sophisticated "guard hypothesis" model (Soosaar et al., 2005). Many NBS-LRR proteins conferring resistance against viruses have been identified so far, and are classified on the basis of their N-terminal structure, that carries either a Toll–interleukin-1 receptor (TIR) domain or a coiled-coil (CC) domain (Moffett, 2009; De Ronde et al., 2014).

One of the best characterized example is the potato Rx1 gene that encodes a typical CC-NBS-LRR protein and mediates resistance to Potato potexvirus X (PVX) through the recognition of the PVX capsid (CP). Rx_1 CC domain has been shown to form a heterodimer with the cellular ranGTPase-activating protein 2 ($ranGAP_2$), interaction required for Rx_1 function (Rairdan et al., 2008; Tameling et al., 2010). Although direct interaction between $ranGAP_2$ and PVX CP has not been detected, it has been proposed that $ranGAP_2$ might nonetheless interact with the CP, causing a conformational change perceived by the rest of the protein and leading Rx to switch to an active state (Hao et al., 2013). In another well-characterized R gene example, the tobacco N gene product, that displays a TIR-NBS-LRR structure, interacts directly with the helicase domain of the replicase ofTobacco mosaic tobamovirus (TMV), in an ATP dependent manner (Figure 2A; Ueda et al., 2006). Full resistance to TMV requires the N receptor-interacting protein 1 ($NRIP_1$), which is recruited from the chloroplast to the cytoplasm and nucleus and interacts with both viral replicase and N factor (Caplan et al., 2008). In both Rx_1 and N cases, the R gene product is activated in the cytoplasmic compartment, while its nucleocytoplasmic distribution is required for full functionality (Slootweg et al., 2010; Tameling et al., 2010). The study of N- and Rx-mediated resistances have led to characterize the R signaling cascade in plant–virus interactions, that include rapid activation of MAP kinases and the action of a molecular chaperone complex composed of SGT_1 (Suppressor of G_2 allele of SKP_1), HSP90 (Heat shock protein), and RAR_1 (Required for Mla resistance 1; Azevedo et al., 2006; Botër et al., 2007; Kobayashi et al., 2010; Hoser et al., 2013), whose role seems to both stabilize R factors and mediate their degradation, thereby maintaining a tight cellular balance between defense signaling and attenuation (Kadota and Shirasu, 2012).

Immune events downstream of R protein activation are frequently associated with calcium ion influx, MAPK-mediated signaling, reactive oxygen species (ROS) production, salicylic acid (SA) accumulation, and extensive transcriptional reprogramming. In addition, the activation of R genes is most of the time associated with a hypersensitive response (HR), a phenomenon involving the programmed death of the infected and neighboring cells, with the consequences to confine the pathogen in the hypersensitive lesion and to prevent any further pathogen spread in the plant. A functional module that mediates HR against viruses (as well as non-viral pathogens) requires the interaction of two lipases, EDS_1 (Enhanced disease susceptibility 1) and PAD_4 (Phytoalexin deficient 4) with the protein SAG_{101} (Senescence-associated gene 101; Liu et al., 2002; Marathe et al., 2002; Wagner et al., 2013). In Arabidopsis, the $EDS_1/PAD_4/SAG_{101}$ complex regulates HRT-mediated resistance against Turnip crinkle carmovirus (Zhu et al., 2011). Recent insights into R protein downstream signaling indicate that the resistance process and the HR-programmed cell death are distinct physiological pathways, although both can work in concert (Bendahmane et al., 1999; Bai et al., 2012). Meanwhile R gene mediated defense is taking place locally, it also induces defense signaling in distally located tissues, called systemic acquired resistance (SAR), a mechanism that has been demonstrated for both N and Rx_1 genes (Delaney et al., 1994; Liu et al., 2010) and that is described in the Section "Plant Hormone-Mediated Resistance" of this review.

In contrast with the structure of the classical R genes, the RTM genes (for Restricted TEV Movement) were the first cloned non-NBS-LRR dominant resistance genes against viruses. Genetic characterization of natural Arabidopsis accessions and mutants showed that at least five dominantRTM genes are involved in resistance to Tobacco etch potyvirus (TEV), Lettuce mosaic potyvirus(LMV), and PPV (Cosson et al., 2012). It has been recently proposed that RTM members may form a phloem-resident multiprotein complex involved in the resistance mechanisms to block the long-distance movement of potyviruses (Cosson et al., 2010a, b). No induction of HR or production of SA has been detected, in contrast with NBS-LRR mediated resistance responses (Mahajan et al., 1998). Another example of a non-NLS-LRR dominant resistance conferring resistance to TMV comes from the tomato Tm-1 gene, which encodes a protein with a TIM-barrel like structure that interacts directly with the viral replicase strongly impairing the viral genome replication (Ishibashi and Ishikawa, 2013).

Resistance Recessives

Around one-half of the approximately 200 known resistance genes that target plant viruses are recessively inherited (Diaz-Pendon et al., 2004), suggesting that this form of resistance is more common for viruses than for other plant pathogens. The use of such genes is therefore a very important tool in breeding programs to control plant diseases caused by pathogenic viruses.

Recessive resistances are often achieved through the absence of appropriate host factors that are required for the virus to complete its biological cycle. Over the last decade, a large number of recessive resistance genes have been cloned from crop species and shown to encode eukaryotic initiation factors (eIFs) belonging to the eIF_4E and eIF_4G families (Figure 2B; Truniger and Aranda, 2009; Wang and Krishnaswamy, 2012; Julio et al., 2014; Revers and Nicaise, 2014). These factors belong to the cellular $_{43}S$ complex, that recruit both mRNAs and ribosomes before the process of translation (Pestova et al., 2001) and are hypothesized to act as susceptibility factors recruited during the replication/translation steps of the viral genome. Firstly highlighted for successful potyvirus infection, the role of eIF_4E and eIF_4G have been quickly extended to other plant virus families, involving bymoviruses, cucumoviruses, ipomoviruses, sobemoviruses, carmoviruses, and waikiviruses (Revers and Nicaise, 2014), suggesting that they contribute to a broad mechanism of plant susceptibility to viruses. An intriguing aspect of eIF4E- and eIF4G-mediated resistances is that they cover a diverse range of resistance phenotypes. Although in many investigated cases they govern complete qualitative resistance, they have also been shown in some instances to provide partial resistance or to be components of partial or polygenic resistances (Caranta et al., 1997; Nicaise et al., 2003; Acosta-Leal and Xiong, 2008; Charron et al., 2008). Interestingly, ectopic overexpression of an eIF_4E resistance gene generates dominant potyvirus resistance in tomato and potato crops (Kang et al., 2007; Cavatorta et al., 2011), probably through the fact that the abundance of the "resistant eIF_4E" makes the endogenous "susceptible eIF_4E" inaccessible for its recruitment by viruses. Other components belonging to the translational machinery were demonstrated to be required for virus multiplication in Arabidopsis thaliana and Nicotiana benthamiana, including the translation initiation factor 4B (eIF4B), the translation elongation factors 1A and 1B (eEF1A and eEF1B) and the poly (A)-binding proteins (PABPs) $_2$, 4, and 8 (Sasvari et al., 2011; Patarroyo et al., 2012; Hwang et al., 2013; Li et al., 2014b). These results suggest that crop genes encoding these translation factors may lead to new resistance sources that need to be explored for viral disease control.

Other susceptible recessive resistance genes which are not encoding translation initiation factors have been identified through the analysis of mutant's collections and/or natural cultivated or wild species. A positional cloning strategy exploring barley natural variability revealed recently the key role of the PDI_{5-1} (Protein disulfide isomerase like $_{5-1}$) protein in the recessive resistance to bymoviruses (Yang et al., 2014). Another recessive resistance gene named ra blocking vascular transport of Potato potyvirus A (PVA) was genetically characterized in potato (Hämäläinen et al., 2000). In addition, from the perspective of identifying new resistance sources, exploringArabidopsis genetic diversity has provided original insights into the genes involved in plant–virus interactions, which could be used as potentially resistance sources against viruses. For example, the recessive resistance genes rlm1 and rpv1, conferring resistance to LMV and PPV respectively, map in a genomic region containing no translation factor genes (Revers et al., 2003; Decroocq et al., 2006). The lack of co-segregation with eIF_4E or eIF_4G genes is also true for dstm1 mediating TMV resistance (Serrano et al., 2008) and sha3, a major quantitative trait locus (QTL) contributing to systemic resistance against PPV (Pagny et al., 2012). Very recently, data revealed that the gene rwm1conferring resistance to Watermelon mosaic potyvirus in Arabidopsis acts at an early stage of infection by impairing viral accumulation in initially infected leaf tissues and encodes a conserved nucleus-encoded chloroplast phosphoglycerate kinase (Ouibrahim et al., 2014). In the case of tom1and tom2A Arabidopsis mutants, TMV accumulation is suppressed in single cells. After further characterization, it appears that both genes encode transmembrane proteins localized in the tonoplast and are required for tobamovirus replication (Ishibashi et al., 2012). Consistent with the fact that viral replication complexes associate with host intracellular membranes (Hull, 2013), prospective studies identified many membrane-associated components as key factors required for plant infection success, providing new candidates for novel genetic sources of crop resistances (Diaz et al., 2010; Hyodo et al., 2013; Barajas et al., 2014a,b).

Rna Interference Mediated-Resistance

Over the past decades, RNA interference (RNAi; also called gene silencing) has been recognized as an evolutionarily conserved process in most eukaryotes that is triggered by double-stranded RNAs (dsRNAs). These dsRNAs are processed by ribonuclease III-type DICER-like (DCL) enzymes into small RNAs (sRNAs), 21–24 nucleotides in length, that are incorporated into an RNA-induced cytoplasmic silencing complex (RISC), whose key catalytic component corresponds to one member of the ARGONAUTE (AGO) protein family. Once integrated into the RISC, sRNAs base-pair to their target-mRNA and induce their cleavage (Bologna and Voinnet, 2014). In plants, silencing pathways are particularly diverse and partially overlapping. At least, three basic processes can be distinguished: cytoplasmic RNA silencing (or post-transcriptional gene silencing; PTGS) mediated by small interfering

RNAs (siRNAs), silencing-mediated by plant-encoded microRNAs (miRNAs) and transcriptional gene silencing (TGS) mediated by siRNA-directed methylation of DNA and histone proteins. Key components of these RNA silencing pathways have been shown to have an important protective role against invading viral pathogens (Bologna and Voinnet, 2014).

Most plant viruses have RNA genome that commonly contain double-stranded secondary structure elements and/or produce dsRNA intermediates via the action of viral RNA-dependent RNA polymerases (RDRs) during the replication step. These molecules are targeted by the RNA-silencing machinery to produce virus-derived small RNAs (vsRNAs; Pumplin and Voinnet, 2013; Figure 2C). What initially appeared counter-intuitive, infection from DNA viruses produce also dsRNAs, most likely via bidirectional convergent transcription (Aregger et al., 2012). The integration of vsRNAs in the RISC leads to the sequence-specific degradation of viral nucleic acids, as well as the generation of a mobile-silencing signal, which spreads between cells through plasmodesmata and over long-distances via the phloem, through a relay-amplification process involving host RDRs (Pumplin and Voinnet, 2013). This process activates RNA silencing in non-infected cells and is notably responsible for the plant recovery phenomenon. Given gene silencing induces immune mechanisms highly specific to the pathogen, it is commonly accepted that RNAi is classified into plant adaptive immunity (Voinnet, 2001; Waterhouse et al., 2001).

Encoded by multigene families, plant DCLs, RDRs and AGOs are often specialized in the production and function of the distinct sRNA classes. Despite its universality, current knowledge remains incompletely understood, as it relies mostly on reverse genetic studies conducted in Arabidopsis. This notwithstanding, it appears clear that specific RNAi-associated components are involved in silencing antiviral functions. Thus, DCL_4 is the major enzyme for generating RNA virus-derived vsRNAs, even if DCL_2 can substitute for DCL_4 to some extent (Blevins et al., 2006; Deleris et al., 2006; Qu et al., 2008; Jakubiec et al., 2012). In contrast, all four DCLs (DCL_{1-4}) produce DNA virus-derived vsRNAs (Akbergenov et al., 2006; Moissiard and Voinnet, 2006; Blevins et al., 2011). So far, only AGO_1, AGO_2, and AGO7 seem to contribute to antiviral RNAi (Morel et al., 2002; Qu et al., 2008; Shivaprasad et al., 2008; Azevedo et al., 2010; Harvey et al., 2011; Jaubert et al., 2011; Wang et al., 2011; Zhang et al., 2012). In particular, RNA viruses-derived vsRNAs seem to be strongly recruited by AGO_1 and AGO_2 proteins, an observation consistent with the hyper-susceptibility ofago1 ago2 double mutants (Wang et al., 2011). Antiviral RNAi has also been shown to be dependent on one or more of RDR_1, RDR_2, and RDR_6 for signal initiation and/or amplification (Diaz-Pendon et al., 2007; Qu et al., 2008; Qi et al., 2009; Vaistij and Jones, 2009; Garcia-Ruiz et al., 2010; Wang et al., 2011; Jiang et al., 2012). An increase of susceptibility against RNA viruses is also observed inArabidopsis plants defected for the gene HEN_1 (for HUA ENHANCER $_1$) encoding a methyltransferase that protects siRNA and miRNA duplexes from degradation (Vogler et al., 2007;Zhang et al., 2012). Interestingly, the natural resistance genes Ty-1 and Ty-3 conferring resistance toTomato yellow leaf curl begomovirus (TYLCV) has been recently shown to encode a tomato RDR. Although resistant plants do not show symptoms upon a challenge with TYLCV, low levels of virus are still detectable, a phenomenon characteristic of a virus tolerance more than a real resistance (Verlaan et al., 2013).

Viruses have evolved diverse mechanisms to avoid silencing-mediated resistance, most notably through silencing suppressor activities. Identified for almost all types of plant viruses, silencing suppressors target RNAi pathways at different points and through diverse mechanisms, including the impair of siRNA biogenesis, the defect of siRNA incorporation into the RISC, the degradation of AGOs, the trapping of sRNAs and the suppression of RNAi amplification (reviewed in Burgyán and Havelda, 2011; Pumplin and Voinnet, 2013; Bologna and Voinnet, 2014; Li et al., 2014a). In turn, increasing evidence suggest that plants have evolved by establishing specific defenses against RNA-silencing suppression by pathogens, providing yet another illustration of the never-ending molecular arms race between plant pathogens and their hosts (Pumplin and Voinnet, 2013; Sansregret et al., 2013).

Plant Hormone-Mediated Resistance

Plant hormones play important roles in regulating signaling networks involved in plant defenses. Upon pathogen attack, the quantity, composition and timing of the plant hormonal blend produced by the plant depends greatly on the lifestyle and infection strategy of the invading attacker. In the last decades, significant progress has been made in identifying the key components and understanding the role of phytohormones in plant responses to biotic stresses (Robert-Seilaniantz et al., 2011).

During the R-mediated resistance activation (see the Section "Dominant Resistances" in this review), cellular responses elicited at the infection site are emitted to distant non-infected tissues, resulting in a resistance or in reduced susceptibility state that can remain efficient during several weeks (Fu and Dong, 2013). This phenomenon is referred as the SAR (Figure 2D). In the case of TMV-triggered HR, the response persists up to 3 weeks during the time plants are protected against not only TMV but also other pathogens (Ross, 1961). How SAR can be sustained for so long is not clear but epigenetic modifications, such as DNA methylation and chromatin remodeling, seem critical to maintain a SAR signal (Spoel and Dong, 2012). Moreover, during a viral infection (in a manner similar to non-viral infections), this long lasting and broad-spectrum disease resistance requires endogenous accumulation of SA, resulting in transcriptional reprogramming of a battery of genes encoding pathogenesis-related (PR) proteins (Tsuda et al., 2008; Yi et al., 2014). The signal emitted by the infection spot to protect the uninfected tissues against pathogen invasion may circulate as a heterocomplex, where methyl-SA binds to lipid derivatives and lipid-transport proteins and moves through the phloem to the rest of the plant. Thus, studies on TMV-infected tobacco plants revealed that MeSA participates to the perpetuation of SAR defense (Park et al., 2007; Dempsey and Klessig, 2012). Recent data suggest that the composition of the mobile immune signal in SAR relies on a complex network of cross-interacting signals (e.g., MeSA, glycerol-3-phosphate, the lipid-transfer protein DIR_1 and the amino acid-derivative pipecolic acid), that differs depending on the plant species and the type of plant–pathogen interaction (Vlot et al., 2008; Dempsey and Klessig, 2012; Spoel and Dong, 2012). Jasmonic acid (JA) is also strongly involved in plant defense against viruses and the HR response initiated by Avr-R protein interactions results on a modulation of SA and JA. Although SA clearly acts as a positive regulator of plant resistance to viruses, the role of JA is controversial and remains to be fully elucidated. For example, JA seems to regulate negatively the local resistance to TMV in tobacco (Oka et al., 2013) but is essential for systemic resistance to TMV inN. benthamiana (Zhu et al., 2014). It is likely that a balance between endogenous JA and SA play a key role for determining the degree of resistance, in a similar way to pathosystems involving non-viral plant pathogens (Thaler et al., 2012). Interestingly, plant viruses have evolved targeting hormone pathways, often exploiting the antagonistic interactions between SA and JA pathways (Kazan and Lyons, 2014). Although their function in plant–virus interactions remains poorly understood, recent studies indicate that other plant hormones modulate antiviral resistance mechanisms, especially abscisic acid (Chen et al., 2013; Alazem et al., 2014; Seo et al., 2014), ethylene (Fischer and Dröge-Laser, 2004; Love et al., 2007; Zhang et al., 2009; Chen et al., 2013), and brassinosteroids (Ali et al., 2014).

CROP IMPROVEMENT AND VIRAL PEST MANAGEMENT: MODERN TECHNOLOGY TO THE RESCUE

Ultimately, the main objective of research on plant–virus interactions consists on the implementation of efficient antiviral resistances in crop plants. For long, genetics-mediated resistance strategies have involved exploiting plant natural variability by introgression of resistance genes through a classical breeding process. However, the onset of evolved pathogens able to overcome these resistance, sometimes very rapidly after resistance deployment (García-Arenal and McDonald, 2003), raised the problem of crop resistance durability and thus the urge to develop new breeding strategies. The last decades saw the emergence of new approaches combining modern technology and state-of-the-art knowledge on plant–virus interactions toward crop improvement programs, triggering a new green revolution in agriculture.

Technical Assets from Marker-Assisted Selection

The conventional breeding represents a laborious and time consuming process. In this context, the advent of DNA marker techniques such as random amplified polymorphic DNA (RAPD), restriction fragment length polymorphism (RFLP), amplified fragment length polymorphism (AFLP), and simple sequence repeats (SSRs) contributed, through what is called "Marker-assisted selection (MAS)," to make substantially easier, quicker, and more accurate the selection of resistant genotypes during the introgression steps (Collard and Mackill, 2008). Importantly, this progress has recently been driven by next generation sequencing-(NGS-) based technologies successfully employed for de novo whole genome shotgun (WGS) sequencing of reference genotypes and whole genome resequencing (WGRS) of several cultivars, land races, and wild relatives (Mascher and Stein, 2014). Despite the fact that the creation of improved varieties via the MAS strategy is still at its infancy (Collard and Mackill, 2008), MAS has an enormous potential and represents a great challenge of molecular breeding in the 21st century.

Exploiting Resistance-Associated QTLs

It is probable that only a small proportion of the natural biodiversity available for disease resistance has been exploited so far. Although more than 80% of reported plant resistances to viruses are monogenically controlled, most agronomic traits in crop plants do not segregate as single defined qualitative monogenic characters but as quantitative and polygenically controlled traits (Maule et al., 2007). Mapping QTL for quantitative resistance requires large sized progenies, nearly saturated genetic maps, as well as reliable and quantitative phenotyping procedures. To date, relatively few QTLs analyses have been performed in plant–virus interactions compared to other pathogens (Maule et al., 2007; Palloix and Ordon, 2011). It is a general assumption that the greater the number of mutations required for virus virulence, the more durable is the resistance. Consequently, the higher durability of polygenic resistances in plant–virus interactions is commonly hypothesized and has recently been validated in the case of pepper resistance to Potato potyvirus Y (Palloix et al., 2009; Quenouille et al., 2013). Notwithstanding their clear relevance, QTLs present particular technical challenges for their characterization as well as their incorporation into crops, leading them to be mostly dropped in favor of approaches using monogenic resistances or combining several known major genes.

Gene Pyramiding Strategy

The concept of transferring several characterized resistance genes into one plant is called "Gene Pyramiding." The dogma behind this strategy is that the probability of a pathogen mutating to "virulence against all resistance genes in the pyramid would be the product of the probabilities for each gene singly" (Mundt, 2014), thus making the probability of a virulent pathotype arising highly unlikely. Hence, pyramiding relies on resistance genes that have been previously characterized singly and whose functions are combined within the same plant. This strategy has been successfully applied to plant–virus interactions (Werner et al., 2005; Palloix et al., 2009; Shi et al., 2009) and could bring the opportunity to associate different types of antiviral plant resistances targeting different virus-derived molecules (proteins or nucleic acids) and various processes during virus cycle (replication/translation, plasmodesmata crossing, systemic colonization), with the possible consequence to reduce significantly the probability of resistance breaking by new virus variants (Quenouille et al., 2013). Despite arduous population sizes required for multiple resistance genes along with other agronomical traits, gene pyramiding is gaining considerable importance in plant–pathogen interactions and thus is representing an ambitious challenge for crop improvement programs.

Engineering Resistances

Genetic engineering directly manipulates the genome of a plant by inducing the expression of novel proteins or by modulating the expression of targeted genes. The genetically modified (GM) plants resistant to viral attacks are grouped into three major categories. First of all, resistance genes with a dominant determinism can be introduced to another plant species. The best examples in plant–virus interactions involve the successful transfer of R genes into solanaceous related plants (Whitham et al., 1996; Bendahmane et al., 1999; Baurès et al., 2008; Candresse et al., 2010) but it seems also possible across plant families (Seo et al., 2006). The opportunity to introduce R genes from sexually incompatible species is, however, weakened by the fact that most R genes display a limited-spectrum resistance. Alternative transgenic approaches have also been developed based on the integration of a viral protein or genomic region into the host plant, through a phenomenon called pathogen-derived resistance (PDR). Initially reported in transgenic tobacco plants expressing the TMV CP (Abel et al., 1986), the resistance observed was ultimately ascribed to the action of the transgene-encoded CP, which disturbs the disassembly of incoming TMV particles. In many other studies, resistance was obtained through the expression of partial or non-coding viral sequences. This kind of resistance, more efficient that the protein-mediated resistance, was later shown to be based on the RNAi mechanism (Tenllado et al., 2004). These observations led to a plethora of RNAi-mediated engineered resistances, using virus-derived double-stranded, hairpin RNAs, or artificial miRNAs (e.g., Qu et al., 2007; Hashmi et al., 2011; Shimizu et al., 2011; Yadav et al., 2011; Zhang et al., 2011a,b; Lin et al., 2012; Shekhawat et al., 2012; Lemgo et al., 2013; Odipio et al., 2013; Zhao and Song, 2014). In the cases where virus-resistant GM varieties have been deployed and/or commercialized (including transgenic squash, papaya, plum, potato, and bean), this strategy has so far proven to be remarkably efficient (Tricoli et al., 1995; Lius et al., 1997; Ferreira et al., 2002;Malinowski et al., 2006; Zagrai et al., 2008; Bravo-Almonacid et al., 2012; Aragão et al., 2013). In spite of engineering resistance success (especially RNAi-based resistant crops), public concerns over the potential ecological impact of GM crops and GM organisms overall have so far significantly limited their use, in particular in Europe.

The Tilling and Eco Tilling Revolution

The early 2000s have seen the emergence of the TILLING (Targeting Induced Local Lesions IN Genomes) method, that consists of a classical mutagenesis step followed by the targeted search for plants carrying a mutation in a gene of interest (McCallum et al., 2000). By creating artificial polymorphism directly into crops, this technique allows potentially: (i) to replace a resistance allele overcome by viral strains, (ii) to generate novel resistances with a broader spectrum, or (iii) to create a new resistance gene based on knowledge acquired in heterologous systems. Such a strategy increase the natural allelic diversity by the identification of novel artificial alleles. Although this strategy requires the previous characterization of the gene conferring the resistance, the main advantage of TILLING is that it can be applied to any plant species, regardless of its genome size, ploidy level or method of propagation, and without introducing heterologous DNA as for GM plants. Similarly, the TILLING natural alternative (using natural germplasms collections instead of EMS mutants collections) is calling ecoTILLING and consists of exploiting the whole natural variability of a plant species (including wild-related and cultivated genotypes; Comai et al., 2004). Initially developed in A. thaliana, both TILLING and ecoTILLING have spread rapidly to other model plants (Medicago truncatula, Lotus japonicus, Brachypodium distachyon) and major crops (e.g., maize, soybean, sorghum, tomato, pepper, cucumber, pea, wheat, banana, bean, rice, barley, Brassica napus). It seems now clear that these strategies are emerging as major crop improvement tools, with especially successful examples of recent applications to antiviral protection (Nieto et al., 2007; Piron et al., 2010).

Chemical Treatments

Curative treatments for virus control in fields is impossible. However, chemical applications can prime plant defense, i.e., resistance mechanisms are switched on prior to future infections. Based on the findings that exogenous applications of plant hormones trigger an efficient and broad-spectrum resistance to viral and non-viral pathogens, many SAR-priming molecules have been characterized. Among a series of SA analogs (Gozzo and Faoro, 2013), benzothiadiazole (BTH) has been identified as the safest and most efficient SAR activator and was brought to the market with the common name of acibenzolar-S-methyl, that leads to plant protection against many pathogens including viruses (Friedrich et al., 1996; Lawton et al., 1996; Mandal et al., 2008; Takeshita et al., 2013; Trejo-Saavedra et al., 2013). This priming approach represents an environment-friendly and efficient way to control plant diseases by exploiting a natural phenomenon. Despite these positive aspects, so far it has not yet been met with enthusiastic favor by the farmers, as the treatment efficacy with BTH (as well as other SAR-priming molecules) depends on many variables: dose, plant species and cultivar, growth stage of plant, pathogen pressure, climatic conditions and timing of chemical applications (Gozzo and Faoro, 2013). Nevertheless, coupled to genetic strategies, chemical SAR priming is still a valuable method to increase plant resistance against viruses for pest management programs in fields.

Plant Vaccination

Plant vaccination gambles on cross-protection, a phenomenon whereby the inoculation of a virus into a host plant prevents the multiplication of a subsequent challenge virus. This strategy mainly relies on the manipulation of the primary virus, whose infection is weak (symptomless with low viral load), and that triggers virus-induced gene silencing (VIGS) targeting both primary and challenge viruses (Ziebell and Carr, 2010; Nishiguchi and Kobayashi, 2011). Primary viruses act as vaccines and are classified into two categories: the attenuated and the engineering viruses. An attenuated virus corresponds to a weak isolate that triggers cross-protection against virulent isolates of the same virus or closed related viruses (Ziebell and Carr, 2010; Nishiguchi and Kobayashi, 2011).. Many examples of cross-protection have been identified since the discovery of this phenomenon by McKinney in 1920s (McKinney, 1926; Crowdy and Posnette, 1947; Fletcher, 1978; Wang, 1991; Wen et al., 1991;Hugues and Ollennu, 1994; Valkonen et al., 2002; Kurihara and Watanabe, 2003; Ichiki et al., 2005;You et al., 2005; Kosaka et al., 2006; Yoon et al., 2006; Ziebell et al., 2007; Nakazono-Nagaoka et al., 2009; Kurth et al., 2012). A great illustration of this method involves an attenuated isolate ofZucchini yellow mosaic potyvirus that has since been registered as the pesticide CUBIO ZY-02 and successfully employed in virus disease control in Japan (Kosaka and Fukunishi, 1997; Kosaka et al., 2009). Moreover, tristeza disease caused by CTV is currently controlled by mild CTV isolates, which when inoculated into existing field trees, extend the productive life of orchards and enable a more graduate replanting of trees on CTV-tolerant rootstocks (Lee and Keremane, 2013). However, identifying effective attenuated viruses for each virus of economic importance might be very arduous. A possible solution came recently from the development of engineering vaccines based on viral vectors carrying a genomic fragment of the virulent virus of interest (Culver, 1996; Bazzini et al., 2006; Tamura et al., 2013; Taki et al., 2013). Given Apple latent spherical cheravirus (ALSV) is able to infect a broad range of herbaceous crops (e.g., tomato, lettuce, zucchini, spinach, soybean, tobacco) and fruit trees (e.g., apple, cherry, peach, plum, pear, and citrus) without causing any symptoms, this virus was identified as an excellent candidate for VIGS-mediated vaccination, that could be simply applied widely by replacing the insert with a sequence derived from a different virus. This strategy has been recently illustrated (Tamura et al., 2013; Taki et al., 2013) and represents a very promising technology in virus epidemics management in fields.

NEW RESEARCH LEAD: LET'S TALK ABOUT ANTIVIRAL PTI IN PLANTS

In an attempt to increase resistance durability in fields, crop improvement requires a continuous pipeline of new resistance genes. Given the importance of PAMP-triggered immunity (PTI) in the field of pathology over the last decades, PTI against plant viruses represents an unexplored question that needs to be addressed.

Plant Innate Immunity

Over the last decades, a concept revolutionizing the understanding of immunity emerged in plant pathology. This new concept stems from the ability for each organism to discriminate between self and non-self-molecules through the action of pattern recognition receptors (PRRs) perceiving specific microbial molecular signatures, named pathogen-associated molecular patterns (PAMPs). Perception of PAMPs by these immune receptors induces a downstream signaling cascade including PRR association with positive regulators, phosphorylation events, successive activation of cytoplasmic kinases (including the MAP kinases) and defense-related transcription factors, as well as specific defense genes expression. This range of fast, efficient and multi-layered defense responses is referred as PTI (Nicaise et al., 2009; Macho and Zipfel, 2014). To counteract this defense strategy, successful pathogens deploy effectors proteins, the primary function of which is to evade/interfere with PTI. In turn, some plant cultivars have evolved R proteins to block these effectors, leading to effector-triggered immunity (ETI; Jones and Dangl, 2006). Because both PTI and ETI defense layers rely on cellular actors already present primary to the infection, they are commonly classified into plant innate immunity, in opposition to the concept of adaptive immunity (e.g., gene silencing), where the defense responses are acquired following an infection and are adapted to the pathogen.

It's All about Good Questions

To date, most current models about plant innate immunity exclude viruses, as they are not generally viewed as encoding typical PAMPs and effectors that would trigger PTI and ETI responses (Jones and Dangl, 2006; Boller and Felix, 2009; Schwessinger and Ronald, 2012; Dangl et al., 2013). The fact that virus biology differs greatly from other pathogens raises several questions of importance: could plants defend themselves against viral attacks through similar defense mechanisms employed against non-viral pathogens? And then, could the zig-zag model from Jones and Dangl (2006) be applied to plant–virus interactions?

Given that R gene-based immunity in plant–virus interactions involves specific recognition of viral avirulence factors by host-encoded NBS-LRR proteins and leads to HR and SAR (described previously in this review), recent hypotheses advance that it may correspond to antiviral ETI mechanisms (Moffett, 2009; Mandadi and Scholthof, 2013; De Ronde et al., 2014; Nakahara and Masuta, 2014). If so, it may suggest that viral avirulent factors perceived by R proteins correspond to pathogenic effectors, the primary function being to target PTI. Therefore, the concept of antiviral innate immunity as a whole lies mainly in whether it exists virus-associated PRR pathways and viral-encoded effectors interfering with PTI responses. Therefore, the question to assess in priority is whether plant viruses possess molecular features meeting the definition of PAMPs.

Do Viruses Encode Pamps?

Pathogen-associated molecular patterns are defined as highly conserved molecular signatures, characteristic of entire classes of microbes and with an essential function for these microbes (Medzhitov and Janeway, 1997), which implies that they are under strong evolutive pressure and thus, not easily modified by mutations. Many PAMPs from plant pathogens have been identified and so far, they are comprised of proteins, lipids, or carbohydrates (Boller and Felix, 2009). Virus genomes are highly variable even within the same viral family and display high mutation rates during the infection process (≥1 mutation per genome per round of replication), making the expression "viral populations" more suitable than "viruses" when speaking of infected hosts. This elevated mutation rate allows viruses to quickly adapt to changes in their host or environment. Therefore, the PAMP definition referring to small molecular motifs conserved within a class of pathogen is hardly applicable to viral protein motifs. By contrast, virus genomes and their replicative forms (such as dsRNAs) display structural characteristics distinguishing them from the nucleic acid sequences from prokaryotes and eukaryotes. Consequently, these much conserved nucleic acids-composed features could be perceived as non-self-signals in plants, acting thus as viral PAMPs.

What do We know about PTI in Animal–Virus Interactions?

In the animal field, the most prominent group of PRRs comprises the Toll-like receptors (TLRs), a family of a dozen type I-transmembrane proteins containing a domain with LRRs and a tail that contains a conserved region called Toll/IL_{-1} receptor (TIR) domain (Beutler, 2004), and sharing common recognition strategies and structural features with plant PRRs (Zipfel and Felix, 2005; Ronald and Beutler, 2010). The TLRs involved in antiviral PTI are localized in endosomes and perceive pathogen nucleic acids: TLR_3 binds virus-derived dsRNA, TLR7 and TLR8 recognize viral single-stranded RNAs (ssRNA) and TLR9 mediates the perception of unmethylated CpG DNA from viruses (as well as bacteria and protozoa; Beutler, 2004). After PAMP perception, the activated receptor seems to form homo- and/or hetero-dimers with other TLRs. The TIR domains of these complexes recruit cytoplasmic adaptors, triggering a signaling cascade that mainly consist on the successive activation of cytoplasmic kinases including the MAP kinases, defense-related transcription factors and the expression of specific defense genes (Beutler, 2004; Kawai and Akira, 2008; Sandig and Bulfone-Paus, 2012). In addition, subsequent studies revealed that TLR-independent recognition of viruses can be also accomplished by cytosolic PRRs, including the two RNA helicases RIG-I (Retinoic acid-inducible gene I) and MDA_5 (Melanoma differentiation-associated gene 5), both highly activated by dsRNAs, as well as 5′ triphosphate-RNAs for RIG-I (Kawai and Akira, 2008). In the arm race between host and microbes, successful animal viruses have evolved to suppress and/or hijack PTI responses, leading to host colonization. This overcoming of defense mechanisms is orchestrated by virulence effectors encoded by virus genomes. Among the strategies employed by viruses to suppress host immunity, inactivation of cytoplasmic kinases, inhibition of PRR complexes, stabilization of immunity-related negative regulators and inactivation of transcription factors are particularly noteworthy (Schröder and Bowie, 2007; Yokota et al., 2010).

On the Trail of PTI Against Plant Viruses

The existence of animal PRRs specific to viral PAMPs raises the question of the existence of PRRs perceiving plant viruses. Although no PTI pathways against plant viruses has been formally characterized to date, typical PTI cellular responses are observed during plant–virus interactions, including ion fluxes (Shabala et al., 2010, 2011; Otulak and Garbaczewska, 2011), ROS production (Allan et al., 2001; Love et al., 2005; Díaz-Vivancos et al., 2008), ethylene signaling (de Laat and van Loon, 1982; Love et al., 2005), and callose deposition (Iglesias and Meins, 2000; Li et al., 2012;Zavaliev et al., 2013). Key components of plant PTI pathways are also involved in antiviral defense mechanisms. SA, known to regulate innate immunity including PTI responses (Tsuda et al., 2008; Yi et al., 2014), plays an important role in plant–virus interactions (Carr et al., 2010). MAPK$_4$ that regulates negatively plant PTI mechanisms (GAO Et Al., 2008; Kong et al., 2012), represses soybean defense against Bean pod mottle comovirus (Liu et al., 2011). The receptor-like kinases (RLKs) BAK$_1$ (BRI$_1$-Associated receptor Kinase $_1$) and BKK$_1$ (BAK$_1$-like Kinase $_1$), key activators of plant PRRs after PAMP perception (Kim et al., 2013), are required for Arabidopsis resistance to plant viruses (Yang et al., 2010; Kørner et al., 2013). Moreover, the protein NIK$_1$, another RLK highly similar to BAK$_1$ and BKK$_1$, has been strongly associated with Arabidopsis defense against the Cabbage leaf curl geminivirus (CaLCuV). Strikingly, NIK$_1$ is cleaved by the viral protein NSP during the infection, suppressing its kinase activity and thus enhancing susceptibility to CaLCuV infection (Fontes et al., 2004; Carvalho et al., 2008). Overall, these data support the hypothesis that plants may defend themselves against viruses through similar PTI pathways employed against non-viral pathogens.

Accumulating data indicate that: (i) colonization strategies employed by plant and animal viruses inside host cells are highly similar (Thivierge et al., 2005; Taliansky et al., 2010; Nicholson and White, 2014), (ii) plant and animal PTI mechanisms display impressive structural and functional similarities (Zipfel and Felix, 2005; Dardick and Ronald, 2006; Ronald and Beutler, 2010), (iii) effectors from animal and plant pathogens target similar PTI components (Schröder and Bowie, 2007; Yokota et al., 2010; Deslandes and Rivas, 2012) and, (iv) plant defense against viruses involves many components belonging to PTI signaling (described above). Therefore, the apparent universality of PTI strategies identified so far in eukaryotic organisms enable an extrapolation of antiviral pathways from animal toward plant models.

Thus, strongly based on the understanding of plant immune pathways and the knowledge on antiviral mechanisms from the animal field, I propose here a model of plant immunity against viruses (Figure 3), that does not invalidate previous models but instead tries (i) to bring the concept of PTI as a novel plant antiviral mechanism, and (ii) to integrate the concept of innate (PTI and ETI) and adaptive (gene silencing) antiviral immunities, both triggered by similar virus-derived elicitor molecules, in order to address the important question of plant defense against viruses in a more holistic way.

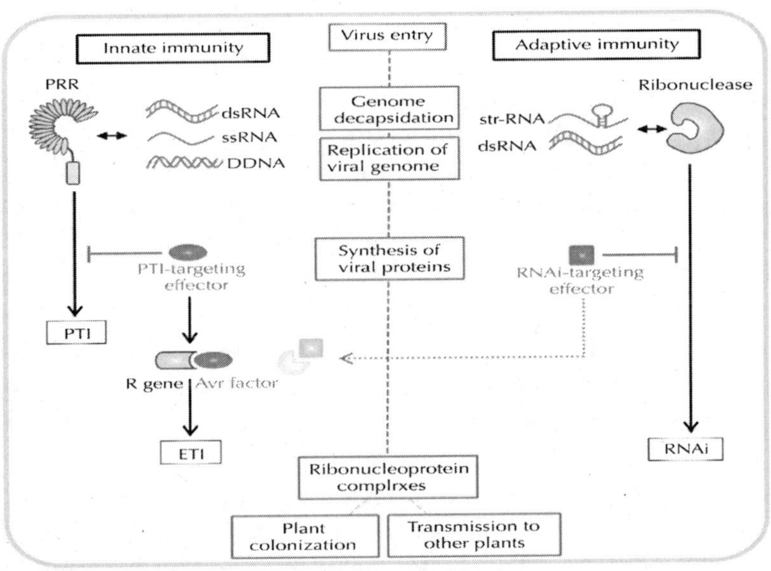

Figure 3: Plant innate immunity vs. Plant adaptive immunity. Following entry into a host cell and genome decapsidation, the virus genome is liberated

and replicated, leading to the accumulation of pathogenic nucleic acids that are perceived as viral PAMPs by specific intracellular PRRs. This recognition triggers a downstream cascade leading to PTI responses. Virus genome translation (occurring simultaneous to replication) leads to the synthesis of the virus-encoded proteins, among which viral pathogenic effectors enable the suppression of PTI signaling. In turn, specific plant Rgenes interact (directly or indirectly) with these effectors (that are then called avirulence factors) to trigger ETI. In the case of plant adaptive immunity illustrated by RNA interference (RNAi), the virus-derived elicitor molecules, corresponding to replicative dsRNAs or structured (str-RNA) viral genomes, are recognized by DCLs, key component of the silencing machinery that leads to virus degradation at the nucleic acid level. Viral proteins acting as RNAi-suppressing effectors interfere with this defense pathway. A recent publication (Sansregret et al., 2013) suggests silencing suppressors may be targeted by ETI-like mechanisms, restoring plant resistance. The main steps of the virus cycle into the first infected cell have been mentioned into gray boxes.

Thus, following the virus entry into a plant cell, viral nucleic acids are perceived as PAMPs by specific intracellular PRRs triggering PTI pathways (Figure 3). Successful viruses encode specialized effectors that are able to suppress this defense layer. In turn, specific R gene products interact with these effectors (playing then the role of avirulence factors), leading to ETI. In the case of RNAi-based adaptive immunity, the viral elicitor molecules (mostly dsRNAs) are recognized by DCLs. This leads to virus degradation at the nucleic acid level, a process that can be suppressed by a second class of virus-encoded effectors, commonly named silencing suppressors. Interestingly, Sansregret et al. (2013) recently highlighted a resistance mechanism antagonizing the silencing suppressor P19, through a phenomenon of extreme resistance characterized by a SA- and ethylene-dependent process without microscopic cell death. Although the reliance of this mechanism on one or several R proteins remains to be established, this data suggest that the action of RNAi-suppressing effectors might be countered by ETI-like mechanisms.

It is important to note here that recent reviews refer DCLs themselves as PRRs perceiving viral nucleic acids and triggering immune responses equivalent to the zig-zag model first layer (Pumplin and Voinnet, 2013; Sansregret et al., 2013; Nakahara and Masuta, 2014). Of course, PTI and RNAi mechanisms are both induced by highly similar virus-derived molecules that both fit the definition of PAMPs, suggesting that these virus-derived molecules trigger two different pathways in parallel (PTI and RNAi), similarly as in the animal field (Yan and Chen, 2012). In line with this, DCL-mediated immunity should maybe be called PTI too. However, this idea faces a conceptual and semantic conflict, as the classical view of PTI lies on innate immunity, while silencing triggers an immunity that is classified as adaptive. This issue could be partially solved by raising the concepts of innate and adaptive PTIs, that both would be targeted by plant ETI. This notwithstanding, this situation reveals that these processes represents two distinct but tight and intricate branches of the whole plant immune arsenal against viruses, raising important conceptual questions that will need to be answered.

Indeed, future investigations will need first to bring the clear proof of the existence of PTI pathways triggered by the recognition of viral PAMPs by plant PRRs. To this perspective, the identification of pathogenic effectors from successful plant viruses that target specifically PTI responses will be definitely an asset, as pathogenic effectors represent fantastic molecular tools commonly used to characterize innate immunity signaling.

Which Advantage PTI could bring into Crop Improvement Programs?

So far, the only powerful strategy to control viral epidemics in field relies on the use of genetic resistances. Unfortunately, viruses evolve very quickly the ability to overcome the resistances employed by breeders. Hence, handling the extreme genetic plasticity of viruses represents a major challenge for the coming years. The durability of plant resistances, even more with viruses than any other plant pathogen, depends mostly on the resistance timing/efficiency and on the elicitor nature/conservation within pathogen families, in order to enable an efficient, sustainable, and broad-spectrum resistance.

Given their importance for pathogen survival, PAMPs are under a strong evolutive pressure and are highly conserved within pathogen families, this definition fitting perfectly with the characteristics of viral nucleic acids, whose features (e.g., dsRNAs replicative forms, highly structured genomes) are highly conserved within virus families and differ radically from prokaryotic or eukaryotic nucleic acids. It is these typical viral features that are perceived by specialized PRRs in animals (e.g., TLRs), triggering very fast and very efficient immune responses. Because the innate pathway is solely programmed by structural and nucleotide sequence genomic features, plant antiviral PTI should be a remarkably versatile mechanism because it could respond virtually to any plant virus. Moreover, timeliness of defense responses is a key parameter in resistance success, especially in the case of viruses, given their tremendous adaptation abilities. And yet, the first measurable PTI responses are within minutes after PAMP perception (Nicaise et al., 2009), making PTI the quickest plant defense identified so far. Hence, translating fundamental knowledge about PTI mechanisms for new crop resistance deployment must bring considerable advances in the control of viral epidemics in fields. Moreover, what is increasingly clear is that plant antiviral arsenal ensure a highly robust plant immune system, through different plant defense layers (R genes, SAR, RNAi, PTI, and ETI) that emerge to be finely coordinated (Voinnet, 2005; Alamillo et al., 2006; Boccara et al., 2014;Padmanabhan and Dinesh-Kumar, 2014). Consequently, addressing plant–virus interactions holistically by combining PTI-mediated resistance with other defense pathways (such as silencing) could provide highly robust resistance phenotypes targeting virus cycle as a whole.

CONCLUSIONS

In the world agriculture context, intensification of cultural practices, climate alterations, and extensive exchanges affecting global markets are associated with an increased incidence of plant viral diseases. In consequence, management of viral epidemics is a matter of global food security. Nowadays, antiviral measures employed in fields benefits greatly from the latest knowledge highlighted by plant virologists, geneticists and molecular biologists, as well as engineers and breeders, whose coordinated effort enables a deployment in crop improvement programs. This notwithstanding, the situation in fields remains worrying, given the emergence of numerous crop diseases ascribed to viruses, a statement that could be aggravated in the future through climate changes that might affect strongly virus and organism-vectors populations in terms of magnitude, locations and dynamics (Anderson et al., 2004; Gautam et al., 2013). In a coordinated manner with technological advances, fundamental research needs to explore new scientific leads, deciphering more and more thoroughly the intricate molecular dialog between a plant virus and its host. Hence, future challenges associated to the management of crop viral diseases will rely mainly on integrated research actions with a view to translating fundamental understanding toward applied programs, and thus reducing the gap existing between the laboratory and the field.

ACKNOWLEDGMENTS

This work was supported by the European Framework Programme 7, through a Marie-Curie Intra-European Fellowship (Grant number 327341).

REFERENCES

1. Abel, P. P., Nelson, R. S., De, B., Hoffmann, N., Rogers, S. G., Fraley, R. T.,et al. (1986). Delay of disease development in transgenic plants that express the tobacco mosaic virus coat protein gene. Science 232, 738–743. doi: 10.1126/science.3457472
2. Abo, M. E., and Sy, A. A. (1998). Rice virus diseases: epidemiology and management strategies. J. Sustain. Agric. 11, 113–134. doi: 10.1300/J064v11n02_09
3. Acosta-Leal, R., and Xiong, Z. (2008). Complementary functions of two recessive R-genes determine resistance durability of tobacco "Virgin A Mutant" (VAM) to potato virus Y. Virology 379, 275–283. doi: 10.1016/j.virol.2008.06.026
4. Akbergenov, R., Si-Ammour, A., Blevins, T., Amin, I., Kutter, C., Vanderschuren, H.,et al. (2006). Molecular characterization of geminivirus-derived small RNAs in different plant species. Nucleic Acids Res. 34, 462–471. doi: 10.1093/nar/gkj447
5. Alamillo, J. M., Saénz, P., and García, J. A. (2006). Salicylic acid-mediated and RNA-silencing defense mechanisms cooperate in the restriction of systemic spread of plum pox virus in tobacco. Plant J. 48, 217–227. doi: 10.1111/j.1365-313X.2006.02861.x
6. Alazem, M., Lin, K.-Y., and Lin, N.-S. (2014). The abscisic acid pathway has multifaceted effects on the accumulation of bamboo

mosaic virus. Mol. Plant Microbe Interact. 27, 177–189. doi: 10.1094/MPMI-08-13-0216-R

7. Ali, S. S., Gunupuru, L. R., Kumar, G. B., Khan, M., Scofield, S., Nicholson, P.,et al. (2014). Plant disease resistance is augmented in uzu barley lines modified in the brassinosteroid receptor BRI1. BMC Plant Biol. 14:227. doi: 10.1186/s12870-014-0227-1
8. Allan, A. C., Lapidot, M., Culver, J. N., and Fluhr, R. (2001). An early tobacco mosaic virus-induced oxidative burst in tobacco indicates extracellular perception of the virus coat protein. Plant Physiol. 126, 97–108. doi: 10.1104/pp.126.1.97
9. Anderson, P. K., Cunningham, A. A., Patel, N. G., Morales, F. J., Epstein, P. R., and Daszak, P. (2004). Emerging infectious diseases of plants: pathogen pollution, climate change and agrotechnology drivers. Trends Ecol. Evol. 19, 535–544. doi: 10.1016/j.tree.2004.07.021
10. Aragão, F. J. L., Nogueira, E. O., Tinoco, M. L., and Faria, J. C. (2013). Molecular characterization of the first commercial transgenic common bean immune to the bean golden mosaic virus. J. Biotechnol. 166, 42–50. doi: 10.1016/j.jbiotec.2013.04.009
11. Aregger, M., Borah, B. K., Seguin, J., Rajeswaran, R., Gubaeva, E. G., Zvereva, A. S.,et al. (2012). Primary and secondary siRNAs in geminivirus-induced gene silencing. PLoS Pathog. 8:e1002941. doi: 10.1371/journal.ppat.1002941
12. Azevedo, C., Betsuyaku, S., Peart, J., Takahashi, A., Noël, L., Sadanandom, A.,et al. (2006). Role of SGT1 in resistance protein accumulation in plant immunity. EMBO J. 25, 2007–2016. doi: 10.1038/sj.emboj.7601084
13. Azevedo, J., Garcia, D., Pontier, D., Ohnesorge, S., Yu, A., Garcia, S.,et al. (2010). Argonaute quenching and global changes in Dicer homeostasis caused by a pathogen-encoded GW repeat protein. Genes Dev. 24, 904–915. doi: 10.1101/gad.1908710
14. Bai, S., Liu, J., Chang, C., Zhang, L., Maekawa, T., Wang, Q.,et al. (2012). Structure-function analysis of barley NLR immune receptor MLA10 reveals its cell compartment specific activity in cell death and disease resistance. PLoS Pathog.8:e1002752. doi: 10.1371/journal.ppat.1002752

15. Barajas, D., Martín, I. F., Pogany, J., Risco, C., and Nagy, P. D. (2014a). Noncanonical role for the host Vps4 AAA+ ATPase ESCRT protein in the formation of tomato bushy stunt virus replicase. PLoS Pathog. 10:e1004087. doi: 10.1371/journal.ppat.1004087

16. Barajas, D., Xu, K., de Castro Martín, I. F., Sasvari, Z., Brandizzi, F., Risco, C.,et al. (2014b). Co-opted oxysterol-binding ORP and VAP proteins channel sterols to RNA virus replication sites via membrane contact sites. PLoS Pathog.10:e1004388. doi: 10.1371/journal.ppat.1004388

17. Baurès, I., Candresse, T., Leveau, A., Bendahmane, A., and Sturbois, B. (2008). The Rx gene confers resistance to a range of potexviruses in transgenic Nicotiana plants. Mol. Plant Microbe Interact. 21, 1154–1164. doi: 10.1094/MPMI-21-9-1154

18. Bazzini, A. A., Asurmendi, S., Hopp, H. E., and Beachy, R. N. (2006). Tobacco mosaic virus (TMV) and potato virus X (PVX) coat proteins confer heterologous interference to PVX and TMV infection, respectively. J. Gen. Virol. 87, 1005–1012. doi: 10.1099/vir.0.81396-0

19. Bendahmane, A., Kanyuka, K., and Baulcombe, D. C. (1999). The Rx gene from potato controls separate virus resistance and cell death responses. Plant Cell 11, 781–792. doi: 10.1105/tpc.11.5.781

20. Beutler, B. (2004). Innate immunity: an overview. Mol. Immunol. 40, 845–859. doi: 10.1016/j.molimm.2003.10.005

21. Blevins,T.,Rajeswaran,R.,Aregger,M.,Borah,B.K.,Schepetilnikov, M., Baerlocher, L.,et al. (2011). Massive production of small RNAs from a non-coding region of cauliflower mosaic virus in plant defense and viral counter-defense. Nucleic Acids Res. 39, 5003–5014. doi: 10.1093/nar/gkr119

22. Blevins, T., Rajeswaran, R., Shivaprasad, P. V., Beknazariants, D., Si-Ammour, A., Park, H. S.,et al. (2006). Four plant Dicers mediate viral small RNA biogenesis and DNA virus induced silencing. Nucleic Acids Res. 34, 6233–6246. doi: 10.1093/nar/gkl886

23. Boccara, M., Sarazin, A., Thiébeauld, O., Jay, F., Voinnet, O., Navarro, L.,et al. (2014). The Arabidopsis miR472-RDR6 silencing pathway modulates PAMP-and effector-triggered immunity through the post-transcriptional control of disease

resistance genes. PLoS Pathog. 10:e1003883. doi: 10.1371/journal.ppat.1003883

24. Boller, T., and Felix, G. (2009). A renaissance of elicitors: perception of microbe-associated molecular patterns and danger signals by pattern-recognition receptors. Annu. Rev. Plant Biol. 60, 379–406. doi: 10.1146/annurev.arplant.57.032905.105346
25. Bologna, N. G., and Voinnet, O. (2014). The diversity, biogenesis, and activities of endogenous silencing small RNAs inArabidopsis. Annu. Rev. Plant Biol. 65, 473–503. doi: 10.1146/annurev-arplant-050213-035728
26. Boonham, N., Kreuze, J., Winter, S., van derVlugt, R., Bergervoet, J., Tomlinson, J.,et al. (2014). Methods in virus diagnostics: from ELISA to next generation sequencing. Virus Res. 186, 20–31. doi: 10.1016/j.virusres.2013.12.007
27. Botër, M., Amigues, B., Peart, J., Breuer, C., Kadota, Y., Casais, C.,et al. (2007). Structural and functional analysis of SGT1 reveals that its interaction with HSP90 is required for the accumulation of Rx, an R protein involved in plant immunity.Plant Cell 19, 3791–3804. doi: 10.1105/tpc.107.050427
28. Bragard, C., Caciagli, P., Lemaire, O., Lopez-Moya, J. J., MacFarlane, S., Peters, D.,et al. (2013). Status and prospects of plant virus control through interference with vector transmission. Annu. Rev. Phytopathol. 51, 177–201. doi: 10.1146/annurev-phyto-082712-102346
29. Bravo-Almonacid, F., Rudoy, V., Welin, B., Segretin, M. E., Bedogni, M. C., Stolowicz, F.,et al. (2012). Field testing, gene flow assessment and pre-commercial studies on transgenic Solanum tuberosum spp. tuberosum (cv. Spunta) selected for PVY resistance in Argentina. Transgenic Res. 21, 967–982. doi: 10.1007/s11248-011-9584-9
30. Burgyán, J., and Havelda, Z. (2011). Viral suppressors of RNA silencing. Trends Plant Sci. 16, 265–272. doi: 10.1016/j.tplants.2011.02.010
31. Calvert, L. A., and Tresh, J. M. (2002). "The viruses and virus diseases of cassava," in Cassava: Biology, Production and Utilization, eds R. J. Hillocks, J. M. Thresh, and A. C. Bellotti (Wallingford: CABI Publishing).

32. Cambra, M., Capote, N., Myrta, A., and Llácer, G. (2006). Plum pox virus and the estimated costs associated with sharka disease. EPPO Bull. 36, 202–204. doi: 10.1111/j.1365-2338.2006.01027.x
33. Candresse, T., Marais, A., Faure, C., Dubrana, M. P., Gombert, J., and Bendahmane, A. (2010). Multiple coat protein mutations abolish recognition of Pepino mosaic potexvirus (PepMV) by the potato rx resistance gene in transgenic tomatoes.Mol. Plant Microbe Interact. 23, 376–383. doi: 10.1094/MPMI-23-4-0376
34. Caplan, J. L., Mamillapalli, P., Burch-Smith, T. M., Czymmek, K., and Dinesh-Kumar, S. P. (2008). Chloroplastic protein NRIP1 mediates innate immune receptor recognition of a viral effector. Cell 132, 449–462. doi: 10.1016/j.cell.2007.12.031
35. Caranta, C., Lefebvre, V., and Palloix, A. (1997). Polygenic resistance of pepper to potyviruses consists of a combination of isolate-specific and broad-spectrum quantitative trait loci. Mol. Plant Microbe Interact. 10, 872–878. doi: 10.1094/MPMI.1997.10.7.872
36. Carr, J. P., Lewsey, M. G., and Palukaitis, P. (2010). Signaling in induced resistance. Adv. Virus Res. 76, 57–121. doi: 10.1016/S0065-3527(10)76003-6
37. Carvalho, C. M., Santos, A. A., Pires, S. R., Rocha, C. S., Saraiva, D. I., Machado, J. P. B.,et al. (2008). Regulated nuclear trafficking of rpL10A mediated by NIK1 represents a defense strategy of plant cells against virus. PLoS Pathog. 4:e1000247. doi: 10.1371/journal.ppat.1000247
38. Cavatorta, J., Perez, K. W., Gray, S. M., Van Eck, J., Yeam, I., and Jahn, M. (2011). Engineering virus resistance using a modified potato gene. Plant Biotechnol. J. 9, 1014–1021. doi: 10.1111/j.1467-7652.2011.00622.x
39. Charron, C., Nicolaï, M., Gallois, J. L., Robaglia, C., Moury, B., Palloix, A.,et al. (2008). Natural variation and functional analyses provide evidence for co-evolution between plant eIF4E and potyviral VPg. Plant J. 54, 56–68. doi: 10.1111/j.1365-313X.2008.03407.x
40. Chen, L., Zhang, L., Li, D., Wang, F., and Yu, D. (2013). WRKY8 transcription factor functions in the TMV-cg defense response by mediating both abscisic acid and ethylene signaling in

Arabidopsis. Proc. Natl. Acad. Sci. U.S.A. 110, E1963–E1971. doi: 10.1073/pnas.1221347110

41. Collard, B. C. Y., and Mackill, D. J. (2008). Marker-assisted selection: an approach for precision plant breeding in the twenty-first century. Philos. Trans. R. Soc. Lond. B Biol. Sci. 363, 557–572. doi: 10.1098/rstb.2007.2170
42. Comai, L., Young, K., Till, B. J., Reynolds, S. H., Greene, E. A., Codomo, C. A.,et al. (2004). Efficient discovery of DNA polymorphisms in natural populations by Ecotilling. Plant J. 37, 778–786. doi: 10.1111/j.0960-7412.2003.01999.x
43. Cosson, P., Schurdi-Levraud, V., Le, Q. H., Sicard, O., Caballero, M., Roux, F.,et al. (2012). The RTM resistance to potyviruses in Arabidopsis thaliana: natural variation of the RTM genes and evidence for the implication of additional genes. PLoS ONE 7:e39169. doi: 10.1371/journal.pone.0039169
44. Cosson, P., Sofer, L., Le, Q. H., Léger, V., Schurdi-Levraud, V., Whitham, S. A.,et al. (2010a). RTM3, which controls long-distance movement of potyviruses, is a member of a new plant gene family encoding a meprin and TRAF homology domain-containing protein. Plant Physiol. 154, 222–232. doi: 10.1104/pp.110.155754
45. Cosson, P., Sofer, L., Schurdi-Levraud, V., and Revers, F. (2010b). A member of a new plant gene family encoding a meprin and TRAF homology (MATH) domain-containing protein is involved in restriction of long distance movement of plant viruses. Plant Signal. Behav. 5, 1321–1323. doi: 10.4161/psb.5.10.13244
46. Crowdy, S. H., and Posnette, A. F. (1947). Virus diseases of cacao in West Africa II. Cross immunity experiments with viruses 1A, 1B and 1C. Ann. Appl. Biol. 34, 403–411. doi: 10.1111/j.1744-7348.1947.tb06373.x
47. Culver, J. N. (1996). Tobamovirus cross protection using a potexvirus vector. Virology 226, 228–235. doi: 10.1006/viro.1996.0650
48. Dangl, J. L., Horvath, D. M., and Staskawicz, B. J. (2013). Pivoting the plant immune system from dissection to deployment.Science 341, 746–751. doi: 10.1126/science.1236011

49. Dardick, C., and Ronald, P. (2006). Plant and animal pathogen recognition receptors signal through non-RD kinases. PLoS Pathog. 2:e2. doi: 10.1371/journal.ppat.0020002
50. Decroocq, V., Sicard, O., Alamillo, J. M., Lansac, M., Eyquard, J. P., García, J. A.,et al. (2006). Multiple resistance traits control plum pox virus infection in Arabidopsis thaliana. Mol. Plant Microbe Interact. 19, 541–549. doi: 10.1094/MPMI-19-0541
51. De Laat, A. M. M., and van Loon, L. C. (1982). Regulation of ethylene biosynthesis in virus-infected tobacco leaves 1. Plant Physiol. 69, 240–245. doi: 10.1104/pp.69.1.240
52. Delaney, T. P., Uknes, S., Vernooij, B., Friedrich, L., Weymann, K., Negrotto, D.,et al. (1994). A central role of salicylic acid in plant disease resistance. Science 266, 1247–1250. doi: 10.1126/science.266.5188.1247
53. Deleris, A., Gallego-Bartolome, J., Bao, J., Kasschau, K. D., Carrington, J. C., and Voinnet, O. (2006). Hierarchical action and inhibition of plant Dicer-like proteins in antiviral defense. Science 313, 68–71. doi: 10.1126/science.1128214
54. Dempsey, D. A., and Klessig, D. F. (2012). SOS – too many signals for systemic acquired resistance? Trends Plant Sci. 17, 538–545. doi: 10.1016/j.tplants.2012.05.011
55. De Ronde, D., Butterbach, P., and Kormelink, R. (2014). Dominant resistance against plant viruses. Front. Plant Sci. 5:307. doi: 10.3389/fpls.2014.00307
56. Deslandes, L., and Rivas, S. (2012). Catch me if you can: bacterial effectors and plant targets. Trends Plant Sci. 17, 644–655. doi: 10.1016/j.tplants.2012.06.011
57. Diaz, A., Wang, X., and Ahlquist, P. (2010). Membrane-shaping host reticulon proteins play crucial roles in viral RNA replication compartment formation and function. Proc. Natl. Acad. Sci. U.S.A. 107, 16291–16296. doi: 10.1073/pnas.1011105107
58. Diaz-Pendon, J. A., Li, F., Li, W.-X., and Ding, S.-W. (2007). Suppression of antiviral silencing by cucumber mosaic virus 2b protein in Arabidopsis is associated with drastically reduced accumulation of three classes of viral small interfering RNAs. Plant Cell 19, 2053–2063. doi: 10.1105/tpc.106.047449

59. Diaz-Pendon, J. A., Truniger, V., Nieto, C., Garcia-Mas, J., Bendahmane, A., and Aranda, M. A. (2004). Advances in understanding recessive resistance to plant viruses. Mol. Plant Pathol. 5, 223–233. doi: 10.1111/J.1364-3703.2004.00223.X

60. Díaz-Vivancos, P., Clemente-Moreno, M. J., Rubio, M., Olmos, E., García, J. A., Martínez-Gómez, P.,et al. (2008). Alteration in the chloroplastic metabolism leads to ROS accumulation in pea plants in response to plum pox virus. J. Exp. Bot. 59, 2147–2160. doi: 10.1093/jxb/ern082

61. Dzahini-Obiatey, H., Domfeh, O., and Amoah, F. M. (2010). Over seventy years of a viral disease of cocoa in Ghana: from researchers' perspective. Afr. J. Agric. Res. 5, 476–485. doi: 10.5897/AJAR09.625

62. Elena, S. F., Bedhomme, S., Carrasco, P., Cuevas, J. M., de la Iglesia, F., Lafforgue, G.,et al. (2011). The evolutionary genetics of emerging plant RNA viruses. Mol. Plant Microbe Interact. 24, 287–293. doi: 10.1094/MPMI-09-10-0214

63. Fabre, F., Rousseau, E., Mailleret, L., and Moury, B. (2012). Durable strategies to deploy plant resistance in agricultural landscapes. New Phytol. 193, 1064–1075. doi: 10.1111/j.1469-8137.2011.04019.x

64. Ferreira, S. A., Pitz, K. Y., Manshardt, R., Zee, F., Fitch, M., and Gonsalves, D. (2002). Virus coat protein transgenic papaya provides practical control of papaya ringspot virus in Hawaii. Plant Dis. 86, 101–105. doi: 10.1094/PDIS.2002.86.2.101

65. Fischer, U., and Dröge-Laser, W. (2004). Overexpression of NtERF5, a new member of the tobacco ethylene response transcription factor family enhances resistance to tobacco mosaic virus. Mol. Plant Microbe Interact. 17, 1162–1171. doi: 10.1094/MPMI.2004.17.10.1162

66. Fletcher, J. T. (1978). The use of avirulent virus strain to protect plants against the effects of virulent strains. Ann. Appl. Biol. 89, 110–114. doi: 10.1111/j.1744-7348.1978.tb02581.x

67. Fontes, E. P. B., Santos, A. A., Luz, D. F., Waclawovsky, A. J., and Chory, J. (2004). The geminivirus nuclear shuttle protein is a virulence factor that suppresses transmembrane receptor kinase activity. Genes Dev. 18, 2545–2556. doi: 10.1101/gad.1245904

68. Friedrich, L., Lawton, K., Ruess, W., Masner, P., Specker, N., Rella, M. G.,et al. (1996). A benzothiadiazole derivative induces systemic acquired resistance in tobacco. Plant J. 10, 61–70. doi: 10.1046/j.1365-313X.1996.10010061.x

69. Fu, Z. Q., and Dong, X. (2013). Systemic acquired resistance: turning local infection into global defense. Annu. Rev. Plant Biol. 64, 839–863. doi: 10.1146/annurev-arplant-042811-105606

70. Gao, M., Liu, J., Bi, D., Zhang, Z., Cheng, F., Chen, S.,et al. (2008). MEKK1, MKK1/MKK2 and MPK4 function together in a mitogen-activated protein kinase cascade to regulate innate immunity in plants. Cell Res. 18, 1190–1198. doi: 10.1038/cr.2008.300

71. García-Arenal, F., and McDonald, B. A. (2003). An analysis of the durability of resistance to plant viruses. Phytopathology93, 941–952. doi: 10.1094/PHYTO.2003.93.8.941

72. Garcia-Ruiz, H., Takeda, A., Chapman, E. J., Sullivan, C. M., Fahlgren, N., Brempelis, K. J.,et al. (2010). ArabidopsisRNA-dependent RNA polymerases and dicer-like proteins in antiviral defense and small interfering RNA biogenesis during turnip mosaic virus infection. Plant Cell 22, 481–496. doi: 10.1105/tpc.109.073056

73. Gautam, H. R., Bhardwaj, M. L., and Kumar, R. (2013). Climate change and its impact on plant diseases. Curr. Sci. 105, 1685–1691.

74. Gozzo, F., and Faoro, F. (2013). Systemic acquired resistance (50 years after discovery): moving from the lab to the field. J. Agric. Food Chem. 61, 12473–12491. doi: 10.1021/jf404156x

75. Hämäläinen, J. H., Kekarainen, T., Gebhardt, C., Watanabe, K. N., and Valkonen, J. P. (2000). Recessive and dominant genes interfere with the vascular transport of potato virus a in diploid potatoes. Mol. Plant Microbe Interact. 13, 402–412. doi: 10.1094/MPMI.2000.13.4.402

76. Hao, W., Collier, S. M., Moffett, P., and Chai, J. (2013). Structural basis for the interaction between the potato virus X resistance protein (Rx) and its cofactor Ran GTPase-activating protein 2 (RanGAP2). J. Biol. Chem. 288, 35868–35876. doi: 10.1074/jbc.M113.517417

77. Harper, S. J. (2013). Citrus tristeza virus: evolution of complex and varied genotypic groups. Front. Microbiol. 4:93. doi: 10.3389/fmicb.2013.00093
78. Harvey, J. J. W., Lewsey, M. G., Patel, K., Westwood, J., Heimstädt, S., Carr, J. P.,et al. (2011). An antiviral defense role of AGO2 in plants. PLoS ONE 6:e14639. doi: 10.1371/journal.pone.0014639
79. Hashmi, J. A., Zafar, Y., Arshad, M., Mansoor, S., and Asad, S. (2011). Engineering cotton (Gossypium hirsutum L.) for resistance to cotton leaf curl disease using viral truncated AC1 DNA sequences. Virus Genes 42, 286–296. doi: 10.1007/s11262-011-0569-9
80. Hoser, R., Zurczak, M., Lichocka, M., Zuzga, S., Dadlez, M., Samuel, M. A.,et al. (2013). Nucleocytoplasmic partitioning of tobacco N receptor is modulated by SGT1. New Phytol. 200, 158–171. doi: 10.1111/nph.12347
81. Hugues, J. A., and Ollennu, L. A. A. (1994). Mild strain protection of cocoa in Ghana against cocoa swollen shoot virus—a review. Plant Pathol. 43, 442–457. doi: 10.1111/j.1365-3059.1994.tb01578.x
82. Hull, R. (2013). Plant Virology, 5th Edn. New York: Academic Press.
83. Hwang, J., Oh, C. S., and Kang, B. C. (2013). Translation elongation factor 1B (eEF1B) is an essential host factor for tobacco mosaic virus infection in plants. Virology 439, 105–114. doi: 10.1016/j.virol.2013.02.004
84. Hyodo, K., Mine, A., Taniguchi, T., Kaido, M., Mise, K., Taniguchi, H.,et al. (2013). ADP ribosylation factor 1 plays an essential role in the replication of a plant RNA virus. J. Virol. 87, 163–176. doi: 10.1128/JVI.02383-12
85. Ichiki, T. U., Nagaoka, E. N., Hagiwara, K., Uchikawa, K., Tsuda, S., and Omura, T. (2005). Integration of mutations responsible for the attenuated phenotype of pepper mild mottle virus strains results in a symptomless cross-protecting strain. Arch. Virol. 150, 2009–2020. doi: 10.1007/s00705-005-0551-8
86. Iglesias, V. A., and Meins, F. (2000). Movement of plant viruses is delayed in a beta-1,3-glucanase-deficient mutant showing a reduced plasmodesmatal size exclusion limit and enhanced

callose deposition. Plant J. 21, 157–166. doi: 10.1046/j.1365-313X.2000.00658.x

87. Ishibashi, K., and Ishikawa, M. (2013). The resistance protein Tm-1 inhibits formation of a tomato mosaic virus replication protein-host membrane protein complex. J. Virol. 87, 7933–7939. doi: 10.1128/JVI.00743-13
88. Ishibashi, K., Miyashita, S., Katoh, E., and Ishikawa, M. (2012). Host membrane proteins involved in the replication of tobamovirus RNA. Curr. Opin. Virol. 2, 693–698. doi: 10.1016/j.coviro.2012.09.011
89. Jakubiec, A., Yang, S. W., and Chua, N. H. (2012). Arabidopsis DRB4 protein in antiviral defense against turnip yellow mosaic virus infection. Plant J. 69, 14–25. doi: 10.1111/j.1365-313X.2011.04765.x
90. Jaubert, M., Bhattacharjee, S., Mello, A. F. S., Perry, K. L., and Moffett, P. (2011). ARGONAUTE2 mediates RNA-silencing antiviral defenses against potato virus X in Arabidopsis. Plant Physiol. 156, 1556–1564. doi: 10.1104/pp.111.178012
91. Jiang, L., Qian, D., Zheng, H., Meng, L. Y., Chen, J., Le, W. J.,et al. (2012). RNA-dependent RNA polymerase 6 of rice (Oryza sativa) plays role in host defense against negative-strand RNA virus, rice stripe virus. Virus Res. 163, 512–519. doi: 10.1016/j.virusres.2011.11.016
92. Jones, J. D. G., and Dangl, J. L. (2006). The plant immune system. Nature 444, 323–329. doi: 10.1038/nature05286
93. Julio, E., Cotucheau, J., Decorps, C., Volpatti, R., Sentenac, C., Candresse, T.,et al. (2014). A eukaryotic translation initiation factor 4E (eIF4E) is responsible for the "va" tobacco recessive resistance to potyviruses. Plant Mol. Biol. Rep. doi: 10.1007/s11105-014-0775-4
94. Kadota, Y., and Shirasu, K. (2012). The HSP90 complex of plants. Biochim. Biophys. Acta 1823, 689–97. doi: 10.1016/j.bbamcr.2011.09.016
95. Kang, B. C., Yeam, I., Li, H., Perez, K. W., and Jahn, M. M. (2007). Ectopic expression of a recessive resistance gene generates dominant potyvirus resistance in plants. Plant Biotechnol. J. 5, 526–536. doi: 10.1111/j.1467-7652.2007.00262.x

96. Kawai, T., and Akira, S. (2008). Toll-like receptor and RIG-I-like receptor signaling. Ann. N. Y. Acad. Sci. 1143, 1–20. doi: 10.1196/annals.1443.020

97. Kazan, K., and Lyons, R. (2014). Intervention of phytohormone pathways by pathogen effectors. Plant Cell 26, 2285–2309. doi: 10.1105/tpc.114.125419

98. Kim, B. H., Kim, S. Y., and Nam, K. H. (2013). Assessing the diverse functions of BAK1 and its homologs in Arabidopsis, beyond BR signaling and PTI responses. Mol. Cells 35, 7–16. doi: 10.1007/s10059-013-2255-3

99. King, A. M. Q., Adams, M. J., Carstens, E. B., and Lefkowitz, E. (2012). Virus Taxonomy: Classification and Nomenclature of Viruses: Ninth Report of the International Committee on Taxonomy of Viruses. San Diego: Elsevier Academic Press.

100. Kobayashi, M., Seo, S., Hirai, K., Yamamoto-Katou, A., Katou, S., Seto, H.,et al. (2010). Silencing of WIPK and SIPK mitogen-activated protein kinases reduces tobacco mosaic virus accumulation but permits systemic viral movement in tobacco possessing the N resistance gene. Mol. Plant Microbe Interact. 23, 1032–1041. doi: 10.1094/MPMI-23-8-1032

101. Kong, Q., Qu, N., Gao, M., Zhang, Z., Ding, X., Yang, F.,et al. (2012). The MEKK1-MKK1/MKK2-MPK4 kinase cascade negatively regulates immunity mediated by a mitogen-activated protein kinase kinase kinase in Arabidopsis. Plant Cell 24, 2225–2236. doi: 10.1105/tpc.112.097253

102. Kørner, C. J., Klauser, D., Niehl, A., Domínguez-Ferreras, A., Chinchilla, D., Boller, T.,et al. (2013). The immunity regulator BAK1 contributes to resistance against diverse RNA viruses. Mol. Plant Microbe Interact. 26, 1271–1280. doi: 10.1094/MPMI-06-13-0179-R

103. Kosaka, Y., and Fukunishi, T. (1997). Multiple inoculation with three attenuated viruses for the control of cucumber virus disease. Plant Dis. 81, 733–738. doi: 10.1094/PDIS.1997.81.7.733

104. Kosaka, Y., Ryang, B.-S., Katagiri, N., and Yasuhara, H. (2009). "Development and diffusion of CUBIO ZY-02, a water-soluble formulation of an attenuated zucchini yellow mosaic virus isolate, for the control of cucumber mosaic disease," inProceedings of PSJ Biocontrol Workshop, Vol. 11, Yamagata, 63–72.

105. Kosaka, Y., Ryang, B.-S., Kobori, T., Shiomi, H., Yasuhara, H., and Kataoka, M. (2006). Effectiveness of an attenuatedZucchini yellow mosaic virus isolate for cross-protecting cucumber. Plant Dis. 90, 67–72. doi: 10.1094/PD-90-0067

106. Kurihara, Y., and Watanabe, Y. (2003). Cross-protection in Arabidopsis against crucifer tobamovirus Cg by an attenuated strain of the virus. Mol. Plant Pathol. 4, 259–269. doi: 10.1046/j.1364-3703.2003.00174.x

107. Kurth, E. G., Peremyslov, V. V., Prokhnevsky, A. I., Kasschau, K. D., Miller, M., Carrington, J. C.,et al. (2012). Virus-derived gene expression and RNA interference vector for grapevine. J. Virol. 86, 6002–6009. doi: 10.1128/JVI.00436-12

108. Lawton, K. A., Friedrich, L., Hunt, M., Weymann, K., Delaney, T., Kessmann, H.,et al. (1996). Benzothiadiazole induces disease resistance in Arabidopsis by activation of the systemic acquired resistance signal transduction pathway. Plant J. 10, 71–82. doi: 10.1046/j.1365-313X.1996.10010071.x

109. Lee, R. F., and Keremane, M. L. (2013). Mild strain cross protection of tristeza: a review of research to protect against decline on sour orange in Florida. Front. Microbiol. 4:259. doi: 10.3389/fmicb.2013.00259

110. Legg, J. P. (1999). Emergence, spread and strategies for controlling the pandemic of cassava mosaic virus disease in east and central Africa. Crop Prot. 18, 627–637. doi: 10.1016/S0261-2194(99)00062-9

111. Legg, J. P., and Thresh, J. M. (2000). Cassava mosaic virus disease in East Africa: a dynamic disease in a changing environment. Virus Res. 71, 135–149. doi: 10.1016/S0168-1702(00)00194-5

112. Lemgo, G. N. Y., Sabbadini, S., Pandolfini, T., and Mezzetti, B. (2013). Biosafety considerations of RNAi-mediated virus resistance in fruit-tree cultivars and in rootstock. Transgenic Res. 22, 1073–1088. doi: 10.1007/s11248-013-9728-1

113. Li, F., Huang, C., Li, Z., and Zhou, X. (2014a). Suppression of RNA silencing by a plant DNA virus satellite requires a host calmodulin-like protein to repress RDR6 expression. PLoS Pathog. 10:e1003921. doi: 10.1371/journal.ppat.1003921

114. Li, Z., Gonzalez, P. A., Sasvari, Z., Kinzy, T. G., and Nagy, P. D. (2014b). Methylation of translation elongation factor 1A by the METTL10-like See1 methyltransferase facilitates tombusvirus replication in yeast and plants. Virology 448, 43–54. doi: 10.1016/j.virol.2013.09.012

115. Li, W., Zhao, Y., Liu, C., Yao, G., Wu, S., Hou, C.,et al. (2012). Callose deposition at plasmodesmata is a critical factor in restricting the cell-to-cell movement of soybean mosaic virus. Plant Cell Rep. 31, 905–916. doi: 10.1007/s00299-011-1211-y

116. Lin, C. Y., Ku, H. M., Chiang, Y. H., Ho, H. Y., Yu, T. A., and Jan, F. J. (2012). Development of transgenic watermelon resistant to cucumber mosaic virus and watermelon mosaic virus by using a single chimeric transgene construct. Transgenic Res. 21, 983–993. doi: 10.1007/s11248-011-9585-8

117. Liu, J.-Z., Horstman, H. D., Braun, E., Graham, M. A., Zhang, C., Navarre, D.,et al. (2011). Soybean homologs of MPK4 negatively regulate defense responses and positively regulate growth and development. Plant Physiol. 157, 1363–1378. doi: 10.1104/pp.111.185686

118. Liu, P.-P., Yang, Y., Pichersky, E., and Klessig, D. F. (2010). Altering expression of benzoic acid/salicylic acid carboxyl methyltransferase 1 compromises systemic acquired resistance and PAMP-triggered immunity in Arabidopsis. Mol. Plant Microbe Interact. 23, 82–90. doi: 10.1094/MPMI-23-1-0082

119. Liu, Y., Schiff, M., Marathe, R., and Dinesh-Kumar, S. P. (2002). Tobacco Rar1, EDS1 and NPR1/NIM1 like genes are required for N-mediated resistance to tobacco mosaic virus. Plant J. 30, 415–429. doi: 10.1046/j.1365-313X.2002.01297.x

120. Lius, S., Manshardt, R. M., Fitch, M. M. M., Slightom, J. L., Sanford, J. C., and Gonsalves, D. (1997). Pathogen-derived resistance provides papaya with effective protection against papaya ringspot virus. Mol. Breed. 3, 161–168. doi: 10.1023/A:1009614508659

121. Love, A. J., Laval, V., Geri, C., Laird, J., Tomos, A. D., Hooks, M. A.,et al. (2007). Components of Arabidopsis defense- and ethylene-signaling pathways regulate susceptibility to cauliflower mosaic virus by restricting long-distance movement. Mol. Plant Microbe Interact. 20, 659–670. doi: 10.1094/MPMI-20-6-0659

122. Love, A. J., Yun, B. W., Laval, V., Loake, G. J., and Milner, J. J. (2005). Cauliflower mosaic virus, a compatible pathogen ofArabidopsis, engages three distinct defense-signaling pathways and activates rapid systemic generation of reactive oxygen species. Plant Physiol. 139, 935–948. doi: 10.1104/pp.105.066803

123. Macho, A. P., and Zipfel, C. (2014). Plant PRRs and the activation of innate immune signaling. Mol. Cell 54, 263–272. doi: 10.1016/j.molcel.2014.03.028

124. Mahajan, S. K., Chisholm, S. T., Whitham, S. A., and Carrington, J. C. (1998). Identification and characterization of a locus (RTM1) that restricts long-distance movement of tobacco etch virus in Arabidopsis thaliana. Plant J. 14, 177–186. doi: 10.1046/j.1365-313X.1998.00105.x

125. Malinowski, T., Cambra, M., Capote, N., Zawadzka, B., Gorris, M. T., Scorza, R.,et al. (2006). Field Trials of plum clones transformed with the plum pox virus coat protein (PPV-CP) gene. Plant Dis. 90, 1012–1018. doi: 10.1094/PD-90-1012

126. Mandadi, K. K., and Scholthof, K.-B. G. (2013). Plant immune responses against viruses: how does a virus cause disease?Plant Cell 25, 1489–1505. doi: 10.1105/tpc.113.111658

127. Mandal, B., Mandal, S., Csinos, A. S., Martinez, N., Culbreath, A. K., and Pappu, H. R. (2008). Biological and molecular analyses of the acibenzolar-S-methyl- induced systemic acquired resistance in flue-cured tobacco against tomato spotted wilt virus. Phytopathology 98, 196–204. doi: 10.1094/PHYTO-98-2-0196

128. Marathe, R., Anandalakshmi, R., Liu, Y., and Dinesh-Kumar, S. P. (2002). The tobacco mosaic virus resistance gene, N.Mol. Plant Pathol. 3, 167–172. doi: 10.1046/j.1364-3703.2002.00110.x

129. Mascher, M., and Stein, N. (2014). Genetic anchoring of whole-genome shotgun assemblies. Front. Genet. 5:208. doi: 10.3389/fgene.2014.00208

130. Maule, A. J., Caranta, C., and Boulton, M. I. (2007). Sources of natural resistance to plant viruses: status and prospects. Mol. Plant Pathol. 8, 223–231. doi: 10.1111/j.1364-3703.2007.00386.x

131. McCallum, C. M., Comai, L., Greene, E. A., and Henikoff, S. (2000). Targeting Induced Local Lesions IN Genomes (TILLING) for plant functional genomics. Plant Physiol. 123, 437–442. doi: 10.1104/pp.123.2.439

132. McKinney, H. H. (1926). Virus mixtures that may not be detected in young tobacco plants. Phytopathology 16, 883.
133. Medzhitov, R., and Janeway, C. A. (1997). Innate immunity: the virtues of a nonclonal system of recognition. Cell 91, 295–298. doi: 10.1016/S0092-8674(00)80412-2
134. Moffett, P. (2009). Mechanisms of recognition in dominant R gene mediated resistance. Adv. Virus Res. 75, 1–33. doi: 10.1016/S0065-3527(09)07501-0
135. Moissiard, G., and Voinnet, O. (2006). RNA silencing of host transcripts by cauliflower mosaic virus requires coordinated action of the four Arabidopsis Dicer-like proteins. Proc. Natl. Acad. Sci. U.S.A. 103, 19593–19598. doi: 10.1073/pnas.0604627103
136. Morel, J.-B., Godon, C., Mourrain, P., Béclin, C., Boutet, S., Feuerbach, F.,et al. (2002). Fertile hypomorphic ARGONAUTE (ago1) mutants impaired in post-transcriptional gene silencing and virus resistance. Plant Cell 14, 629–639. doi: 10.1105/tpc.010358
137. Moreno, P., Ambrós, S., Albiach-Martí, M. R., Guerri, J., and Peña, L. (2008). Citrus tristeza virus: a pathogen that changed the course of the citrus industry. Mol. Plant Pathol. 9, 251–268. doi: 10.1111/j.1364-3703.2007.00455.x
138. Mundt, C. C. (2014). Durable resistance: a key to sustainable management of pathogens and pests. Infect. Genet. Evol. 27C, 446–455. doi: 10.1016/j.meegid.2014.01.011
139. Nakahara, K. S., and Masuta, C. (2014). Interaction between viral RNA silencing suppressors and host factors in plant immunity. Curr. Opin. Plant Biol. 20, 88–95. doi: 10.1016/j.pbi.2014.05.004
140. Nakazono-Nagaoka, E., Takahashi, T., Shimizu, T., Kosaka, Y., Natsuaki, T., Omura, T.,et al. (2009). Cross-protection against bean yellow mosaic virus (BYMV) and clover yellow vein virus by attenuated BYMV isolate M11. Phytopathology 99, 251–257. doi: 10.1094/PHYTO-99-3-0251
141. Nicaise, V., German-Retana, S., Sanjuán, R., Dubrana, M., Mazier, M., Maisonneuve, B.,et al. (2003). The eukaryotic translation initiation factor 4E controls lettuce susceptibility to the potyvirus lettuce mosaic virus. Plant Physiol. 132, 1272–1282. doi: 10.1104/pp.102.017855.1

142. Nicaise, V., Roux, M., and Zipfel, C. (2009). Recent advances in PAMP-triggered immunity against bacteria: pattern recognition receptors watch over and raise the alarm. Plant Physiol. 150, 1638–1647. doi: 10.1104/pp.109.139709

143. Nicholson, B. L., and White, K. A. (2014). Functional long-range RNA–RNA interactions in positive-strand RNA viruses.Nat. Rev. Microbiol. 12, 493–504. doi: 10.1038/nrmicro3288

144. Nieto, C., Piron, F., Dalmais, M., Marco, C. F., Moriones, E., Gómez-Guillamón, M. L.,et al. (2007). EcoTILLING for the identification of allelic variants of melon eIF4E, a factor that controls virus susceptibility. BMC Plant Biol. 7:34. doi: 10.1186/1471-2229-7-34

145. Nishiguchi, M., and Kobayashi, K. (2011). Attenuated plant viruses: preventing virus diseases and understanding the molecular mechanism. J. Gen. Plant Pathol. 77, 221–229. doi: 10.1007/s10327-011-0318-x

146. Odipio, J., Ogwok, E., Taylor, N. J., Halsey, M., Bua, A., Fauquet, C. M.,et al. (2013). RNAi-derived field resistance to cassava brown streak disease persists across the vegetative cropping cycle. GM Crops Food 5, 1–4.

147. Oka, K., Kobayashi, M., Mitsuhara, I., and Seo, S. (2013). Jasmonic acid negatively regulates resistance to tobacco mosaic virus in tobacco. Plant Cell Physiol. 54, 1999–2010. doi: 10.1093/pcp/pct137

148. Ordon, F., Habekuss, A., Kastirr, U., Rabenstein, F., and Kühne, T. (2009). Virus resistance in cereals: sources of resistance, genetics and breeding. J. Phytopathol. 157, 535–545. doi: 10.1111/j.1439-0434.2009.01540.x

149. Otulak, K., and Garbaczewska, G. (2011). Cellular localisation of calcium ions during potato hypersensitive response to potato virus Y. Micron 42, 381–391. doi: 10.1016/j.micron.2010.11.001

150. Ouibrahim, L., Mazier, M., Estevan, J., Pagny, G., Decroocq, V., Desbiez, C.,et al. (2014). Cloning of the Arabidopsis rwm1 gene for resistance to watermelon mosaic virus points to a new function for natural virus resistance genes. Plant J. 79, 705–716. doi: 10.1111/tpj.12586

151. Padmanabhan, M. S., and Dinesh-Kumar, S. P. (2014). The conformational and subcellular compartmental dance of plant NLRs during viral recognition and defense signaling. Curr. Opin. Microbiol. 20, 55–61. doi: 10.1016/j.mib.2014.05.003

152. Pagny, G., Paulstephenraj, P. S., Poque, S., Sicard, O., Cosson, P., Eyquard, J.-P.,et al. (2012). Family-based linkage and association mapping reveals novel genes affecting plum pox virus infection in Arabidopsis thaliana. New Phytol. 196, 873–86. doi: 10.1111/j.1469-8137.2012.04289.x

153. Palloix, A., Ayme, V., and Moury, B. (2009). Durability of plant major resistance genes to pathogens depends on the genetic background, experimental evidence and consequences for breeding strategies. New Phytol. 183, 190–199. doi: 10.1111/j.1469-8137.2009.02827.x

154. Palloix, A., and Ordon, F. (2011). "Advanced breeding for virus resistance in plants," in Recent Advances in Plant Virology, eds C. Caranta, M. A. Aranda, M. Tepfer, and J. J. Lopez-Moya (Norwich: Caister Academic Press), 195–218.

155. Park, S. W., Kaimoyo, E., Kumar, D., Mosher, S., and Klessig, D. F. (2007). Methyl salicylate is a critical mobile signal for plant systemic acquired resistance. Science 318, 113–116. doi: 10.1126/science.1147113

156. Patarroyo, C., Laliberté, J.-F., and Zheng, H. (2012). Hijack it, change it: how do plant viruses utilize the host secretory pathway for efficient viral replication and spread? Front. Plant Sci. 3:308. doi: 10.3389/fpls.2012.00308

157. Pestova, T. V, Kolupaeva, V. G., Lomakin, I. B., Pilipenko, E. V, Shatsky, I. N., Agol, V. I.,et al. (2001). Molecular mechanisms of translation initiation in eukaryotes. Proc. Natl. Acad. Sci. U.S.A. 98, 7029–7036. doi: 10.1073/pnas.111145798

158. Piron, F., Nicolaï, M., Minoïa, S., Piednoir, E., Moretti, A., Salgues, A.,et al. (2010). An induced mutation in tomato eiF4E leads to immunity to two potyviruses. PLoS ONE 5:e11313. doi: 10.1371/journal.pone.0011313

159. Pumplin, N., and Voinnet, O. (2013). RNA silencing suppression by plant pathogens: defence, counter-defence and counter-counter-defence. Nat. Rev. Microbiol. 11, 745–760. doi: 10.1038/nrmicro3120

160. Qi, X., Bao, F. S., and Xie, Z. (2009). Small RNA deep sequencing reveals role for Arabidopsis thaliana RNA-dependent RNA polymerases in viral siRNA biogenesis. PLoS ONE 4:e4971. doi: 10.1371/journal.pone.0004971

161. Qu, F., Ye, X., and Morris, T. J. (2008). Arabidopsis DRB4, AGO1, AGO7, and RDR6 participate in a DCL4-initiated antiviral RNA silencing pathway negatively regulated by DCL1. Proc. Natl. Acad. Sci. U.S.A. 105, 14732–14737. doi: 10.1073/pnas.0805760105

162. Qu, J., Ye, J., and Fang, R. (2007). Artificial microRNA-mediated virus resistance in plants. J. Virol. 81, 6690–6699. doi: 10.1128/JVI.02457-06

163. Quenouille, J., Montarry, J., Palloix, A., and Moury, B. (2013). Farther, slower, stronger: how the plant genetic background protects a major resistance gene from breakdown. Mol. Plant Pathol. 14, 109–118. doi: 10.1111/j.1364-3703.2012.00834.x

164. Rairdan, G. J., Collier, S. M., Sacco, M. A., Baldwin, T. T., Boettrich, T., and Moffett, P. (2008). The coiled-coil and nucleotide binding domains of the potato Rx disease resistance protein function in pathogen recognition and signaling.Plant Cell 20, 739–751. doi: 10.1105/tpc.107.056036

165. Revers, F., Guiraud, T., Houvenaghel, M.-C., Mauduit, T., Le Gall, O., and Candresse, T. (2003). Multiple resistance phenotypes to lettuce mosaic virus among Arabidopsis thaliana accessions. Mol. Plant Microbe Interact. 16, 608–616. doi: 10.1094/MPMI.2003.16.7.608

166. Revers, F., and Nicaise, V. (2014). "Plant resistance to infection by viruses," in Encyclopedia of Life Sciences, ed. Wiley-Blackwell (Chichester: John Wiley & Sons Ltd). doi: 10.1002/9780470015902.a0000757.pub3

167. Robert-Seilaniantz, A., Grant, M., and Jones, J. D. G. (2011). Hormone crosstalk in plant disease and defense: more than just JASMONATE-SALICYLATE antagonism. Annu. Rev. Phytopathol. 49, 317–343. doi: 10.1146/annurev-phyto-073009-114447

168. Ronald, P. C., and Beutler, B. (2010). Plant and animal sensors of conserved microbial signatures. Science 330, 1061–1064. doi: 10.1126/science.1189468

169. Ross, F. A. (1961). Systemic acquired resistance induced by localized virus infections in plants. Virology 14, 340–358. doi: 10.1016/0042-6822(61)90319-1

170. Sandig, H., and Bulfone-Paus, S. (2012). TLR signaling in mast cells: common and unique features. Front. Immunol. 3:185. doi: 10.3389/fimmu.2012.00185

171. Sansregret, R., Dufour, V., Langlois, M., Daayf, F., Dunoyer, P., Voinnet, O.,et al. (2013). Extreme resistance as a host counter-counter defense against viral suppression of RNA silencing. PLoS Pathog. 9:e1003435. doi: 10.1371/journal.ppat.1003435

172. Sasaya, T., Nakazono-Nagaoka, E., Saika, H., Aoki, H., Hiraguri, A., Netsu, O.,et al. (2013). Transgenic strategies to confer resistance against viruses in rice plants. Front. Microbiol. 4:409. doi: 10.3389/fmicb.2013.00409

173. Sastry, S. K., and Zitter, T. A. (2014). "Management of virus and viroid diseases of crops in the tropics," in Plant Virus and Viroid Diseases in the Tropics, Vol. 2, Epidemiology and Management (Dordrecht: Springer), 149–480. doi: 10.1007/978-94-007-7820-7_

174. Sasvari, Z., Izotova, L., Kinzy, T. G., and Nagy, P. D. (2011). Synergistic roles of eukaryotic translation elongation factors 1Bγ and 1A in stimulation of tombusvirus minus-strand synthesis. PLoS Pathog. 7:e1002438. doi: 10.1371/journal.ppat.1002438

175. Schröder, M., and Bowie, A. G. (2007). An arms race: innate antiviral responses and counteracting viral strategies. Biochem. Soc. Trans. 35, 1512–1514. doi: 10.1042/BST0351512

176. Schwessinger, B., and Ronald, P. C. (2012). Plant innate immunity: perception of conserved microbial signatures. Annu. Rev. Plant Biol. 63, 451–482. doi: 10.1146/annurev-arplant-042811-105518

177. Seo, J.-K., Kwon, S.-J., Cho, W. K., Choi, H.-S., and Kim, K.-H. (2014). Type 2C protein phosphatase is a key regulator of antiviral extreme resistance limiting virus spread. Sci. Rep. 4, 5905. doi: 10.1038/srep05905

178. Seo, Y.-S., Rojas, M. R., Lee, J.-Y., Lee, S.-W., Jeon, J.-S., Ronald, P.,et al. (2006). A viral resistance gene from common bean functions across plant families and is up-regulated in a non-

virus-specific manner. Proc. Natl. Acad. Sci. U.S.A. 103, 11856–11861. doi: 10.1073/pnas.0604815103

179. Serrano, C., González-Cruz, J., Jauregui, F., Medina, C., Mancilla, P., Matus, J. T.,et al. (2008). Genetic and histological studies on the delayed systemic movement of tobacco mosaic virus in Arabidopsis thaliana. BMC Genet. 9:59. doi: 10.1186/1471-2156-9-59
180. Shabala, S., Babourina, O., Rengel, Z., and Nemchinov, L. G. (2010). Non-invasive microelectrode potassium flux measurements as a potential tool for early recognition of virus-host compatibility in plants. Planta 232, 807–815. doi: 10.1007/s00425-010-1213-y
181. Shabala, S., Baekgaard, L., Shabala, L., Fuglsang, A., Babourina, O., Palmgren, M. G.,et al. (2011). Plasma membrane Ca^{2+}transporters mediate virus-induced acquired resistance to oxidative stress. Plant. Cell Environ. 34, 406–417. doi: 10.1111/j.1365-3040.2010.02251.x
182. Shekhawat, U. K. S., Ganapathi, T. R., and Hadapad, A. B. (2012). Transgenic banana plants expressing small interfering RNAs targeted against viral replication initiation gene display high-level resistance to banana bunchy top virus infection. J. Gen. Virol. 93, 1804–1813. doi: 10.1099/vir.0.041871-0
183. Shi, A., Chen, P., Li, D., Zheng, C., Hou, A., and Zhang, B. (2009). Gene pyramiding of genes of soybean mosaic virus resistance. Mol. Breed. 23, 113–124. doi: 10.1007/s11032-008-9219-x
184. Shimizu, T., Nakazono-Nagaoka, E., Akita, F., Uehara-Ichiki, T., Omura, T., and Sasaya, T. (2011). Immunity to rice black streaked dwarf virus, a plant reovirus, can be achieved in rice plants by RNA silencing against the gene for the viroplasm component protein. Virus Res. 160, 400–403. doi: 10.1016/j.virusres.2011.05.011
185. Shivaprasad, P. V., Rajeswaran, R., Blevins, T., Schoelz, J., Meins, F., Hohn, T.,et al. (2008). The CaMV transactivator/viroplasmin interferes with RDR6-dependent trans-acting and secondary siRNA pathways in Arabidopsis.Nucleic Acids Res. 36, 5896–5909. doi: 10.1093/nar/gkn590
186. Slootweg, E., Roosien, J., Spiridon, L. N., Petrescu, A.-J., Tameling, W., Joosten, M.,et al. (2010). Nucleocytoplasmic distribution is required for activation of resistance by the potato NB-LRR

receptor Rx1 and is balanced by its functional domains. Plant Cell 22, 4195–4215. doi: 10.1105/tpc.110.077537

187. Soosaar, J. L. M., Burch-Smith, T. M., and Dinesh-Kumar, S. P. (2005). Mechanisms of plant resistance to viruses. Nat. Rev. Microbiol. 3, 789–798. doi: 10.1038/nrmicro1239

188. Spoel, S. H., and Dong, X. (2012). How do plants achieve immunity? Defence without specialized immune cells. Nat. Rev. Immunol. 12, 89–100. doi: 10.1038/nri3141

189. Takeshita, M., Okuda, M., Okuda, S., Hyodo, A., Hamano, K., Furuya, N.,et al. (2013). Induction of antiviral responses by acibenzolar-s-methyl against cucurbit chlorotic yellows virus in Melon. Phytopathology 103, 960–965. doi: 10.1094/PHYTO-08-12-0188-R

190. Taki, A., Yamagishi, N., and Yoshikawa, N. (2013). Development of apple latent spherical virus-based vaccines against three tospoviruses. Virus Res. 176, 251–258. doi: 10.1016/j.virusres.2013.06.015

191. Taliansky, M. E., Brown, J. W. S., Rajamäki, M. L., Valkonen, J. P. T., and Kalinina, N. O. (2010). Involvement of the plant nucleolus in virus and viroid infections. Parallels with animal pathosystems. Adv. Virus Res. 77, 119–158. doi: 10.1016/B978-0-12-385034-8.00005-3

192. Tameling, W. I. L., Nooijen, C., Ludwig, N., Boter, M., Slootweg, E., Goverse, A.,et al. (2010). RanGAP2 mediates nucleocytoplasmic partitioning of the NB-LRR immune receptor Rx in the Solanaceae, thereby dictating Rx function. Plant Cell 22, 4176–4194. doi: 10.1105/tpc.110.077461

193. Tamura, A., Kato, T., Taki, A., Sone, M., Satoh, N., Yamagishi, N.,et al. (2013). Preventive and curative effects of Apple latent spherical virus vectors harboring part of the target virus genome against potyvirus and cucumovirus infections.Virology 446, 314–324. doi: 10.1016/j.virol.2013.08.019

194. Tenllado, F., Llave, C., and Díaz-Ruíz, J. R. (2004). RNA interference as a new biotechnological tool for the control of virus diseases in plants. Virus Res. 102, 85–96. doi: 10.1016/j.virusres.2004.01.019

195. Thaler, J. S., Humphrey, P. T., and Whiteman, N. K. (2012). Evolution of jasmonate and salicylate signal crosstalk. Trends Plant Sci. 17, 260–270. doi: 10.1016/j.tplants.2012.02.010

196. Thivierge, K., Nicaise, V., Dufresne, P. J., Cotton, S., Laliberté, J.-F., Le Gall, O.,et al. (2005). Plant virus RNAs. Coordinated recruitment of conserved host functions by (+) ssRNA viruses during early infection events. Plant Physiol. 138, 1822–1827. doi: 10.1104/pp.105.064105.1822

197. Thresh, J. M., and Cooter, R. J. (2005). Strategies for controlling cassava mosaic virus disease in Africa. Plant Pathol. 54, 587–614. doi: 10.1111/j.1365-3059.2005.01282.x

198. Trejo-Saavedra, D. L., García-Neria, M. A., and Rivera-Bustamante, R. F. (2013). Benzothiadiazole (BTH) induces resistance to pepper golden mosaic virus (PepGMV) in pepper (Capsicum annuum L.). Biol. Res. 46, 333–340. doi: 10.4067/S0716-97602013000400004

199. Tricoli, D. M., Carney, K. J., Russell, P. F., McMaster, J. R., and Groff, D. W. (1995). Field-evaluation of transgenic squash containing single or multiple virus coat protein gene constructs for resistance to cucumber mosaic-virus. Artif. Cells Blood Substit. Immobil. Biotechnol. 13, 1458–1465. doi: 10.1038/nbt1295-1458

200. Truniger, V., and Aranda, M. A. (2009). Recessive resistance to plant viruses. Adv. Virus Res. 119–159. doi: 10.1016/S0065-3527(09)07504-6

201. Tsuda, K., Sato, M., Glazebrook, J., Cohen, J. D., and Katagiri, F. (2008). Interplay between MAMP-triggered and SA-mediated defense responses. Plant J. 53, 763–775. doi: 10.1111/j.1365-313X.2007.03369.x

202. Ueda, H., Yamaguchi, Y., and Sano, H. (2006). Direct interaction between the tobacco mosaic virus helicase domain and the ATP-bound resistance protein, N factor during the hypersensitive response in tobacco plants. Plant Mol. Biol. 61, 31–45. doi: 10.1007/s11103-005-5817-8

203. Vaistij, F. E., and Jones, L. (2009). Compromised virus-induced gene silencing in RDR6-deficient plants. Plant Physiol.149, 1399–1407. doi: 10.1104/pp.108.132688

204. Valkonen, J. P. T., Rajamäki, M.-L., and Kekarainen, T. (2002). Mapping of viral genomic regions important in cross-protection between strains of a potyvirus. Mol. Plant Microbe Interact. 15, 683–692. doi: 10.1094/MPMI.2002.15.7.683

205. Verlaan, M. G., Hutton, S. F., Ibrahem, R. M., Kormelink, R., Visser, R. G. F., Scott, J. W.,et al. (2013). The tomato yellow leaf curl virus resistance genes Ty-1 and Ty-3 are allelic and code for DFDGD-Class RNA-dependent RNA polymerases.PLoS Genet. 9:e1003399. doi: 10.1371/journal.pgen.1003399

206. Vlot, A. C., Klessig, D. F., and Park, S. W. (2008). Systemic acquired resistance: the elusive signal(s). Curr. Opin. Plant Biol.11, 436–442. doi: 10.1016/j.pbi.2008.05.003

207. Vogler, H., Akbergenov, R., Shivaprasad, P. V, Dang, V., Fasler, M., Kwon, M.-O.,et al. (2007). Modification of small RNAs associated with suppression of RNA silencing by tobamovirus replicase protein. J. Virol. 81, 10379–10388. doi: 10.1128/JVI.00727-07

208. Voinnet, O. (2001). RNA silencing as a plant immune system against viruses. Trends Genet. 17, 449–459. doi: 10.1016/S0168-9525(01)02367-8

209. Voinnet, O. (2005). Induction and suppression of RNA silencing: insights from viral infections. Nat. Rev. Genet. 6, 206–220. doi: 10.1038/nrg1555

210. Wagner, S., Stuttmann, J., Rietz, S., Guerois, R., Brunstein, E., Bautor, J.,et al. (2013). Structural basis for signaling by exclusive EDS1 heteromeric complexes with SAG101 or PAD4 in plant innate immunity. Cell Host Microbe 14, 619–630. doi: 10.1016/j.chom.2013.11.006

211. Wale, S., Platt, B., and Cattlin, N. D. (2008). Diseases, Pests and Disorders of Potatoes: A Colour Handbook. London: Manson Publishing Ltd.

212. Wang, A., and Krishnaswamy, S. (2012). Eukaryotic translation initiation factor 4E-mediated recessive resistance to plant viruses and its utility in crop improvement. Mol. Plant Pathol. 13, 795–803. doi: 10.1111/j.1364-3703.2012.00791.x

213. Wang, H. L. (1991). Effectiveness of cross protection by a mild strain of zucchini yellow mosaic virus in cucumber, melon, and squash. Plant Dis. 75, 203. doi: 10.1094/PD-75-0203

214. Wang, X.-B., Jovel, J., Udomporn, P., Wang, Y., Wu, Q., Li, W.-X.,et al. (2011). The 21-nucleotide, but not 22-nucleotide, viral secondary small interfering RNAs direct potent antiviral defense by two cooperative argonautes in Arabidopsis thaliana. Plant Cell 23, 1625–1638. doi: 10.1105/tpc.110.082305

215. Waterhouse, P. M., Wang, M. B., and Lough, T. (2001). Gene silencing as an adaptive defence against viruses. Nature 411, 834–842. doi: 10.1038/35081168

216. Wen, F., Lister, R. M., and Fattouh, F. A. (1991). Cross-protection among strains of barley yellow dwarf virus. J. Gen. Virol.72, 791–799. doi: 10.1099/0022-1317-72-4-791

217. Werner, K., Friedt, W., and Ordon, F. (2005). Strategies for pyramiding resistance genes against the barley yellow mosaic virus complex (BaMMV, BaYMV, BaYMV-2). Mol. Breed. 16, 45–55. doi: 10.1007/s11032-005-3445-2

218. Whitham, S., McCormick, S., and Baker, B. (1996). The N gene of tobacco confers resistance to tobacco mosaic virus in transgenic tomato. Proc. Natl. Acad. Sci. U.S.A. 93, 8776–8781. doi: 10.1073/pnas.93.16.8776

219. Yadav, J. S., Ogwok, E., Wagaba, H., Patil, B. L., Bagewadi, B., Alicai, T.,et al. (2011). RNAi-mediated resistance to cassava brown streak uganda virus in transgenic cassava. Mol. Plant Pathol. 12, 677–687. doi: 10.1111/j.1364-3703.2010.00700.x

220. Yan, N., and Chen, Z. J. (2012). Intrinsic antiviral immunity. Nat. Immunol. 13, 214–222. doi: 10.1038/ni.2229

221. Yang, H., Gou, X., He, K., Xi, D., Du, J., Lin, H.,et al. (2010). BAK1 and BKK1 in Arabidopsis thaliana confer reduced susceptibility to turnip crinkle virus. Eur. J. Plant Pathol. 127, 149–156. doi: 10.1007/s10658-010-9581-5

222. Yang, P., Lüpken, T., Habekuss, A., Hensel, G., Steuernagel, B., Kilian, B.,et al. (2014). PROTEIN DISULFIDE ISOMERASE LIKE 5-1 is a susceptibility factor to plant viruses. Proc. Natl. Acad. Sci. U.S.A. 111, 2104–2109. doi: 10.1073/pnas.1320362111

223. Yi, S.Y., Shirasu, K., Moon, J. S., Lee, S. G., and Kwon, S. Y. (2014). The activated SA and JA signaling pathways have an influence on flg22-triggered oxidative burst and callose deposition. PLoS ONE 9:e88951. doi: 10.1371/journal.pone.0088951

224. Yokota, S.-I., Okabayashi, T., and Fujii, N. (2010). The battle between virus and host: modulation of Toll-like receptor signaling pathways by virus infection. Mediators Inflamm. 2010, 184328. doi: 10.1155/2010/184328

225. Yoon, J. Y., Ahn, H. II, Kim, M., Tsuda, S., and Ryu, K. H. (2006). Pepper mild mottle virus pathogenicity determinants and cross protection effect of attenuated mutants in pepper. Virus Res. 118, 23–30. doi: 10.1016/j.virusres.2005.11.004

226. You, B.-J., Chiang, C.-H., Chen, L.-F., Su, W.-C., and Yeh, S.-D. (2005). Engineered mild strains of papaya ringspot virus for broader cross protection in cucurbits. Phytopathology 95, 533–540. doi: 10.1094/PHYTO-95-0533

227. Zagrai, I., Capote, N., Ravelonandro, M., Cambra, M., Zagrai, L., and Scorza, R. (2008). Plum pox virus silencing of C5 transgenic plums is stable under challenge inoculation with heterologous viruses. J. Plant Pathol. 90, S1–S63.

228. Zavaliev, R., Levy, A., Gera, A., and Epel, B. L. (2013). Subcellular dynamics and role of Arabidopsis β-1, 3-glucanases in cell-to-cell movement of tobamoviruses. Mol. Plant Microbe Interact. 26, 1016–1030. doi: 10.1094/MPMI-03-13-0062-R

229. Zhang, G., Chen, M., Li, L., Xu, Z., Chen, X., Guo, J.,et al. (2009). Overexpression of the soybean GmERF3 gene, an AP2/ERF type transcription factor for increased tolerances to salt, drought, and diseases in transgenic tobacco. J. Exp. Bot.60, 3781–3796. doi: 10.1093/jxb/erp214

230. Zhang, X., Li, H., Zhang, J., Zhang, C., Gong, P., Ziaf, K.,et al. (2011a). Expression of artificial microRNAs in tomato confers efficient and stable virus resistance in a cell-autonomous manner. Transgenic Res. 20, 569–581. doi: 10.1007/s11248-010-9440-3

231. Zhang, X., Sato, S., Ye, X., Dorrance, A. E., Morris, T. J., Clemente, T. E.,et al. (2011b). Robust RNAi-based resistance to mixed infection of three viruses in soybean plants expressing separate short hairpins from a single transgene.Phytopathology 101, 1264–1269. doi: 10.1094/PHYTO-02-11-0056

232. Zhang, X., Singh, J., Li, D., and Qu, F. (2012). Temperature-dependent survival of turnip crinkle virus-infectedArabidopsis plants relies on an RNA silencing-based defense that requires

DCL2, AGO2, and HEN1. J. Virol. 86, 6847–6854. doi: 10.1128/JVI.00497-12

233. Zhao, D., and Song, G. Q. (2014). Rootstock-to-scion transfer of transgene-derived small interfering RNAs and their effect on virus resistance in nontransgenic sweet cherry. Plant Biotechnol. J. doi: 10.1111/pbi.12243 [Epub ahead of print].
234. Zhu, F., Xi, D.-H., Yuan, S., Xu, F., Zhang, D.-W., and Lin, H.-H. (2014). Salicylic acid and jasmonic acid are essential for systemic resistance against tobacco mosaic virus in Nicotiana benthamiana. Mol. Plant Microbe Interact. 27, 567–577. doi: 10.1094/MPMI-11-13-0349-R
235. Zhu, S., Jeong, R. D., Venugopal, S. C., Lapchyk, L., Navarre, D., Kachroo, A.,et al. (2011). Sag101 forms a ternary complex with EDS1 and PAD4 and is required for resistance signaling against turnip crinkle virus. PLoS Pathog. 7:e1002318. doi: 10.1371/journal.ppat.1002318
236. Ziebell, H., and Carr, J. P. (2010). Cross-protection: a century of mystery. Adv. Virus Res. 76, 211–264. doi: 10.1016/S0065-3527(10)76006-1
237. Ziebell, H., Payne, T., Berry, J. O., Walsh, J. A., and Carr, J. P. (2007). A cucumber mosaic virus mutant lacking the 2b counter-defence protein gene provides protection against wild-type strains. J. Gen. Virol. 88, 2862–2871. doi: 10.1099/vir.0.83138-0
238. Zipfel, C., and Felix, G. (2005). Plants and animals: a different taste for microbes? Curr. Opin. Plant Biol. 8, 353–360. doi: 10.1016/j.pbi.2005.05.004.

Chapter 4

Activity of the Novel Fungicide SYP-Z048 against Plant Pathogens

Fengping Chen[1,2], Ping Han[2,3], Pengfei Liu[2], Naiguo Si[4], Junli Liu[4], and Xili Liu[2]

[1]Key Laboratory of Plant Virology of Fujian Province, Institute of Plant Virology, Fujian Agriculture and Forestry University, Fuzhou, 350002, China,

[2]Department of Plant Pathology, China Agricultural University, Beijing 100193, China, [3]Beijing Research Center for Agricultural Standards and Testing, Beijing 100097, China, [4]State Key Laboratory of the Discovery and Development of Novel Pesticide, China ShenYang Research Institute of Chemical Industry, Shenyang 110021, China.

ABSTRACT

In *in vitro* tests with 18 plant pathogens, the fungicide 3-[5-(4-chl orophenyl)-2,3-dimethyl-3-isoxazolidinyl] pyridine (SYP-Z048) was highly effective on inhibiting mycelial growth of various ascomycota and basidiomycota, with EC_{50} values ranging from 0.008 to 1.140 μg/ml. SYP-Z048 had much weaker activity against growth of oomycota with EC_{50} values > 100 μg/ml. In a second *in vitro* test with *Monilinia fructicola* isolates, SYP-Z048 inhibited mycelial growth (EC_{50} = 0.013 μg/ml), germ tube elongation (EC_{50} = 0.007 μg/ml), and sporulation but did not affect spore germination. In a detached pear fruit assay inoculated with *M. fructicola* isolates, SYP-Z048 showed protective and curative activity. Field tests indicated that SYP-Z048 was an efficacious fungicide for control of brown rot disease in two peach orchards. When applied to a single spot on a tomato leaflet in a compound leaf, SYP-Z048 suppressed the growth of *Botrytis cinerea* isolates on the rest 4 leaflets, indicating that the fungicide has systemic movement in plant tissues. These results indicate that SYP-Z048 has potential for management of brown rot causing by *M. fructicola* and other diseases caused by ascomycota and basidiomycota.

INTRODUCTION

Fungi are among the crop pests that reduce crop yield and food quality, and they also can produce toxic compounds that make the produce harmful for consumers. Damage caused by plant-pathogenic fungi is often controlled by the application of fungicides. Currently more than 80% of the fruit and vegetable crops grown in the U.S. receive at least one fungicide application every season; if the crops were left untreated, the yields of most fruit and vegetables would reduce 50 to 95 percent[1]. Fungicides are estimated to increase farm income in the U.S. by nearly $13 billion annually[1]. In China, pesticide application has been increasing annually. Fungicides have become an integral part of efficient food production in China and other countries.

Many fungicides that target different pathogens and diseases have been reported[2, 3, 4, 5], but because of problems relating to toxic residues, disease resurgence and pathogen resistance[6, 7, 8,9, 10], the use of many fungicides is restricted. Therefore, researchers have been searching for fungicides with low toxicity, high selectivity, and high activity against fungal strains that are resistant to other fungicides.

SYP-Z048, 3-[5-(4-chlorophenyl)-2,3-dimethyl-3-isoxazolidinyl] pyridine, a mixture of two diasteromers (Fig. 1), is a novel fungicide developed for control of plant-pathogenic fungi by China Shenyang Research Institute of Chemical Industry in 1997. SYP-Z048 have been registered for control of grey mold caused by *Botrytis cinerea* on tomato in China as Junsiqi (25% EC, China Shenyang Research Institute of Chemical Industry, Shenyang, China)[11]. SYP-Z048 inhibits ergosterol synthesis in *B. cinerea*[12] and has been demonstrated to be a sterol 14α-demethylase inhibitor (DMI)[13].

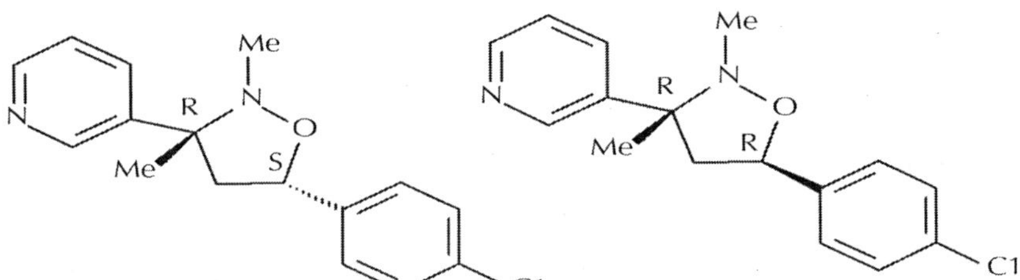

Figure 1: Structure of SYP-Z048 used in the experiments, a mixture of two stereoisomers: 3-[(3R, 5S)-5-(4-chloro-phenyl)-2,3-dimethyl-3-isoxazolidinyl] pyridine (1) and its (3R, 5R)-isomer (2), drawn by Fengping Chen.

DMI fungicides have been playing an important role in chemical control of many diseases since they were developed in 1970s, providing high efficacy against a broad spectrum of fungal pathogens. DMIs consist of many structurally heterogeneous compounds, belonging to different subclasses. The subclasses of imidazole and triazole are among the most widely used fungicides, as well as pyridines, such as pyrifenox[14] and buthiobate[15]. SYP-Z048 is a pyridine derivative that includes a isoxazolidine ring.

It has been documented SYP-Z048 has highly activity on *M. fructicola*, the causal agent of brown rot. The mean EC_{50} values for baseline population of *M. fructicola* is 0.017 μg/ml[13], which is similar to the values for propiconazole (0.03 μg/ml)[16] and tebuconazole (0.016 μg/ml)[17]. The antifungal activity of SYP-Z048 against other plant-pathogenic fungi or on *M. fructicola* causing brown rot in field, however, is unknown.

The objectives of this study were to determine: (a) the *in vitro* antifungal activity of SYP-Z048 against 18 plant-pathogenic fungi; (b) the fungicidal activities of SYP-Z048 on different developmental stages of *M. fructicola*; (c) the preventative and curative activities of SYP-Z048 by detach fruits assay and control efficacy in field on brown rot and (d) systemic activity in tomato leaves.

RESULTS

In Vitro Activity of SYP-Z048 against 18 Plant Pathogens

SYP-Z048 inhibited the *in vitro* mycelial growth of many plant-pathogenic fungi, including ascomycota and basidiomycotac (Table 1) with EC_{50} values of 0.008–1.140 µg/ml). In contrast, SYP-Z048 was much less effective against oomycota with EC_{50} values > 100 µg/ml (Table 1).

Table 1: *In vitro* inhibition of plant pathogens by SYP-Z048

Organism	Species	Correlation* coefficient	EC_{50} (µg/ml)
Ascomycota	*Gibberella zeae*	0.976	0.047
	G.fujikuroi	0.991	0.044
	Fusariumoxysporum	0.994	0.315
	Valsamali	0.978	0.067
	Physalosporapiricola	0.996	0.033
	Moniliniafructicola	0.989	0.013
	Sclerotiniasclerotiorum	0.995	0.009
	Gaeumannomycesgraminis	0.995	0.081
	Cladosporiumcucumerium	0.988	0.046
	Cochlioboluslunatus	0.986	0.008
	Botrytiscinerea	0.976	0.306
	Magnaportheoryzae	0.989	1.140
	Passalorafulva	0.986	0.126
	Alternariasolani	0.992	0.418
Basidiomycota	*Ceratobasidiumcerealis*	0.988	0.573
	Sphacelothecareiliana	0.969	0.079

Oomycota	*Pythiumaphanidermatum*	-	>100
	Phytophthoracapsici	-	>100

Effect of SYP-Z048 on the Different Developmental Stages of *M. Fructicola*

SYP-Z048 substantially inhibited mycelial growth, germ tube elongation, and sporulation of *M. fructicola* (Table 2). The EC_{50} values were 0.013 and 0.001 μg/ml for mycelial growth and germ tube elongation, respectively The inhibition activity of SYP-Z048 was greater for initial germ tube elongation than for mycelial growth. Spore production was significantly inhibited by SYP-Z048 when the concentration was >0.02 μg/ml. Spore germination was not affected by SYP-Z048 (Table 2).

Table 2: The *in vitro* effect of SYP-Z048 on different developmental stages of *M. fructicola*

SYP-Z048 concentration (μg/ml)	Inhibition of mycelial growth (%)	Inhibition of germ tube elongation (%)	Spore production 10^4 spores/ cm^2	Spore germination (%)
0.00	-	-	15.94a*	85.6a
0.005	27.7	65.5	14.91ab	85.6a
0.01	48.9	68.7	12.99ab	85.5a
0.02	56.8	73.1	7.12bc	83.0a
0.03	71.7	74.3	1.15c	85.0a
0.06	81.6	81.1	0.33c	84.8a
EC_{50}	0.013	0.001	-	-

Protective and Curative Activity of SYP-Z048 on Detached Pear Fruit

Whether applied 24 h before inoculation (protective activity) or 24 h after inoculation (curative activity), SYP-Z048 provided substantial control of *M. fructicola* on pear fruit (Table 3). The efficacy for protective and curative activity was 89.3% and 85.2%, respectively (Table 3). Comparison of disease incidence and lesion area after inoculation of *M. fructicola* for 3 days indicated that the fungicide was slowing down disease progression.

Table 3: Protective and curative effect of SYP-Z048 on brown rot in detached pears after inoculation of *M. fructicola* for 3 days

Treatment	**Disease Incidence (%)**	**Lesion area (mm^2)**	**Control Efficacy (%)**
Water spray with inoculation	100a*	193.7a	-
Curative treatment	65.1b	28.7b	85.2
Protective treatment	66.3b	20.8b	89.3
Water spray without inoculation	0c	0c	-

Control Efficacy of SYP-Z048 on Brown Rot in Field

Significant differences among fungicides treatments were observed in both peach orchards, (Fig. 2). All fungicides treatments were statistically superior to water control ($P < 0.05$). SYP-Z048 treatment controlled brown rot significantly better than thiophonate-methyl and propineb when the concentration was > 100 g/L in both locations. Compared to fenbuconazole at same dose, lower efficacy was observed with SYP-Z048 application at the Liaoning location whereas no difference among these treatments was observed at the Shandong location.

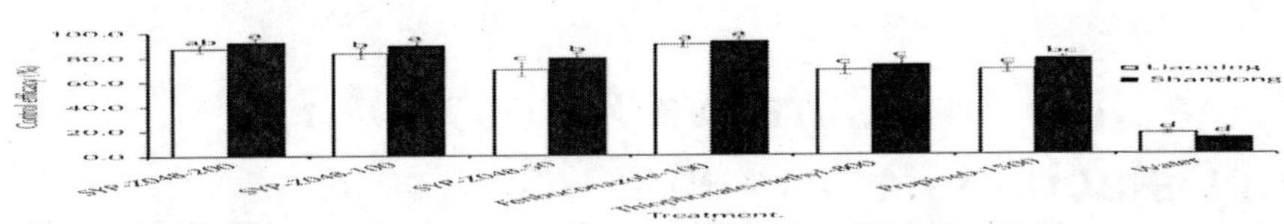

Figure 2: Efficacy of SYP-Z048 and other fungicides for brown rot disease control in peach orchards in Liaoning and Shandong province in China in 2013.

Systemic Activity of SYP-Z048 in Tomato Leaves

In the first experiment concerning systemic activity of SYP-Z048, the fungicide SYP-Z048 inhibited*B cinerea* on the leaflet where the fungicide was applied (the fourth leaflet) (Fig. 3) and suppressed lesion developing on the other four leaflets, although the fungicide was less effective on the untreated leaflets than on the treated leaflet (Fig. 4).

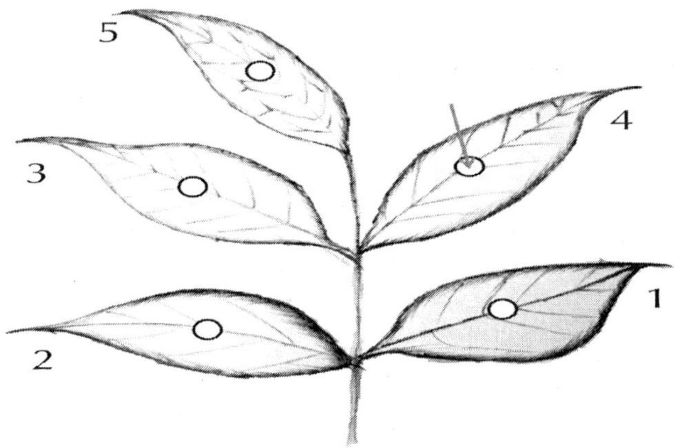

Figure 3: Schematic of a tomato leaf (with five numbered leaflets, created by Fengping Chen) used in the experiments to determine systemic activity of SYP-Z048.

A drop of fungicide was added to the fourth leaflet (arrow) before leaflets were inoculated with mycelium plug (circles) colonized by *B. cinerea*. See Methods section for additional details.

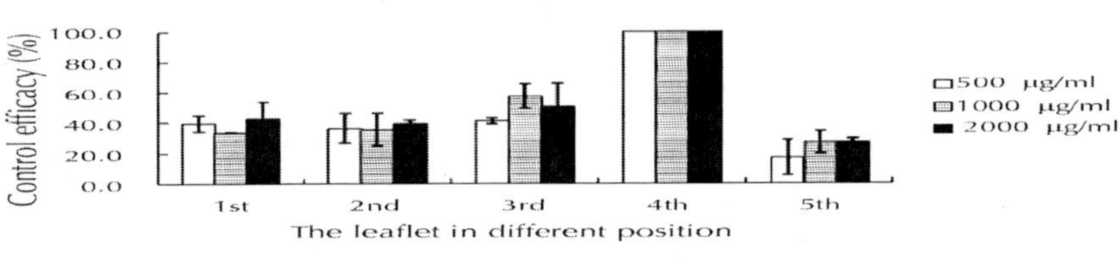

Figure 4: The efficacy of SYP-Z048 against *B. cinerea* after 2 days inoculation when the fungicide was applied to the fourth leaflet and the fungus was

inoculated to all five leaflets on a tomato leaf. In the second experiment concerning movement of SYP-Z048 in tomato leaves, inhibition of *B. cinerea* on the upper epidermis side of the leaflet was similar to inhibition on the lower epidermis side of the leaflet when the fungicide was added on the opposite side of the leaflet (Fig. 5).

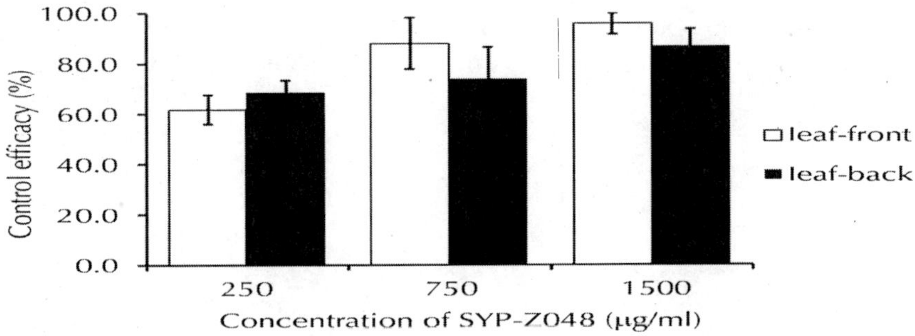

Figure 5: The efficacy of SYP-Z048 against *B. cinerea* after 2 days inoculation when the fungicide was applied to the upper epidermis (leaf-front) or lower epidermis (leaf-back) of a tomato leaflet and the fungus was inoculated to the other side of the same leaflet.

DISCUSSION

SYP-Z048 is a novel fungicide, being both pyridine derivative and also carrying a isoxazolidine ring. Both groups are effective on some pathogens. The combination of two active groups in SYP-Z048 would make this compound behave specially. Isoxazolidine or oxazole derivatives have been synthesized for antitumor, antiamoebic, antimicrobial, herbicidal, and fungicidal activities[18, 19, 20,21, 22, 23]; many are effective against human pathogens. Activity against plant pathogens by other oxazolidinones and oxazolidine-diones involves inhibition of RNA polymerase I and cytochrome b[24, 25]. Pyridine fungicides possess different modes of action according to the specific fungicide. Most of pyridines are respiration inhibitors, for example succinate dehydrate inhibitors boscalid and fluopyram[5] and succinate-cytochrome c reductase inhibitor pyribencarb[26]; some pyridines are 14α-demethylase inhibitors such as buthiobate[15] and pyrifenox[14]; dipyrithione is response to lipopolysaccharide[27] and fluopicolide is involve in modification of the cellular localization of a spectrin-like protein[28]. SYP-Z048 has typical 14α-demethylase inhibitor activity[13], which should be a novel member of the demethylation inhibitors of ergosterol biosynthesis, but whether the activity of SYP-Z048 also results from other mechanism is not clear yet.

The lack of inhibition on *M. fructicola* spore germination supports the modes of action of SYP-Z048 which was implied as a DMI fungicide[13]. In general, DMI fungicides are known to inhibit hyphal growth strongly rather than spore germination[29]. For example, DMIs doesn't affect the urediniospore germination of *Puceinia hemerocallidis*[30] and conidium germination of *Spilocaea oleagina*[31]. But in some pathogens, it is effective on spore germination[32, 33]. The behavior observed for DMIs in pathogens differs due to the stage of starting sterol biosynthesis. Conidia of some species are able to synthesize sterols in the early stage of germination, whereas other initiate sterol synthesis at the very end of the germ tube phase[33].

SYP-Z048 has potential for control of diseases caused by 16 plant-pathogenic ascomycota and basidiomycota fungi but not the oomycota. According to our *in vitro* experiment, the EC_{50} values for SYP-Z048 < 1 μg/ml to all experimental fungi except *M. oryzae*. The data collected under controlled conditions might not be the same as field study, but it reflects the real efficacy to some extent. For example, *M. fructicola* was inhibited strongly by SYP-Z048 *in vitro* (EC_{50} = 0.013 μg/ml), and in consistent, brown rot caused by *M. fructicola* was controlled efficaciously in the field with the application dose of 100 μg/ml (The efficacy > 80%).

SYP-Z048 exhibits substantial activity on brown rot of stone fruit, caused by *M. fructicola*. *M. fructicola* is a world widespread pathogen, including Asia, Europe, Africa, North America, Central America, as well as South America and Oceania[34]. The fungus can cause not only fruit rot but also blossom blight and twig canker, especially causing damage in postharvest fruit[35, 36]. The most effective management approaches are pre-harvest fungicide applications. Although resistance to benzimidazoles[37], dicarboxamides[38], succinate dehydrogenase inhibitors (SDHIs) and quinone outside inhibitor fungicides (QoIs)[39] has been reported in *M. fructicola*, SYP-Z048 is not cross-resistant to compounds belonging to those classes[13]. The comparison of efficacy in two peach orchards showed that SYP-Z048 was more efficacious than thiophonate-methyl (a representative of benzimidazoles) and propineb (a protective fungicide) in this study, even the application concentration was 8 times and 15 times less, respectively. All the facts taken together suggest that SYP-Z048 is a fungicide with practical significance in brown rot management.

SYP-Z048 possesses both protective and curative fungicidal activity. Such fungicides are key components of modern agriculture due to their ability of preventing disease expanding. The detached fruit assay showed SYP-Z048 controlled brown rot well when applied before inoculation, but also showed high efficacy when applied after inoculation which is not significantly different from the pre-treatment. The movement of SYP-Z048 in plant tissue was confirmed using the compound leaf of tomato. The data we obtained indicated that SYP-Z048 could penetration into leaflet and move acropetally, horizontally, and basipetally in plant, and also moved through the leaflet tissues. after application.

In conclusion, SYP-Z048 has substantial *in vitro* activity against a very broad spectrum of plant pathogenic fungi. The fungicide is a systemic fungicide with both protective activity and curative activity against brown rot disease. The novel structure fungicide SYP-Z048 may be expected to be an effective agrochemical agent in plant disease management.

METHODS

Media, Fungicides, and Pathogens

Potato dextrose agar (PDA medium) was used for the routine culture of the fungi and contained 200 g of potato, 15 g of glucose, and 12 g of agar in 1 L of sterile deionized water. The medium was autoclaved for 20 min at 121°C. Water agar was used for the spore germination and tube elongation tests and contained 12 g of agar in 1 L of sterile deionized water, without sterilization. The chemicals were purchased from Syngenta China Ltd. (Beijing). The same batch of medium was used for all replicates within each experiment to reduce variability. The fungicide SYP-Z048 (84%) was kindly provided by China ShenYang Research Institute of Chemical Industry. The chemical was dissolved in methanol and diluted to 10 mg/ml as a stock solution, which was stored at 4°C in the dark to preserve fungicide activity. For each assay on PDA plate, the stock solution was diluted to working concentrations with sterile deionized water. In all cases, the final amount of solvent never exceeded 0.1% in treated and control samples. The formulations of SYP-Z048 (25% Junsiqi EC) was provided by Shenyang Sciencreat ChemicalsCo.,Ltd; fenbuconazole (24% Yingde SC), thiophonate-methyl (70% Jiajiliujunling WP) and propineb (70% bingsenxin WG) were purchased from Dow AgroScience, Jiangsu Rotam Chemistry Co., Ltd., and NantongBaoye Chemical Co., Ltd., respectively.

The activity of SYP-Z048 was tested against the following plant-pathogenic fungi: *Gibberella zeae, G. fujikuroi, Fusarium oxysporum, Valsa mali, Physalospora piricola, M. fructicola, Sclerotinia sclerotiorum, Gaeumannomyces graminis, Cladosporium cucumerium, Cochliobolus lunatus, Botrytis cinerea, Magnaporthe oryzae, Passalora fulva, Alternaria solani, Ceratobasidium cerealis, Sphacelotheca reiliana, Pythium aphanidermatum* and *Phytophthora capsici. S. reiliana* was cultured in Richard's liquid medium (10.0 g of KNO_3, 5.0 g of KH_2PO_4, 2.5 g of $MgSO_4{\cdot}7H_2O$, 0.02 g of $FeCl_3$, 30.0 g of glucose and 1 L of sterile deionized water) at 20°C. The other fungi were cultured on PDA at 25°C.

Inhibitory Activity of SYP-Z048 *in Vitro*

The stock solution of SYP-Z048 in methanol was diluted serially in autoclaved PDA (cooled to around 50°C) to achieve final concentrations according to preliminary test (data not show). Before it solidified, the PDA plus fungicide was added to 9-cm diameter Petri plates, with three replicate plates per concentration. Control plates contained PDA without fungicide but with corresponding concentrations of methanol (≤0.1%). For each of the 18 test fungi listed in Table 1, except *S. reiliana*, a mycelium plug (5 mm) was transferred from the leading edge of an actively growing colony on PDA to the dishes amended with 0 to different concentrations fungicide, with percentage inhibition from 10% to 90%. The dishes were incubated at 25°C in darkness. When the colony in control plates (without fungicide) had covered 75% of the agar surface, the diameter of all colonies of that microorganism were measured. The percentage of inhibition of radial growth was calculated. The tests were replicated three times for each microorganism.

The effect of SYP-Z048 on growth of *S. reiliana*, which does not grow on PDA, was determined with the ELISA plate turbidity method. *S. reiliana* teliospores (10 mg) collected from diseased corn were suspended in 10 ml of 0.03% Tween 20 for 4 h. The suspension was centrifuged at 4000 gfor 10 min. The pelleted teliospores were added to a 0.25% $KMnSO_4$ solution. After 25 min, sterilized water was added and then centrifuged again. The pelleted teliospores were immersed in 0.05 M HCl for 6 h before they were mixed into Richard's medium adjusted to approximately 10^5 teliospores/ml. Liquid containing 75 µl teliospores suspension and 75 µl SYP-Z048 of each concentration were added to each well of ELISA plates. The plates were incubated at 28°C on a rotary shaker (120 r/min) in darkness for 84 h. There were three wells per fungicide concentration, and the experiment was performed twice. The transmittance and optical density at 650 nm (OD_{650}) of the suspension in each well was measured with a microplate reader (Bio-rad model550) and inhibition based on differences in optical densities between fungicide treatments and controls. The median effective concentration value (EC_{50}) was calculated by regressing the percentage inhibition (probability value transformed) against the logarithm value of fungicide concentration (log transformed). Regression analysis was performed with Microsoft Excel 2007.

Effect of SYP-Z048 on different Developmental Stages of *M. Fructicola*

Effect of SYP-Z048 on mycelial growth of *M. fructicola* isolate was measured as described above. The fungicide concentrations were 0, 0.005, 0.01, 0.02, 0.03, and 0.06 mg/l. After colony diameters were measured, the cultures were used to determine the effect of SYP-Z048 on spore production. In cultures with < 0.02 mg/l, six colonized agar plugs, 5 mm in diameter, were excised from each replicate plate; the selected plugs were taken along the radius from the center of the cultureto the margin of the colony. In plates with higher concentrations of fungicide, the colony was too small to collect plugs along the radius, therefore a single plug was collected. The plugs from each plate were put into a 15-ml tube with 2.0 ml of sterile deionized H_2O, agitated with a sterile needle, and shaken for 5 min with a laboratory vortex mixer. Spores were counted using a haemocytometer. Spore production was expressed as the number of spores per mm^2 of colony area. To determine the effect of SYP-Z048 on spore germination and germ tube elongation, spores were harvested from a colony of *M. fructicola* isolate growing on PDA without fungicide by adding sterile deionized water to the plate and agitating. The spore suspension was adjusted to approximately 10^6 spores/ml One hundred µl of the spore suspension was spread on water agar in 9 cm diam petri dishes containing 0, 0.005, 0.01, 0.02, 0.03 and 0.06 ug/ml SYP-Z048 Germination was measured after 6 h at room temperature in the dark by examining 100 spores in three locations on the petri dish at 100× magnification. A spore was scored as germinated if the germ tube had reached at least the full length of the spore. The dishes were then incubated for an additional 4 h under the same conditions and germ tube length was measured on 60 germinated spores per dish at 400× magnification using Image-Pro Plus 5.0 software. Three replicate plates were used for each concentration of SYP-Z048.

Preventive and Curative Activity of SYP-Z048 on Pear Fruit

The methods for determining the preventive and curative activity of *M. fructicola* on pear fruits were similar to those described previously[40, 41]. The pears of "Fengshui" variety were purchased from a supermarket. Before inoculation, pears were kept at 23°C overnight and were rinsed with tap water and air-dried. After the pears were peeled and cut in half, they were placed in a plastic box with a wet filter paper at the bottom, cut side facing down, and were then wounded with a sterile needle on the peeled but uncut surface (six wounds per fruit half). Each wounded site was inoculated with 10 µl of an *M. fructicola* spore suspension (5 × 10^6 conidia/ml) obtained from other pears that had been previously inoculated and infected with *M. fructicola*. SYP-Z048 (200 µg/ml) was applied either 24 h before the fruit was inoculated (protective treatment) or 24 h after the fruit was inoculated (curative treatment); the fungicide was applied to runoff with a fine-mist sprayer. Two control treatments were included. In one control (positive control), fruit were inoculated with *M. fructicola* 24 h after spraying with sterile water. In the other control (negative control), wounded fruit were sprayed with water but not inoculated. The moisture boxes were sealed with clear plastic and arranged randomly in climate control chamber maintained at 23°C, >80% RH, and a 12-h photoperiod. After 4 days, the number of infected wounds and the diameter of the lesions around the wounds were recorded. The experiment was performed twice with 3 replicates per experiment. Each replicate consisted of 12 inoculations.

Control Efficacy of SYP-Z048 on Brown Rot in Field

To assess the efficacy of SYP-Z048 for control of brown rot in the field, the fungicides of SYP-Z048 (25% Junsiqi EC), fenbuconazole (24% Yingde SC), thiophonate-methyl (70% Jiajiliujunling WP) and propineb (70% bingsenxin WG) were applied to trees in two orchards. Three concentrations of SYP-Z048 were tested (50, 100 and 200 g/L) and compared to fenbuconazole, thiophonate-methyl and propineb applied at 100, 800, and 1500 g/L, respectively. Control treatments were trees sprayed with water. Treatments were replicated four times in a randomized block with four trees per replication. All fungicides were applied by sprayer to runoff four times before harvest. 'Jinlu' peach trees were treated on 6 May, 20 May, 4 June and 18 June 2013 in Liaoning province in China. 'Baifeng' peach trees were treated on 29 April, 16 May, 30 May, and 15 June 2013 in Shandong province in China. When fruit were at commercial shipping maturity, a sample consisting of 150 fruit was randomly harvested from each tree and the total number of diseased fruit was counted. Disease incidence was expressed as percentage of diseased fruit and control efficacy was express as: Control efficacy (%) = $(1 - A) \times 100$, where A = disease incidence with fungicide treatment divided by mean disease incidence with water application.

Systemic Activity of SYP-Z048 against *B. Cinerea* on Tomato Leaves

To investigate potential movement and activity of the fungicide in acropetal, horizontal, and basipetal directions, the fungicide was applied in one spot on a tomato leaf with inoculation in the same spot and other locations in two experiments. In the first experiment, pinnately compound leaves (with five leaflets) were collected at the 3rd

leaf position from tomato (*Lycopersicum esculentum* Mill cv. Jiabao) plants growing in the greenhouse. The leaves were rinsed with tap water, air dried, and placed in a plastic box with wet filter paper at the bottom. In the first experiment 10-μL containing 0, 500, 1000, or 2000 μg ml^{-1} SYP-Z048 was applied as a drop to the adaxial side of the fourth leaflet (Fig. 3). The boxes were sealed with clear plastic and arranged randomly in a climate control chamber at 23°C, >80% RH, and a 12-h photoperiod. One day after application, a 3-mm-diam mycelium plug from the margin of an actively growing colony of *B. cinerea* on PDA was placed on the adaxial side of each of the five leaflets. After 2 days the diameter of the *B. cinerea* colony formed around each plug was recorded. The experiment was performed twice with three replicates per treatment. Each replicate consisted of 10 leaves.

In a second experiment, 10 μL containing 0, 250, 750, or 1500 μg ml^{-1} SYP-Z048 was applied to the upper epidermis side of the leaflet or lower epidermis side. After 24 h, the leaflet was incubated as above with a 3-mm-diam mycelium plug placed on the opposite side of the leaflet; i.e., one plug was on the upper epidermis, when the fungicide was treated on the lower epidermis. Colonies were measured and efficacy was calculated as described in the previous paragraph. The experiment was performed twice with 30 replicate leaves per treatment.

Statistical Analysis

All data were analyzed using SAS software ver. 8.0. Difference among means was determined with the Duncan's multiple range test at $P = 0.05$.

ACKNOWLEDGEMENTS

This work was supported by the Special Fund for Agroscientific Research in the Public Interest of China (201303023). This work was also partially supported by National High Technology Research and Development Program of China (2012CB111401). We thank J. L. Starr for his editorial efforts and critique on the manuscript.

REFERENCES

1. Gianessi, L. & Reigner, N. The importance of fungicides in U.S. crop production. *Outlook Pest Manag.* 10, 209–213 (2006).
2. Bartlett, D. W. *et al.* The strobilurin fungicides. *Pest Manag. Sci.* 58, 649–662 (2002).
3. Becher, R. & Wirsel, S. G. R. Fungal cytochrome P450 sterol 14α-demethylase (CYP51) and azole resistance in plant and human pathogens. *Appl. Microbiol. Biotechnol.* 95,825–840 (2012).
4. Kumar, L. *et al.* Imidazole derivatives as possible microbicides with dual protection. *Eur. J. Med. Chem.* 45, 817–824 (2010).
5. Avenot, H. F. & Michailides, T. J. Progress in understanding molecular mechanisms and evolution of resistance to succinate dehydrogenase inhibiting (SDHI) fungicides in phytopathogenic fungi. *Crop Prot.* 29, 643–651 (2010).
6. Bridges, B. A. Mutagenicity of captan and related fungicides. *Mutat. Res.* 32, 3–34 (1975).
7. Hurley, P. M., Hill, R. N. & Whiting, R. J. Mode of carcinogenic action of pesticides inducing thyroid follicular cell tumors in rodents. *Environ. Health Perspect.* 106, 437–445 (1998).
8. Clary, T. & Ritz, B. Pancreatic cancer mortality and organochlorine pesticide exposure in California, 1989-1996. *Am. J. Ind. Med.* 43, 306–313 (2003).
9. Gordon, E. Captan: Transition from 'B2' to 'not likely'. How pesticide registrants affected the EPA Cancer Classification Update. *J. Appl. Toxicol.* 27, 519–526 (2007).
10. FRAC, List of plant pathogenic organisms resistant to disease control agents. (2013) Available from:http://www.frac.info/publication/anhang/List%20of%20resistant%20plant%20pathogenic%20organisms_February%202013%20updated.pdf(Date of access: 10/7/2014).
11. Si, N. G. *et al.* Biological activity and application of a novel fungicide SYP-Z048 (I). *Chin. J. Pestic.* 43, 16–18 (2004).
12. Han, P., Liu, X., Liu, P. & Si, N. Effect of novel fungicide 5-(4-chloro phenyl)-2,3-dimethyl-3-(pyridine-3)-oxazoline on ergosterol biosynthesis in *Botrytis cinerea* by high performance liquid

chromatography. *Chin. J. Anal. Chem.* 34, 1467–1470 (2006).

13. Chen, F. P. *et al.* Baseline sensitivity of *Monilinia fructicola* from China to the DMI fungicide SYP-Z048 and analysis of DMI-resistant mutants. *Plant Dis.* 96, 416–422 (2012).
14. Masner, P. & Kerkenaar, A. Effect of the fungicide pyrifenox on sterol biosynthesis in*Ustilago maydis*. *Pestic. Sci.* 22, 61–69 (1988).
15. Aoyama, Y., Yoshida, Y., Hata, S., Nishino, T. & Katsuki, H. Buthiobate: A potent inhibitor for yeast cytochrome P-450 catalyzing 14α-demethylation of lanosterol. *Biochem. Biophys. Res. Commun.* 115, 642–647 (1983).
16. Zehr, E. I., Luszcz, L. A., Olien, W. C., Newall, W. C. & Toler, J. E. Reduced sensitivity in*Monilinia fructicola* to propiconazole following prolonged exposure in peach orchards. *Plant Dis.* 83, 913–916 (1999).
17. Yoshimura, M. A., Luo, Y., Ma, Z. H. & Michailides, T. J. Sensitivity of *Monilinia fructicola*from stone fruit to thiophanate-methyl, iprodione, and tebuconazole. *Plant Dis.* 88, 373–378(2004).
18. Shaber, S. H., Zhang, L., Szapacs, E. M. & Quinn, J. A. Heterocyclic substituted isoxazolidines and their use as fungicides. U.S. Patent 6313147.B1 (2001).
19. Damodiran, M. *et al.* Antibacterial Activity, Quantitative structure-activity relationship and diastereoselective synthesis of isoxazolidine derivatives via 1,3-dipolar cycloaddition of d-glucose derived nitrone with olefin. *Chem. Biol. Drug Des.* 74, 494–506 (2009).
20. Liu, X.-H., Lv, P.-C., Xue, J.-Y., Song, B.-A. & Zhu, H.-L. Novel 2,4,5-trisubstituted oxazole derivatives: Synthesis and antiproliferative activity. *Eur. J. Med. Chem.* 44, 3930–3935(2009).
21. Turan-Zitouni, G., Kaplancikli, Z. A., Ozdemir, A., Revial, G. & Guven, K. Synthesis and antimicrobial activity of some 2-(benzod oxazol/benzod imidazol-2-ylthio)-N-(9H-fluoren-9-yl)acetamide derivatives. *Phosphorus Sulfur Silicon Relat. Elem.* 182, 639–646 (2007).
22. Vicentini, C. B. *et al.* Synthesis and biological evaluation of a series of substituted pyrazolo 3,4-d -1,2,3-triazoles and pyrazolo 3,4-d oxazoles. *Arch. Pharm.* 331, 269–272 (1998).

23. Bhat, A. R., Athar, F. & Azam, A. New derivatives of 3,5-substituted-1,4,2-dioxazoles: Synthesis and activity against *Entamoeba histolytica*. *Eur. J. Med. Chem.* 44, 926–936(2009).
24. Xiao, K. H. *et al.* Effect of famoxadone on photoinduced electron transfer between the iron-sulfur center and cytochrome c1 in the cytochrome bc1 complex. *J. Biol. Chem.* 278,11419–11426 (2003).
25. Jiang, Y. Effects of oxadixyl treatment on the ultrastructure of *Phytophthora infestans*infecting tomato leaves. *Pestic. Sci.* 35, 153–159 (1992).
26. Kataoka, S., Takagaki, M., Kaku, K. & Shimizu, T. Mechanism of action and selectivity of a novel fungicide, pyribencarb. *J. Pestic. Sci.* 35, 99–106 (2010).
27. Liu, Z. *et al.* Dipyrithione inhibits lipopolysaccharide-induced iNOS and COX-2 up-regulation in macrophages and protects against endotoxic shock in mice. *FEBS Lett.* 582, 1643–1650(2008).
28. Toquin, V. *et al.* A new mode of action for fluopicolide: modification of the cellular localization of a spectrin-like protein. *Pflanzenschutz-Nachrichten Bayer* 59, 171–184 (2006).
29. Sisler, H. D. & Ragsdale, N. N. [Biochemical and cellular aspects of the antifungal action of ergosterol biosynthesis inhibitors]. *Mode of action of antifungal agents* [257–282] (*Cambridge,UK*,1984).
30. Buck, J. W. & Williams-Woodward, J. L. The effect of fungicides on urediniospore germination and disease development of daylily rust. *Crop Prot.* 22, 135–140 (2003).
31. Obanor, F. O., Walter, M., Jones, E. E. & Jaspers, M. V. *In vitro* effects of fungicides on conidium germination of *Spilocaea oleagina*, the cause of olive leaf spot. *N. Z. Plant Prot.*58, 278–282 (2005).
32. Clarkson, J. P., Kennedy, R., Phelps, K., Davies, J. & Bowtell, J. Quantifying the effect of reduced doses of propiconazole (Tilt) and initial disease incidence on leek rust development.*Plant Pathol.* 46, 952–963 (1997).
33. Pontzen, R. & Scheinpflug, H. Effects of triazole fungicides on sterol biosynthesis during spore germination of *Botrytis cinerea*,

Venturia inaequalis and *Puccinia graminis* f. sp. *tritici.Eur. J. Plant Pathol.* 95, 151–160 (1989).

34. EPPO/CABI, Data sheets on quarantine pests:*Monilinia fructicola*. (2010) Available from:https://www.eppo.int/QUARANTINE/fungi/Monilinia_fructicola/MONIFC_ds.pdf(Date of access: 11/7/2014).
35. Hong, C., Holtz, B. A., Morgan, D. P. & Michailides, T. J. Significance of thinned fruit as a source of the secondary inoculum of *Monilinia fructicola* in California nectarine orchards.*Plant Dis.* 81, 519–524 (1997).
36. Luo, Y. & Michailides, T. J. Risk analysis for latent infection of prune by *Monilinia fructicola* in California. *Phytopathology* 91, 1197–1208 (2001).
37. Jones, A. L. & Ehret, G. R. Isolation and characterization of benomyl-tolerant strains of*Monilinia fructicola*. *Plant Dis.* 60, 765–769 (1976).
38. Elmer, P. A. G. & Gaunt, R. E. The biological characteristics of dicarboximide-resistant isolates of *Monilinia fructicola* from New Zealand stone-fruit orchards. *Plant Pathol.* 43,130–137 (1994).
39. Amiri, A., Brannen, P. M. & Schnabel, G. Reduced sensitivity in *Monilinia fructicola* field isolates from South Carolina and Georgia to respiration inhibitor fungicides. *Plant Dis.* 94,737–743 (2010).
40. Harada, Y. *et al. Monilia mumecola,* a new brown rot fungus on *Prunus mume* in Japan. *J. Gen. Plant Pathol.* 70, 297–307 (2004).
41. Valencia-Chamorro, S. A., Perez-Gago, M. B., Del Rio, M. A. & Palou, L. Curative and preventive activity of hydroxypropyl methylcellulose-lipid edible composite coatings containing antifungal food additives to control citrus postharvest green and blue molds. *J. Agric. Food Chem.* 57, 2770–2777 (2009).

Chapter 5

Production of Platinum atom Nanoclusters at one end of Helical Plant Viruses

Yuri Drygin[1], Olga Kondakova[2], and Joseph Atabekov

[1]Belozersky Institute of Physico-Chemical Biology, Lomonosov Moscow State University, Vorob'evy gory 1, Building 40, Moscow 119992, Russia

[2]Department of Biology, Lomonosov Moscow State University, Vorob'evy gory 1, Building 12, Moscow 119992, Russia

ABSTRACT

Platinum atom clusters (Pt nanoparticles, Pt-NPs) were produced selectively at one end of helical plant viruses, tobacco mosaic virus (TMV) and potato virus X (PVX), when platinum coordinate compounds were reduced chemically by borohydrides. Size of the platinum NPs depends on conditions of the electroless deposition of platinum atoms on the virus. Results suggest that the Pt-NPs are bound concurrently to the terminal protein subunits and the 5′ end of encapsidated TMV RNA. Thus, a special structure of tobacco mosaic virus and potato X virus particles with nanoparticles of platinum, which looks like a push-pin with platinum head and virus needle, was obtained. Similar results were obtained with ultrasonically fragmented TMV particles. By contrast, the Pt-NPs fully filled the central axial hole of in vitro assembled RNA-free TMV-like particles. We believe that the results presented here will be valuable in the fundamental understanding of interaction of viral platforms with ionic metals and in a mechanism of nanoparticles formation.

INTRODUCTION

One current goal in nanobiotechnology is to investigate biological nanoplatforms capable of binding metal atoms with the aim of generating new bioinorganic materials for nanoelectronics and medicine. Helical plant viruses, in particular tobacco mosaic virus (TMV) and potato virus X (PVX), have been widely used as templates and scaffolds in nanotechnology [1–12].

TMV particles are rod-like with a diameter of 18 nm and modal length of 300 nm. They consist of 2130 identical 17.5 kDa protein subunits helically arranged around a cylindrical canal and closely packed into a rigid tube. The two-layer cylindrical substructure, each layer consisting of a ring of 17 molecules of coat protein (CP), is known as a "disk," and 16 1/3 molecules are present in each turn of the assembled helix. RNA is introduced between the CP turns and follows the helix of protein subunits [13, 14]. Apparently, the front stereochemical surface of terminal CP molecules of the helical TMV particle is not equivalent to inner surfaces of other CP subunits.

It is known that in the absence of RNA the viral CP may be in vitro assembled into several types of aggregate. In particular, TMV CP can be in vitro assembled into virus-like particles (VLPs) that are structurally similar to native virions [15, 16].

The virions of another helical virus, PVX, are flexuous filaments with modal lengths in the range of 470–580 nm and a diameter of approximately 13 nm [17]. In this work, we found that ions of platinum coordinate compounds reduced by borohydrides could form nucleation centers for the very selective growth of nanoparticles on one pole of helical plant viruses (tobacco mosaic virus and potato virus X).

MATERIALS AND METHODS

Modification of Viruses with (Dien) Platinum, (Chloro(Diethylenetriamine)Platinum(Ii) Chloride or [Pt(Dien)Cl]Cl), and Sodium Borohydride

100 µL of TMV or PVX suspension (0.4 mg/mL) in TDW (deionized double distilled water) was mixed with 2 µL of 25 mM [Pt(dien)Cl]Cl in TDW. The mixture was incubated for 1 h at 30°C, then borate buffer 0.5 M (pH 8.3) was added up to 0.1 M. The reaction mixture was cooled in an ice bath, and sodium borohydride (5mg/mL) was added portion wise to a final concentration of 3 mM or 10 mM. To do Pt-NPs larger and produce short nanowires, concentration of (dien)platinum was increased two fold. The reaction mixture was diluted 4-fold with TDW and samples were examined with transmission electron microscopes Zeis LEO912 AB OMEGA supplied with the energy filter or with TEM Jeol JEM 1011.

Modification of TMV with (Dien) Platinum and Hypophosphite

100 µL of suspension of virus (TMV, PVX, 1 mg/mL) was mixed with 4 µL 20 mM (dien)platinum in water (pH~7), kept for 60 min at 30°C.

Then reducing agent sodium hypophosphite (Fluka) was added to a final concentration of 30 mM, and the mixture was incubated for 20 min at room temperature.

Electron Microscopy of the Virus RNA and Ribonucleoproteins

2% collodion solution (Fluka) strengthened by an evaporated carbon was used for preparation of support films specimens for electron microscopy. Usually, 5 μL of TMV suspension was placed on the grid for 1 min, after which the droplet was removed by a filter paper. While samples were examined with transmission electron microscopes Zeis LEO912 AB OMEGA supplied with the energy filter, no staining was done. In some cases (indicated), virus particles were briefly stained with 2% uranyl acetate for additional positive contrasting.

To observe the RNA, samples were prepared by the protein-free monolayer spreading method described in [18]. Concentrations in the spreading solution were 0.1 mg/mL TMV, 3 M urea, 3.7% formaldehyde, and 0.01% benzyldimethylalkylammonium chloride. The hypophase was 0.15 M sodium acetate and 0.1% formaldehyde. The monolayer was picked up on collodion support, and the sample was rotatory shadowed with tungsten under 7°.

RESULTS AND DISCUSSION

Polar Electroless Deposition of Platinum onto Helical Viruses

In the present work, the structure of bioinorganic nanocomplexes, produced by platinum atoms reduced on a platform of helical plant viruses (TMV, PVX) and on RNA-free TMV VLPs, was studied.

Previously, to synthesize DNA [19] and RNA [20] probes, we used the platinum coordinate compound chloro(diethylenetriamine) platinum(II) chloride (mw 369.15) as a label. [Pt(dien)Cl]Cl was synthesized according to Watt and Cude [21]. It is known that at a ratio of 1 and less per 10 nucleotide residues of double-stranded DNA, (dien)platinum is associated almost exclusively with guanine residues at position N7 [22].

Because it was possible that the (dien)platinum would react with the TMV protein, it was necessary to determine the working concentrations of the reactants and the optimal (dien)platinum/virus ratio to obtain stable complexes and avoid TMV aggregation. As found by sedimentation analysis in an analytical ultracentrifuge, the optimal ratio of (dien)platinum to the virus (in moles of CP) must be 50:1 or less to obtain stable (dien)platinum reacted TMV particles in solution.

Chemical reduction of the complex platinum ions bound to the TMV platform was performed in aqueous sodium borohydride or dimethylamine borane solutions using different reaction times and excesses of the reducing agent. Clusters of platinum atoms and short nanowires formed on the native TMV scaffold were revealed under these experimental conditions (Figure1) by transmission electron microscopy (TEM).

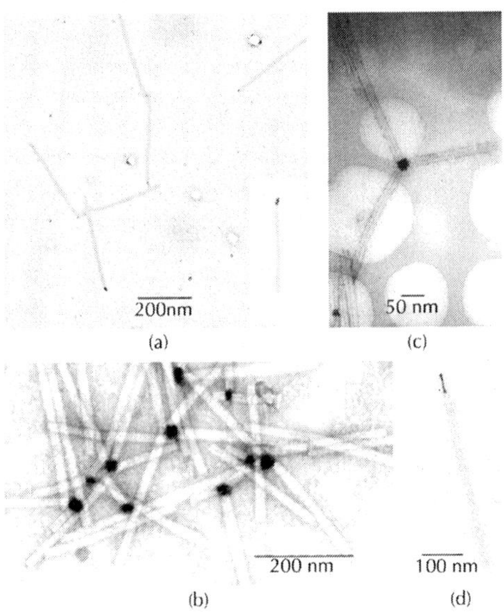

Figure 1: Polar growth of the platinum atom clusters on the TMV platform. Platinum nanoparticles were formed after chemical reduction of (dien) platinum by sodium borohydride to a final concentration of 3 mM (a) or 10 mM

((b)–(d)), see Section 2). Samples were examined with transmission electron microscopes Zeis LEO912 AB OMEGA supplied with the energy filter (specimens ((a), (d)) and Jeol JEM 1011 (specimens (b), (c)). No staining of specimens ((a), (d)) was carried out. Samples ((b), (c)) were stained slightly with 2% uranyl acetate. Arrows in (a) indicate the platinum atom clusters bound to the end of the virus particles.

Figure 1(a) shows that small (~1-2 nm) Pt nanoparticles (Pt-NPs) could be formed on one end of the TMV particles upon exposure to 0.5 mM [Pt(dien)Cl]Cl in 3 mM solution of the reducing agent at pH~8. It is noteworthy that 55% of TMV particles (200 particles were analyzed) contained the Pt nanoparticles on one end, while the number of virus particles carrying Pt-NP clusters on both ends was negligible (less than 0.5%), and no lateral electroless deposition of the platinum on the virus exterior was detected. The formation of Pt-NPs was also observed on one end of the end-to-end aggregates of TMV particles (longer than 300 nm) and TMV fragments shorter than 300 nm. Figure 1(a) (inset) shows that several small discrete Pt-NPs could also be linked to one end of the TMV particle; presumably the primary platinum nanoparticle works here as a nucleation center.

Increasing concentrations of (dien)platinum to 1 mM and borohydride to 10 mM resulted in an increase in the size of Pt-NPs to ≤30 nm as shown in Figures 1(b) and 1(c). Frequently, these TMV particles are aggregated by their terminal "Pt-NP ends" into star-like structures (Figure 1(c)). TMV particles with the central hole partly filled with Pt-nanowire (Figure 1(d)) were also revealed, comprising 20% of the total number of the virus particles. The Pt-nanowires were 10–30 nm in length with a diameter of approximately 3 nm and resembled a small helix.

To prove the platinum origin of the nanoparticles, solution of TMV with bound platinum nanoparticles was concentrated, and diffraction of electron on nanoparticles was examined. Diffraction of electrons for concentrated sample, shown in Figure 1(b) and unstained with uranyl acetate, was obtained by energy-filtering TEM and showed typical picture of face-centered cubic lattice of platinum (Figure 2), with intrinsic interplanar distances within the measurement error.

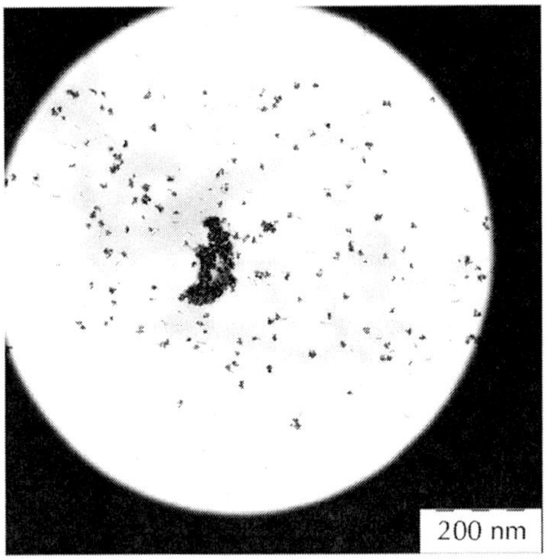

(a)

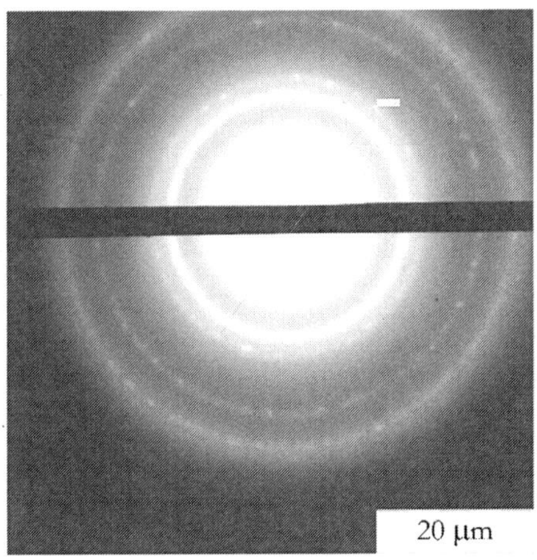

(b)

Figure 2: Electron diffraction of the Pt-NPs-TMV preparation. Large area image of the sample in Figure 1(b) concentrated on the specimen grid and its electron diffraction ((a) and (b), resp.).

Production of Pt-NPs by reduction of the (dien) platinum-pretreated TMV with dimethylamine borane (10 mM, pH 7, 60 min, r.t.) was less efficient than that with sodium borohydride. Presumably, this was because of dimethylamine borane's weaker reducing potential [23].

While potassium tetrachloroplatinate (IV) was reduced by sodium borohydride in weakly alkaline conditions, polar growth of clusters of platinum atoms on virus particles was observed. One hundred of the TMV particles were examined, 55% had platinum atom clusters at one end and 3% at both ends (not shown).

To further elucidate this phenomenon, we examined the interaction of (dien)platinum and its reduction by sodium borohydride with another helical virus, the potexvirus. Figure 3 shows that several discrete small Pt-NPs were associated with one end of filamentous PVX, suggesting that the polar binding of Pt-NPs to helical viruses is a common phenomenon.

Figure 3: Polar binding of the platinum atom clusters to PVX particles. Sample preparation described in Section 2. Specimen was examined under TEM Zeis LEO912 AB OMEGA microscope, no contrasting of specimen.

The (dien)platinum reacts in preference with guanine residues of DNA [22] and, presumably, of RNA [20]. It is known that TMV and PVX RNAs contain only two identical terminal chemical structures, both at the 5′ end, which are guanine residues in the cap group (m^7GpppGp) [24, 25].

It appears that the recognition of one end of the TMV (and PVX) particle by Pt atoms is due to identical 5′-end cap group and peculiarities of the polar geometry of the viral helix. By its selectivity this phenomenon is similar to the specific interaction of monoclonal antibodies with antigenically different terminal surfaces at the virus particle ends [26, 27].

There were three possible mechanisms for platinum atom interaction with the 5′ end of the initial TMV and PVX particles: (i) (dien)platinum interacts with the RNA cap structure and Pt-NPs interact strongly with the 5′ end of the virus RNAs; (ii) (dien) platinum atoms and Pt-NPs recognize and interact strongly with the exposed surface exterior of the 5′-terminal protein or with a disk face structure; (iii) both interaction types of (dien) platinum and Pt-NPs are realized. To select between these mechanisms, the following experiments were carried out.

Pt-NPs Are Deposited on The 5′ End of TMV RNA

It has been shown that the stripping of TMV CP subunits from the 5′ end of RNA is induced by urea, DMSO (dimethyl sulfoxide), alkali, and detergent [28, 29]. Polar stripping results in the production of rods with a tail of RNA protruding from one end. The 5′-proximal 69 nucleotides of TMV RNA lack guanine bases and interact more weakly with CP subunits compared to other regions of RNA [30]. As a consequence, polar 5′ to 3′ uncoating of RNA occurs because of the removal of the terminal CP molecules.

Therefore, (dien)platinum was reduced by sodium borohydride on the tobacco mosaic virus, and the Pt-NPs-containing virus particles were then treated with urea. To fix the structure obtained, formaldehyde was added up to 3.7%. Partly stripped TMV rods with a tail of RNA protruding from one end were produced (Figure 4).

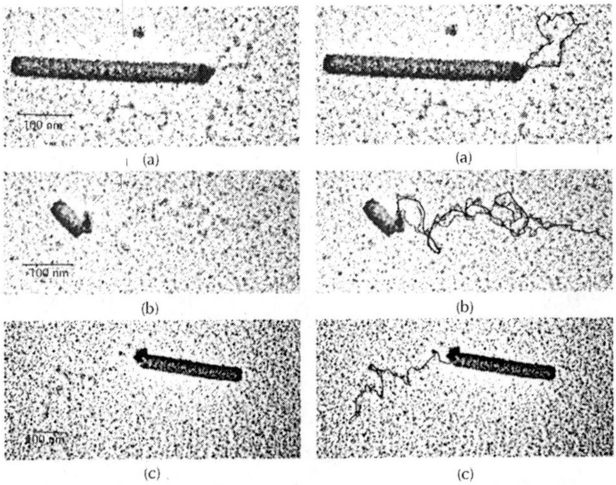

Figure 4: Partial uncoating of TMV and Pt-NPs complexes by 6 M urea. Uncoating was allowed to proceed at 0°C for 5 min in 6 M urea and 60 mM sodium phosphate at a virus concentration of 0.5 mg/mL. Shadowed samples (see Section 2) were examined with a Jeol JEM1011 microscope. Analysis was conducted on 130 uncoated partially TMV particles. Left: ((a), (b)) show a loop of the free RNA released from TMV particle retained by Pt nanoparticle and (c) shows the partially hydrolyzed linear RNA. It is important to note that RNA has secondary structure under spreading conditions. Right: ((a)–(c)) merge images with inked over RNA strand.

Treatment of the Pt-NPs-TMV complex with urea leads to a partial stripping of the viral capsid proteins adjacent to the nanoparticle and to the protruding free RNA (Figure 4(c)). The Pt-NPs were still associated with the shortened virus rods after removal of RNA tails by RNase A (20 μg/mL, 30 min, 30°C) (data not shown). However, Figure 4 shows that, after treatment with ultrasound and urea, the Pt-NP is located on the inner edge of the truncated viral particles. How does this translocation of nanoparticles take place?

There are two possible explanations.

- Since the disassembly and reassembly of TMV are reversible, and the viral proteins have a high propensity toward association, the 5′-end complex of "Pt-NPs-RNA-CP" does not move out and retain contact with the 5′ end of the truncated viral particle via protein-protein interactions. Freed bare RNA is looped out, but bound continuously to the 5′-terminal facing of the viral capsid (Figures 4(a) and 4(b)). It is well known that the purified TMV preparation has some sticky ribonuclease [31]. Free denatured RNA undergoes simple hydrolysis, producing free ends retaining links with the truncated virus particle and leaving Pt-NPs on the edge of the shortened virus rod (Figure 4(c)).

- Pt-NPs bind more strongly to the 5′-terminal end of TMV preeminently because of the more appropriate spatial arrangement of the complex-forming groups of the virus protein and RNA. Therefore, the nanoparticle translocates from the disembodied capsid protein to the truncated (via the urea) end of the viral particle because of linking with bare RNA is weaker (Figure 4(c)).

This observation allows us to hypothesize that the cap groups were not accessible or weakly associated with the Pt-NPs being encapsidated in the virus helix. This also suggests that the Pt-NPs were linked tightly to CP subunits within the 5′-terminal facing or several exterior surfaces of subunits in the TMV helix. If so, the Pt-NPs were linked to the domain(s) of CP subunits located at the terminal surface and adjacent to the central hole of the helical particle. This suggestion is also in line with the ability of Pt-NPs to penetrate the internal canal of TMV tubes, producing nanowires.

Pt-NPs are Bound to the 5′-Terminal Protein Facing of the TMV Virus Particle

Strong binding of the Pt-NPs to the TMV was confirmed by the linkage resistance to the ultrasonication treatment. Results of two series of experiments illustrate this finding.

(i) The virus was first ultrasonicated at 0°C to produce halves of particles and then reacted with (dien)platinum under the same conditions. The size of fragments varied from 25 to 200 nm with the peak (42%) at 100–120 nm. As expected, approximately half (55%, 100 TMV particles were analyzed) of sonication-generated fragments contained Pt-NPs bound to one end (Figures 5(a) and 5(b)).

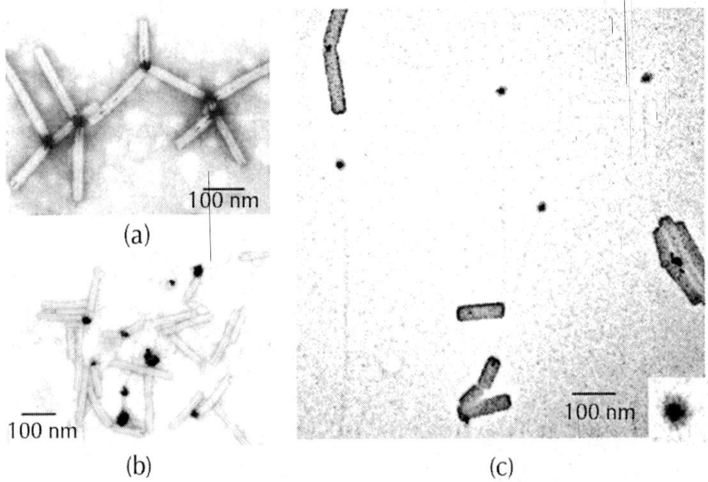

Figure 5: Platinum modification of the ultrasonically treated TMV. (a) chilled TMV samples (1 mg/mL, in ice water) were ultrasonicated at 40 W for five seconds six times with intervals of 60 s using the ultrasonic disintegrator MSE Soniprep 150. TMV was then reacted with 1 mM (dien)platinum and reduced by 10 mM sodium borohydride, (pH~8). (b) TMV was first incubated with 1 mM (dien)platinum and reduced with 10 mM sodium borohydride, then, was exposed to 40 W sonication for 5 s six times with intervals of 60 s at 0°C. Samples (a) and (b) were diluted 10-fold with TDW and stained with 2% uranyl acetate. To observe viral RNA (c), a sample prepared as described in (b) was developed via the monolayer technique (see Section 2). Inset shows 3-fold magnification of the Pt nanoparticle apparently included in the terminal TMV protein facing or disk.

(ii) TMV was first reacted with (dien)platinum and then ultrasonicated. TMV fragments ranged in size from 25 to 200 nm with two peaks: 160 nm (22%) and 120 nm (17%). Only 23% of the TMV particles (100 TMV particles were analyzed) were associated with Pt-NPs; approximately half of the Pt-NPs were detached from the virus rods (20–30% at least, some aggregated) (Figure 5(b)). One can propose that the essential part of free Pt-NPs is just shaken off from the TMV fragments by ultrasonication or they are associated with a low contrasted and weakly visible filamentary material (RNA?).

To gain further insight into the origin of a low contrasted filamentary material with apparently bound Pt-NPs, the fragments generated by sonication of reacted with (dien)platinum virus were fixed with formaldehyde, and benzylalcylammonium chloride was then added. To spread over RNA molecules, urea was added up to a final concentration of 3 M, and the reaction cocktail was loaded onto a hypophase (0.15 M sodium acetate and 0.1% formaldehyde).

Significantly, this technique revealed unusual structures comprising the disk-like element (apparently consisting of CP subunits) with Pt-NPs bound to this disk and the tail of RNA connecting the disk with the remainder of the stripped TMV particle (Figure 5(c) and inset). We assume also that the Pt-NP, by binding to the 5′ end of the RNA and to the 5′-terminal protein (or disk) facing of the TMV virus particle, weakens the connection between this exterior and the rest of the virus particle.

The results obtained suggest that the rigid protein shell is broken into fragments by sonication, while the flexible RNA remains intact (Figures 4 and 5(c)). This finding can be explained by the varying diametrical stiffness of the virion and RNA. It should be stressed that uncoating the TMV modified with (dien)platinum by ultrasonic treatment starts from one end (Figure 4) and is similar to the native TMV uncoating with urea, DMSO, alkali, SDS (sodium dodecyl sulfate), and heat treatment [28, 29]. This confirms, despite its seemingly symmetrical form (see, e.g., the PDB 3D image for TMV), that the virus particle has chemical and physical structural asymmetries at input and output ends. Interestingly, this is logically related in the biological sense to the polarity of TMV RNA-concurrent with uncoating, translation of the parent virion RNA starts at the 5′ end [30].

It appears that the platinum nanoparticle is bound both to the 5′ end of the viral RNA and to the capsid protein, and this association is resistant to the ultrasonic processing and urea treatment. Taken together, these data provide strong evidence that Pt-NPs can be linked directly to the first 5′ surface exterioir of terminal subunits in the TMV CP helix. Concurrent link formation between the Pt-NPs and the 5′ end of encapsidated RNA is also indicated.

It is difficult now to predict with which atoms of amino acid residues of CP and nucleotide residues of RNA at the very 5′-end of the virus RNA interacts (dien)platinum ion without special study of (dien)platinum docking on the pole TMV and PVX exterior surfaces.

In connection with this, it would be interesting to add other ligands to a platinum modified virus particle that are able to enter the metal coordinate sphere, using chemical properties of coordinate metals and clusters of their atoms bound to a virus.

Platinum Modification of TMV-Like Particles

As mentioned above, RNA-free TMV-like particles can be assembled from viral CP in vitro[15–17].

CP was extracted from TMV using an acetic acid according to Fraenkel-Conrat et al. [32]. Then, a sodium phosphate buffer (100 mM, pH 5.6) was added to the suspension of TMV CP to obtain a final concentration of 1 mg/mL, and the protein was incubated at room temperature for 24 h in order to produce VLP. This solution was stored at 4°C.

It was discovered that (dien)platinum and clusters of the reduced platinum atoms do not bind to the VLP via the polar mode; instead, they fill the central hole of the TMV-like particles throughout their length (Figure 6). This finding supports the conclusion on the high sensitivity of (dien) platinum and/or Pt-NPs to spatial organization of functional chemical groups of CP on the surface of VLP and TMV.

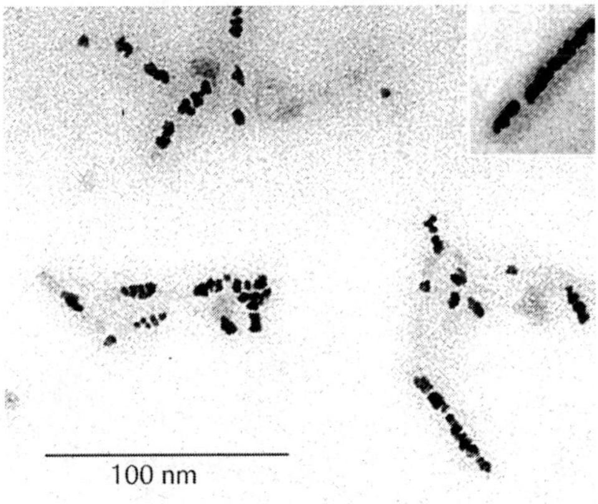

Figure 6: Nanoparticles of platinum fill the inner canal of the TMV VLP. TEM images of Pt nanoparticles in the inner canal of the virus-like rods produced from the coat protein of TMV. Incubation of VLP with (dien)platinum followed

by electroless deposition of Pt atoms took place under the same conditions as with TMV. Inset in the upper corner indicates clearly the coat protein tube around the Pt nanoparticles. No contrasting of specimen was done.

Reduction of (Dien) Platinum with Hypophosphite

Influence of the reduction agent and the reaction conditions were of much importance to understand the mechanism of the (dien)platinum interaction with TMV. Both, the surface and central axial hole of TMV were stained slightly with tiny atom clusters of platinum when (dien) platinum was reduced by 30 mM hypophosphite at pH 6-7 (Figure 7). Thus, modification of the reaction conditions drastically changed character of (dien)platinum-TMV interaction.

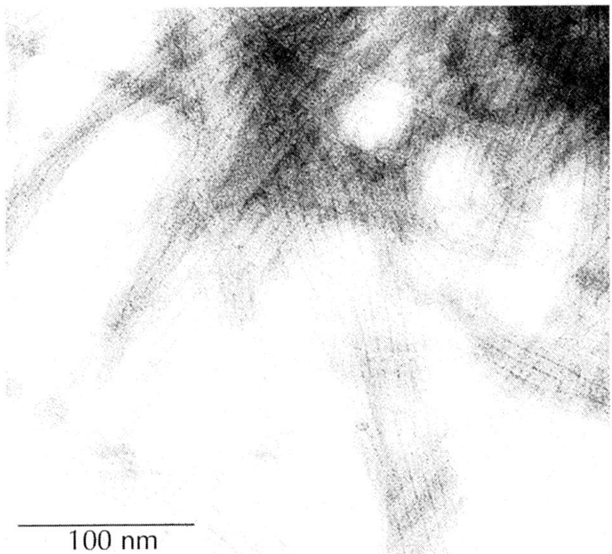

Figure 7: Chemical reduction of (dien)platinum on the TMV platform by hypophosphite. No additional staining was carried out.

Summing up, one can conclude that platinum coordinate compounds (dien)platinum and tetrachloroplatinate(IV) interact very selectively with spiral viruses TMV and PVX and after chemical reduction with borohydrides form nanoparticles via the polar mode with the 5′ end of the viruses (corresponding to RNA), as demonstrated through the polar uncoating of TMV with urea.

Interestingly Balci et al. [4] described the self-assembling of prepared preliminary gold nanoparticles with a diameter of 6 nm and TMV into a metal-virus nanodumbbells. The gold nanoparticles were selectively bound to both ends of the virus rods and can be enlarged by electroless deposition of gold associated with shortening of the virus particles to yield gold-virus-gold dumbbells.

Complexes of TMV and different ionic metals (Ni, Co, Cu, Fe, Ag, and Au) were examined by us as well. We found that copper could only interact with TMV in a polar mode, 100 TMV particles were examined and found 53% one end- and 1% two end- labeled TMV particles (data not shown) after borohydride reduction.

CONCLUSIONS AND PROPOSAL

Nanoparticles of platinum that grows by polar mode on TMV and PVX helical viruses were produced by the borohydrides reduction of platinum coordinate compounds. Size of the platinum NPs depends on conditions of the chemical reduction of platinum atoms on the virus. Electron microscopy examination of polar uncoating of TMV particles showed that the Pt-NPs are bound concurrently to the terminal protein subunits and the 5′ end of encapsidated TMV RNA.

Above and beyond nanorods, nanowires and nanorings produced earlier on the plant viruses as templates, tobacco mosaic virus, and potato X virus particles in a complex with nanoparticles of platinum, which look like a push-pin with platinum head and virus needle, were obtained. Thus, in this work, a novel structure of tobacco mosaic virus, and potato X virus particles in a complex with nanoparticles of platinum, which demonstrates extraordinary specificity of the coordinate metal interaction with a complex polyvalent biological construction, were obtained.

TMV is established as one of the best-investigated models of macromolecular organization in biology. Moreover, partially and completely TMV can be reconstituted from its RNA and coat protein. The process of the TMV reconstitution was studied kinetically [16]. Thus, population of the TMV nucleoproteins of definite size (length) containing 5′ pole (in respect to RNA) can be obtained. On the other hand, 3′-terminal ribose of RNA in these nucleoproteins might be modified (oxidized, e.g.) and fixed covalently on any appropriate surface. One can propose that modified with platinum (or copper) population of these nanosize nucleoproteins of different lengths could serve as comb-like structures, or as strings of different thickness in a stringed instrument that are structurally (and functionally) similar to a musical organ. From our point of view, such wellorganized nanoparticle structures are of high demand in nanoelectronics.

We also cannot exclude that manifold ligands might be linked to the TMV and PVX viruses due to high chemical coordinate potential of platinum atoms.

ACKNOWLEDGMENTS

The authors are grateful to Dr. O. V. Karpova for the purified TMV (strain U1) and PVX (Russian strain) and Dr. M. Arkhipenko for the TMV coat protein preparation. They are thankful to Sergei Ambramchuk and Anatoly Bogdanov for assistance with the electron microscopes operation. They deeply appreciate the help of Dr. Emma Hoyle for English language assistance. This work was supported in parts by the Federal Agency for Science and Innovations contract No. 16.M04.12.0024 from the Russian Federation Ministry of Science and Education and the RBRR grant 12-04-01711-a to Yu.F.D.

REFERENCES

1. E. Dujardin, C. Peet, G. Stubbs, J. N. Culver, and S. Mann, "Organization of metallic nanoparticles using tobacco mosaic virus templates," Nano Letters, vol. 3, no. 3, pp. 413–417, 2003.
2. M. Knez, A. Kadri, C. Wege, U. Gösele, H. Jeske, and K. Nielsch, "Atomic layer deposition on biological macromolecules: metal oxide coating of tobacco mosaic virus and ferritin," Nano Letters, vol. 6, no. 6, pp. 1172–1177, 2006.
3. R. J. Tseng, C. Tsai, L. Ma, J. Ouyang, C. S. Ozkan, and Y. Yang, "Digital memory device based on tobacco mosaic virus

conjugated with nanoparticles," Nature Nanotechnology, vol. 1, no. 1, pp. 72–77, 2006.

4. S. Balci, K. Noda, A. M. Bittner et al., "Self-assembly of metal-virus nanodumbbells,"Angewandte Chemie, vol. 46, no. 17, pp. 3149–3151, 2007.

5. M. Ł. Górzny, A. S. Walton, M. Wn k, P. G. Stockley, and S. D. Evans, "Four-probe electrical characterization of Pt-coated TMV-based nanostructures," Nanotechnology, vol. 19, no. 16, Article ID 165704, 2006.

6. M. Kobayashi, K. Onodera, Y. Watanabe, and I. Yamashita, "Fabrication of 3-nm platinum wires using a tobacco mosaic virus template," Chemistry Letters, vol. 39, no. 6, pp. 616–618, 2010.

7. X. Chen, K. Gerasopoulos, J. Guo et al., "Virus-enabled silicon anode for lithium-ion batteries," ACS Nano, vol. 4, no. 9, pp. 5366–5372, 2010.

8. A. Mueller, F. J. Eber, C. Azucena et al., "Inducible site-selective bottom-up assembly of virus-derived nanotube arrays on RNA-equipped wafers," ACS Nano, vol. 5, no. 6, pp. 4512–4520, 2011.

9. A. K. Manocchi, S. Seifert, B. Lee, and H. Yi, "In situ small-angle X-ray scattering analysis of palladium nanoparticle growth on tobacco mosaic virus nanotemplates,"Langmuir, vol. 27, no. 11, pp. 7052–7058, 2011.

10. J. C. Zhou, C. M. Soto, M. S. Chen et al., "Biotemplating rod-like viruses for the synthesis of copper nanorods and nanowires," Journal of Nanobiotechnology, vol. 10, article 18, 2012.

11. C. Azucena, F. J. Eber, V. Trouillet et al., "New approaches for bottom-up assembly of tobacco mosaic virus-derived nucleoprotein tubes on defined patterns on silica- and polymer-based substrates," Langmuir, vol. 28, no. 42, pp. 14867–14877, 2012.

12. C. Hou, Q. Luo, J. Liu et al., "Construction of GPx active centers on natural protein nanodisk/nanotube: a new way to develop artificial nanoenzyme," ACS Nano, vol. 6, no. 10, pp. 8692–8701, 2012.

13. A. Klug, "The tobacco mosaic virus particle: structure and assembly," Philosophical Transactions of the Royal Society B, vol. 354, no. 1383, pp. 531–535, 1999.
14. F. A. Anderer, "Recent studies on the structure of tobacco mosaic virus," Advances in Protein Chemistry, vol. 18, pp. 1–35, 1964.
15. D. L. D. Caspar, "Assembly and stability of the tobacco mosaic virus particle," Advances in Protein Chemistry, vol. 18, pp. 37–121, 1964.
16. P. J. G. Butler, "Self-assembly of tobacco mosaic virus: the role of an intermediate aggregate in generating both specificity and speed," Philosophical Transactions of the Royal Society B, vol. 354, no. 1383, pp. 537–550, 1999.
17. J. Atabekov, E. Dobrov, O. Karpova, and N. Rodionova, "Potato virus X: structure, disassembly and reconstitution," Molecular Plant Pathology, vol. 8, no. 5, pp. 667–675, 2007.
18. H. J. Vollenweider, J. M. Sogo, and K. T. Koller Th., "A routine method for protein free spreading of double and single stranded nucleic acid molecules," Proceedings of the National Academy of Sciences of the United States of America, vol. 72, no. 1, pp. 83–87, 1975.
19. O. A. Kondakova and Y. F. Drygin, "Diagnostics of potato spindle tubor viroid disease with dien-Pt DNA probe," Biotechnologya, vol. 1, pp. 83–90, 1999 (Russian).
20. Y. F. Drygin, J. G. Atabekov, O. A. Kondakova, S. N. Chirkov, R. A. Zinovkin, and V. I. Kiseleva, "Method for simultaneously detecting a plurality of RNA sequences in a biological sample," Patent WO 123494, 2009.
21. G. W. Watt and W. A. Cude, "Diethylenetriamine complexes of platinum(II) halides,"Inorganic Chemistry, vol. 7, no. 2, pp. 335–338, 1968.
22. N. P. Johnson, J. P. Macquet, J. L. Wiebers, and B. Monsarrat, "Structures of the adducts formed between [Pt(dien)C1]C1 and DNA in vitro," Nucleic Acids Research, vol. 10, no. 17, pp. 5255–5269, 1982.
23. W. Buchner and H. Niederprünt, "Sodium borohydride and amine-boranes, commercially important reducing agents," Pure & Applied Chemistry, vol. 49, no. 6, pp. 733–743, 1977.

24. D. Zimmern, "The 5′ end group of tobacco mosaic virus RNA is m7G5′ ppp5′Gp,"Nucleic Acids Research, vol. 2, no. 7, pp. 1189–1201, 1975.
25. N. Sonenberg, A. J. Shatkin, and R. P. Ricciardi, "Analysis of terminal structures of RNA from potato virus X," Nucleic Acids Research, vol. 5, no. 7, pp. 2501–2512, 1978.
26. M. H. V. van Regenmortel, D. Altschuh, and G. Zeder-Lutz, "Tobacco mosaic virus: a model antigen to study virus-antibody interactions," Biochimie, vol. 75, no. 8, pp. 731–739, 1993.
27. M. H. V. van Regenmortel, "The antigenicity of tobacco mosaic virus," Philosophical Transactions of the Royal Society B, vol. 354, no. 1383, pp. 559–568, 1999.
28. A. Buzzell, "Action of urea on tobacco mosaic virus. II. The bonds between protein subunits," Biophysical Journal, vol. 2, pp. 223–233, 1962.
29. L. E. Blowers and T. M. A. Wilson, "The effect of urea on tobacco mosaic virus: polarity of diassembly," Journal of General Virology, vol. 61, no. 1, pp. 137–141, 1982.
30. K. W. Mundry, P. A. C. Watkins, T. Ashfield, K. A. Plaskitt, S. Eisele-Walter, and T. M. A. Wilson, "Complete uncoating of the 5′ leader sequence of tobacco mosaic virus RNA occurs rapidly and is required to initiate cotranslational virus disassembly in vitro,"Journal of General Virology, vol. 72, no. 4, pp. 769–777, 1991.
31. H. Fraenkel-Conrat, B. Singer, and A. Tsugita, "Purification of viral RNA by means of bentonite," Virology, vol. 14, no. 1, pp. 54–58, 1961.
32. H. Fraenkel-Conrat, "Degradation of tobacco mosaic virus with acetic acid," Virology, vol. 4, no. 1, pp. 1–4, 1957.

Chapter 6

The Induction of Stromule Formation by a Plant DNA-Virus in Epidermal Leaf Tissues Suggests a Novel Intra- and Intercellular Macromolecular Trafficking Route

Björn Krenz[1], Holger Jeske[2], and Tatjana Kleinow[2]

[1]Plant Pathology and Plant-Microbe Biology, Cornell University, Ithaca, NY, USA

[2]Molecular Biology and Plant Virology, Institute of Biology, Universität Stuttgart, Stuttgart, Germany

ABSTRACT

Stromules are dynamic thin protrusions of membrane envelope from plant cell plastids. Despite considerable progress in understanding the importance of certain cytoskeleton elements and motor proteins for stromule maintenance, their function within the cell has yet to be unraveled. Several viruses cause a remodulation of plastid structures and stromule biogenesis within their host plants. For RNA-viruses these interactions were demonstrated to be relevant to the infection process. An involvement of plastids and stromules is assumed in the DNA-virus life cycle as well, but their functional role needs to be determined. Recent findings support a participation of heat shock cognate 70 kDa protein (cpHSC70-1)-containing stromules induced by a DNA-virus infection (*Abutilon* mosaic virus, AbMV, *Geminiviridae*) in intra- and intercellular molecule exchange. The chaperone cpHSC70-1 was shown to interact with the AbMV movement protein (MP). Bimolecular fluorescence complementation confirmed the interaction of cpHSC70-1 and MP, and showed a homo-oligomerization of either protein *in planta*. The complexes were detected at the cellular margin and co-localized with plastids. In healthy plant tissues cpHSC70-1-oligomers occurred in distinct spots at chloroplasts and in small filaments extending from plastids to the cell periphery. AbMV-infection induced a cpHSC70-1-containing stromule network that exhibits elliptical dilations and transverses whole cells. Silencing of the *cpHSC70* gene revealed an impact of cpHSC70 on chloroplast stability and restricted AbMV movement, but not viral DNA accumulation. Based on these data, a model is suggested in which these stromules function in molecule exchange between plastids and other organelles and perhaps other cells. AbMV may utilize cpHSC70-1 for trafficking along plastids and stromules into a neighboring cell or from plastids into the nucleus. Experimental approaches to investigate this hypothesis are discussed.

INTRODUCTION

In plants, transport of endogenous macromolecules such as proteins and nucleic acids over cellular boundaries occurs in a highly selective and regulated manner (Oparka, 2004; Lee and Lu, 2011;Maule et al., 2011; Niehl and Heinlein, 2011; Zavaliev et al., 2011). These controlled intra- and intercellular pathways are exploited by plant viruses for their systemic spread within their hosts; viruses can thus be used as tools to study basic endogenous transport processes within plants (Lee et al., 2003; Lucas, 2006; Benitez-Alfonso et al., 2010; Harries and Ding, 2011; Harries et al., 2011;Niehl and Heinlein, 2011; Schoelz et al.,

2011; Ueki and Citovsky, 2011). There is evidence accumulating that interactions of viruses with the cytoskeleton or the endomembrane system are involved in the targeting of viral nucleoprotein complexes and transport-mediating movement proteins (MPs) to plasmodesmata. However, it is still not possible to generate a complete model of intra- and intercellular movement for any known plant virus. Considering the diverse and sometimes contrasting reports on the roles of various cellular components in viral spread, it is conceivable that viruses use fundamentally different transport mechanisms within their hosts. This seems to be the case for members within one genus, as shown, for example by research into RNA-viruses of the genus *Tobamovirus* [turnip vein-clearing virus (TVCV) and tobacco mosaic virus (TMV); Harries et al., 2009] and the genus *Potexvirus* [*Alternanthera* mosaic virus (AltMV) and potato virus X (PVX);Lim et al., 2010].

TRANSPORT MODELS FOR THE PLANT DNA GEMINIVIRUSES

In contrast to RNA-viruses, plant-infecting DNA geminiviruses (family *Geminiviridae*) replicate within the nucleus, and systemic infection requires the crossing of two cellular barriers, the nuclear envelope via pores and the cell wall via plasmodesmata (Waigmann et al., 2004; Krichevsky et al., 2006; Lucas, 2006; Jeske, 2009). The geminiviruses have relatively small genomes (2.5–3.0 kb per single-stranded DNA circle) and with this limited coding capacity exhibit a strong dependency on host proteins to complete their life cycle. As a consequence, viral-encoded transport-mediating proteins have to interact with a variety of plant factors involved in macromolecular trafficking to overcome cellular boundaries and transfer viral DNA (vDNA) from a nucleus through the cytoplasm and via plasmodesmata into an adjacent cell and into the nucleus of that cell. The genome of bipartite geminiviruses (genus *Begomovirus*) consists of two DNA molecules: DNA A and DNA B. The two DNA B-encoded proteins, nuclear-shuttle protein (NSP) and MP, mediate the viral transport processes (Gafni and Epel, 2002; Rojas et al., 2005; Wege, 2007; Jeske, 2009) and both proteins have an impact on viral pathogenicity (Rojas et al., 2005; Zhou et al., 2007; Jeske, 2009). Previous work showed the C-terminal domain of begomoviral MPs to be important for symptom development and pathogenicity (von Arnim and Stanley, 1992; Ingham and Lazarowitz, 1993; Pascal et al., 1993; Duan et al., 1997; Hou et al., 2000; Saunders et al., 2001; Kleinow et al., 2009a). The DNA A-encoded coat protein (CP) is not essential for systemic infection of bipartite begomoviruses, suggesting that the transport complex is distinct from virions (Rojas et al., 2005; Jeske, 2009). However, CP was able to complement defective NSP mutants, and is therefore regarded as a redundant element in viral movement (Qin et al., 1998). Several studies provide evidence that NSP facilitates trafficking of vDNA into and out of the nucleus, and that MP serves as a membrane adaptor and mediates cell-to-cell transfer via plasmodesmata as well as long-distance spread through the phloem (Rojas et al., 2005; Krichevsky et al., 2006; Wege, 2007; Jeske, 2009).

Two models are currently suggested for the role of NSP and MP during cell-to-cell transport of bipartite geminiviruses: the "couple-skating" and the "relay race" models (Rojas et al., 2005; Jeske, 2009). The "couple-skating" model is based on the experimental data of the phloem-limited begomoviruses squash leaf curl virus (SLCV; Pascal et al., 1994; Sanderfoot and Lazarowitz, 1995;Sanderfoot et al., 1996), cabbage leaf curl virus (CaLCuV; Carvalho et al., 2008a,b), and *Abutilon*mosaic virus (AbMV; Zhang et al., 2001; Aberle et al., 2002; Hehnle et al., 2004; Frischmuth et al., 2007). This model suggests that MP binds the NSP/vDNA complex at the cytoplasmic side of plasma membranes or microsomal vesicles, and transfers the nucleoprotein complex into the next cell either along the plasma membrane or via the endoplasmic reticulum (ER) that spans the plasmodesmata. In contrast, the "relay race" model predicts that after NSP-mediated nuclear export the vDNA is taken over by MP, which then transports the vDNA into the adjacent cell (Noueiry et al., 1994; Rojas et al., 1998, 2005). This model is based on experimental data of the mesophyll-invading begomovirus bean dwarf mosaic virus (BDMV; Levy and Tzfira, 2011). Nevertheless, details of how both proteins co-ordinate vDNA transfer from the nucleus to the cell periphery and further throughout the plant body, are mostly unknown.

For a controlled cycle of geminiviral replication, transcription, encapsidation, and movement, NSP and MP are most likely integrated into a regulatory network consisting of other viral proteins and plant factors. Several studies have characterized a set of interacting host proteins for NSP and MP. NSPs of CaLCuV, tomato golden mosaic virus (TGMV), and tomato crinkle leaf yellows virus (TCrLYV) were found to interact with two classes of receptor-like kinases from *Arabidopsis thaliana*(Fontes et al., 2004; Mariano et al., 2004; Florentino et al., 2006). The further analysis of the NSP/kinase interactions indicated that they play a role in infectivity and symptom development. NSP counters activation of defense signaling mediated by one kinase class via phosphorylation of an immediate downstream target, the ribosomal protein L10/QM (Fontes et al., 2004; Mariano et al., 2004; Florentino et al., 2006; Carvalho et al., 2008c; Rocha et al., 2008; Santos et al., 2010). Additionally, CaLCuV NSP was found to interact with an acetyltransferase (AtNSI; McGarry et al., 2003; Carvalho and Lazarowitz, 2004; Carvalho et al., 2006) and with a small GTPase (Carvalho et al., 2008a,b). AtNSI is proposed to regulate nuclear export of vDNA by acetylating histones and CP.Carvalho et al. (2008a,b) suggest a function for the small GTPase in nuclear export processes, probably as a co-factor of NSP.

Independent of the transport model, the begomoviral MPs have to mediate multiple functions during intra- and intercellular trafficking. The identification of three phosphorylation sites in the AbMV MP, which have an impact on symptom development and/or vDNA accumulation (Kleinow et al., 2009a), indicates a regulation of diverse MP functions by yet unknown host kinases. Currently, three interacting host factors of begomoviral MPs have been identified: a histone H3 (Zhou et al., 2011), a synaptotagmin (SYTA; Lewis and Lazarowitz, 2010), and a chaperone, the heat shock cognate 70 kDa protein cpHSC70-1 (Krenz et al., 2010). Gel overlay assays, and *in vitro* and *in vivo* co-immunoprecipitation (Co-IP) experiments showed an interaction of H3 with NSP and MP of BDMV as well as with CPs of different geminiviruses (Zhou et al., 2011). In *Nicotiana tabacum* protoplasts and *N. benthamiana* leaves, transiently expressed H3 co-localized with NSP in the nucleus and the presence of MP redirected H3 to the cell periphery and plasmodesmata. A complex composed of H3, NSP, MP, and vDNA was recovered by Co-IP from *N. benthamiana* leaves transiently expressing epitope-tagged H3. The data support a model in which histone H3 is a component of a geminiviral movement-competent vDNA complex that assembles in the nucleus and is transferred to the cell periphery and plasmodesmata. SYTA localized to endosomes in *Arabidopsis* cells, and interacted with MPs of the begomoviruses CaLCuV and SLCV as well as with the unrelated MP of the RNA-virus TMV (Lewis and Lazarowitz, 2010). Transgenic *Arabidopsis* lines with either a reduced SYTA level or expressing a dominant-negative SYTA mutant exhibited a delayed systemic infection and an inhibition of cell-to-cell trafficking of the different MPs. Consequently, Lewis and Lazarowitz (2010)proposed that: (i) SYTA regulates endocytosis and (ii) distinct viral MPs transport their cargo to plasmodesmata for cell-to-cell spread via an endocytotic recycling pathway. The chaperone cpHSC70-1 of *Arabidopsis* was shown to specifically interact with the N-terminal domain of AbMV MP in a yeast two-hybrid system (Krenz et al., 2010). Bi-molecular fluorescence complementation (BiFC) analysis provided further evidence for the chaperone/MP interaction, and revealed an MP as well as a cpHSC70-1 self-interaction *in planta* (Krenz et al., 2010). MP/cpHSC70-1 complexes and MP-oligomers were observed at the cell periphery and co-localized with chloroplasts. The detection of MP-homo-oligomers at the cellular margin is in agreement with other localization studies in plant cells (Zhang et al., 2001; Kleinow et al., 2009b) and with earlier yeast two-hybrid assays that showed an MP oligomerization via the C-terminal domain (Frischmuth et al., 2004). MP-oligomer formation has also been detected at chloroplasts (Krenz et al., 2010). It is unknown whether BiFC results from MP imported into plastids or merely associated with the outer envelope of the chloroplast. No BiFC signal was seen in peri-nuclear sites as was previously found for AbMV MP transiently expressed as green fluorescent protein (GFP) fusion in plant cells (Zhang et al., 2001). Thus, MP/MP interaction may be restricted to chloroplasts and the cell periphery.

Bi-molecular fluorescence complementation showed that cpHSC70-1-oligomers were mainly associated with chloroplasts where they accumulated in distinct spots, and occurred to a lower extent in small filaments extending from plastids to the cell periphery and distributed at the periphery (Krenz et al., 2010). The localization of cpHSC70-1 was significantly influenced by AbMV-infection, accumulating in fluorescent foci on long filamental tubular structures reminiscent of plastid stromules, stroma-filled plastid tubules (Natesan et al., 2005; Hanson and Sattarzadeh, 2008). It remains uncertain whether cpHSC70-1 was maintained exclusively within the stroma or whether it was re-located to other structures upon geminiviral infection such as envelope membranes or the intermembrane space. Altogether, AbMV-infection seems to induce a prominent formation of stromules. To our knowledge the geminivirus AbMV is the only plant DNA-virus so far for which stromule biogenesis was documented. Silencing of the *cpHSC70* gene of *N. benthamiana* with the aid of an AbMV DNA A-derived gene silencing vector caused tiny white leaf sectors, which indicated an impact of cpHSC70 on chloroplast stability (Krenz et al., 2010). vDNA accumulated within these small chlorotic areas that were spatially restricted to small sectors adjacent to veins, suggesting a functional relevance of the MP/chaperone interaction for AbMV transport to symptom induction *in planta*.

CELLULAR FUNCTIONS OF HSP70 AND HSC70 AND THEIR PUTATIVE ROLES IN VIRAL INFECTION

The expression of chaperones from the heat shock protein 70 kDa (HSP70) family is induced in response to developmental signals and various abiotic and biotic stress stimuli (Escaler et al., 2000a,b; Maule et al., 2000; Sung et al., 2001; Aparicio et al., 2005; Brizard et al., 2006; Swindell et al., 2007). Some family members exhibit a low constitutive expression level and are therefore named heat shock cognate proteins 70 kDa (HSC70s) (Sung et al., 2001; Swindell et al., 2007). The cellular functions of this chaperone family are quite diverse. They assist newly translated proteins to obtain their active conformation, misfolded or aggregated proteins to refold, assist in membrane translocation of proteins, in assembly and disassembly of macromolecular complexes and in controlling the activity of regulatory factors (Kanzaki et al., 2003; Mayer and Bukau, 2005;Weibezahn et al., 2005; Bukau et al., 2006; Noel et al., 2007; Kampinga and Craig, 2010; Mayer, 2010; Flores-Pérez and Jarvis, 2012). In addition to their intracellular functions in different subcellular compartments, HSP70s play a role in cell-to-cell transport as indicated by two non-cell-autonomous cytoplasmic HSP70s from *Cucurbita maxima* (Aoki et al., 2002) and by closterovirus-encoded homologs of HSP70s which are essential for virus transport and plasmodesmata targeting (Alzhanova et al., 2007; Avisar et al., 2008, and references therein). For HSP70s and HSC70s, substrate binding and release is regulated by a conformational change that is driven by their ATPase activity. Co-chaperones (DNAJ-like/HSP40 type proteins) assist HSP70s and HSC70s functions with their delivery and release of substrates and by enhancing ATP hydrolysis activity.

HSP70s and HSC70s transcript and protein levels are up-regulated in plants upon an infection with RNA- or DNA-viruses (Escaler et al., 2000a,b; Maule et al., 2000; Aparicio et al., 2005; Brizard et al., 2006). Accumulation of viral proteins within the cell during the infection causes stress and might thereby induce the expression of this chaperone family. Several classes of chaperones and co-chaperones including HSP70s/HSC70s and their specific co-chaperones were identified to interact with viral proteins to facilitate the regulation of viral replication, transcription, encapsidation, and intra- and intercellular movement as well as to suppress pathogen responses (Noel et al., 2007;Benitez-Alfonso et al., 2010; Nagy et al., 2011). Recently, silencing of a cytosolic HSC70-1 was found to impair infection by the monopartite geminivirus tomato yellow leaf curl Sardinia virus (TYLCSV) in *N. benthamiana* (Lozano-Duran et al., 2011). However, none of these HSP70s and HSC70s involved in viral life cycles were located in the chloroplast stroma where cpHSC70-1 was identified to interact with the MP of the geminivirus AbMV (Krenz et al., 2010). In addition to the localization of cpHSC70-1 in the chloroplast stroma and stromules, it is also seen in mitochondria and as a nuclear protein in response to cold stress (Sung et al., 2001; Peltier et al., 2002, 2006; Bae et al., 2003; Ito et al., 2006; Su and Li, 2008, 2010; Krenz et al., 2010; Latijnhouwers et al., 2010). An analysis of an*Arabidopsis* knock-out mutant of cpHSC70-1 revealed that its deficiency caused severe developmental defects (Su and Li, 2008, 2010; Latijnhouwers et al., 2010), but the functions of cpHSC70-1 and other stroma-targeted HSP70s/HSC70s are not completely understood. Recent genetic and biochemical analyses indicated that cpHSC70-1 seems to play a role in protein translocation into the plastid stroma in early developmental stages of plants (Su and Li, 2010;Flores-Pérez and Jarvis, 2012). It is well known that HSP70s/HSC70s fulfill multiple functions in chloroplasts (Flores-Pérez and Jarvis, 2012), therefore the participation of cpHSC70-1 in protein transport across membranes might not be the only function it provides. What function of cpHSC70-1 is targeted by AbMV MP? It can be speculated that the virus exploits the ATPase activity of the chaperone as a driving force to mediate transport of the geminiviral nucleoprotein complexes.

PLASTIDS AND STROMULES IN VIRAL INFECTION

Several interactions of RNA-viruses with chloroplasts have been described which were important for the viral infection process (Reinero and Beachy, 1986; Schoelz and Zaitlin, 1989; Prod'homme et al., 2003; Jimenez et al., 2006; Torrance et al., 2006; Xiang et al., 2006; Lin et al., 2007; Lim et al., 2010). Virus–chloroplast interactions most likely facilitate viral replication or movement. The role of chloroplasts in the life cycle of plant DNA-viruses needs to be examined. In studies of cellular alterations induced by various geminiviruses in systemically infected plants, dramatic morphological changes in the ultrastructure of chloroplasts were identified, such as vesiculated entities, reduced starch and chlorophyll content, accumulation of fibrillar inclusions, virus-like particles, and vDNA within plastids (Esau, 1933; Jeske and Werz, 1978, 1980a,b; Schuchalter-Eicke and Jeske, 1983;Jeske and Schuchalter-Eicke, 1984; Jeske, 1986; Gröning et al., 1987, 1990; Rushing et al., 1987;Channarayappa et al., 1992). For AbMV it was shown that the severity of chloroplast structure remodeling was dependent on light intensity, and diurnal and seasonal conditions. Geminivirus-induced plastid alterations have thus far been interpreted to be an indirect result of the interference of viral infection with carbohydrate metabolism, mainly through a disruption in translocation via the phloem (Jeske and Werz, 1978). Nevertheless, the detection of vDNA, fibrillar inclusions, or virus-like particles within chloroplasts, suggests other functions of this interplay. Until now, only AbMV vDNA was detected in purified plastids from infected plants (Gröning et al., 1987, 1990). An artificial co-purification was excluded by thermolysin and DNase I treatment. *In situ* hybridization detected high amounts of AbMV vDNA in a low number of purified plastids, which would not be expected for a non-specific co-purification. However, so far, *in situ* hybridization of infected *Abutilon sellovianum* tissue only revealed AbMV vDNA-specific signals on plastids in rare cases (Horns and Jeske, 1991). Furthermore, the finding that an outer envelope membrane protein (Crumpled leaf) is implicated in the CaLCuV infection process also supports an involvement of chloroplasts in the geminiviral life cycle (Trejo-Saavedra et al., 2009). An interesting plastid modification detected upon AbMV-infection, was the induction of stromule biogenesis (Krenz et al., 2010).

Stromules emanate from the main body of the plastid and are confined by the outer and inner envelope membranes (Natesan et al., 2005; Hanson and Sattarzadeh, 2008, 2011). They represent a highly dynamic structure which extends, retracts, branches, bends, and sometimes releases vesicles from their tip (Gunning, 2005; Natesan et al., 2005; Hanson and Sattarzadeh, 2008). The typical diameter is <1 µm and the length is extremely variable due to their dynamic properties (Gray et al., 2001; Kwok and Hanson, 2004b; Waters et al., 2004). Stromules are distinguished from other irregular shaped plastid protrusions by their specific shape index (Holzinger et al., 2007). The movement of stromules relies on the actin cytoskeleton and the motor protein myosin XI (Kwok and Hanson, 2003; Natesan et al., 2009; Sattarzadeh et al., 2009). Differentially shaped stromules have been identified by using expression of various stroma-targeted fluorescent proteins; these include straight or branched tubules which can exhibit either randomly localized elliptical dilations that transverse the tubule length or triangular areas of expansion (Köhler et al., 2000; Hanson and Sattarzadeh, 2008; Schattat et al., 2011, 2012). For the latter type, branch formation occurs in tandem with dynamic remodeling of contiguous ER tubules (Schattat et al., 2011). Schattat and colleagues suppose that this co-alignment might originate from membrane contact points or by an exploitation of the same cytoskeletal elements for development. Single or multiple stromules may arise in all plastid types present in higher plant tissues, but their frequency varies; for example their abundance is significantly higher for achlorophylic plastids in sink tissues than for chlorophyll-containing plastids in green tissues (Köhler et al., 1997; Köhler and Hanson, 2000; Natesan et al., 2005; Hanson and Sattarzadeh, 2008; Schattat et al., 2012). Analyses of the fruit ripening in tomato showed that the formation of stromules is influenced by the plastid differentiation status and is inversely correlated with the density and size of plastids within a cell (Waters et al., 2004). Consequently, a role of stromules in sensing the number of plastids in a cell is supposed. Various abiotic and biotic stress conditions including heat (Holzinger et al., 2007), subcellular redox stress (Itoh et al., 2010), application of extracellular sucrose or glucose (Schattat and Klösgen, 2011), treatment with abscisic acid (Gray et al., 2012), colonization by an arbuscular mycorrhizal fungus (Fester et al., 2001; Hans et al., 2004; Lohse et al., 2005), and infiltration of agrobacteria (Schattat et al., 2012) were described as inducers of stromules. The formation of a dense plastid network in cells close to the main symbiotic structure during mycorrhizae formation supports a putative correlation between plastid metabolic activity and stromule biogenesis. An induction of stromules was detected as well upon viral infections (Esau, 1944; Shalla, 1964; Caplan et al., 2008; Krenz et al., 2010). RNA-virus infected sugar beets exhibit mosaic disease symptoms including mottling and yellow-green sectoring of leaves. Plastids within these yellow areas showed vesiculation and an amoeboid shape resembling stromules (Esau, 1944). Shalla (1964) described the appearance of "long appendages which extend and contracted within a few seconds" from the plastid body and vesicle formation inside chloroplasts of TMV-infected tomato leaflets. TMV-infected tobacco plants exhibited a strong induction of stromule formation (Caplan et al., 2008) just as for *N. benthamiana*plants locally infected with the DNA-virus AbMV (Krenz et al., 2010).

Although several inducers of stromules have been identified, their functional role remains to be determined. They were proposed to participate in plastid motility and in facilitating transport of various molecules, e.g., proteins, metabolites and signaling components, into and out of a plastid, among plastids and even between plastids and other organelles (Köhler et al., 1997, 2000; Kwok and Hanson, 2004a; Natesan et al., 2005; Hanson and Sattarzadeh, 2008, 2011). Chlorophyll, thylakoid membranes, plastid DNA, and ribosomes have not been detected within stromules (Hanson and Sattarzadeh, 2008, 2011; Newell et al., 2012). Nevertheless, the possibility of a rare movement of plastid DNA and ribosomes or the transfer of much smaller DNA molecules, e.g., plastid transformation vectors, via stromules cannot be completely excluded. Exchange of stroma-targeted GFP between two plastids interconnected by stromules was observed using fluorescence recovery after photobleaching (FRAP) experiments in tobacco and *Arabidopsis* (Köhler et al., 1997; Tirlapur et al., 1999). Moreover, two-photon excitation fluorescence correlation spectroscopy revealed two different transport modes through stromules in tobacco suspension cells (Köhler et al., 2000). A simple diffusion of single stroma-targeted GFP molecules was observed in addition to an active ATP-dependent batch movement of GFP "packets." Köhler and colleagues supposed that these GFP bodies represent an accumulation of GFP in small vesicles. Stromules may carry several of these GFP "packets" leading to a beaded appearance (Köhler and Hanson, 2000; Pyke and Howells, 2002;Hanson and Sattarzadeh, 2008, 2011). The stromules induced by AbMV-infection and containing the MP-interacting chaperone cpHSC70-1 exhibited a related appearance like pearls on a string (Krenz et al., 2010; Figures 1 and 2). It is hypothesized that cpHSC70-1 is present in the same type of "packet" structure as GFP in the preceding experiments, probably associated with vesicles and actively transported.

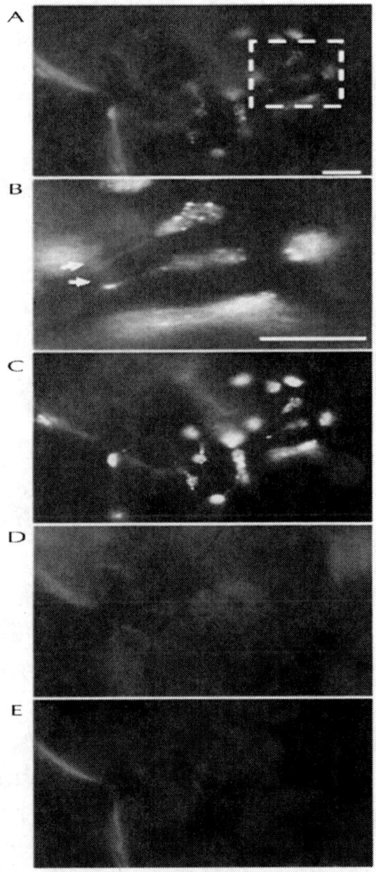

Figure 1: *Abutilon* mosaic virus-induced cpHSC70-1-containing stromules extending from plastids to the cell periphery. Transient co-expression of test constructs in leaf tissues of locally AbMV-infected *N. benthamiana*and epi-fluorescence microscopy were carried out according to Krenz et al.

(2010,2011). AbMV infection was established by simultaneous agro-infiltration of infectious DNA A and DNA B clones with the fluorescent protein expression constructs. (A)Merged image of cells expressing NSP:cyan fluorescent protein (CFP), the two split yellow fluorescent protein (YFP)/BiFC constructs of cpHSC70-1 and the plasmodesmata marker PDCB1:mCherry (callose binding protein 1, Simpson et al., 2009) for 4 days post agro-infiltration (dpai). The square in (A) highlights cpHSC70-1-oligomers at chloroplasts and stromules (yellow, arrows) anchoring at the cell periphery (red: PDCB1:mCherry), and is magnified in (B). The separate fluorescence signals superimposed in (A) are shown in (C) YFP, (D) CFP, and (E) mCherry. Note: The plasmodesmata marker PDCB1:mCherry lost its extracellular localization at the neck region of plasmodesmata upon AbMV-infection (compare Figure 3) and is probably distributed throughout the apoplast. NSP:CFP is redirected from the nucleus to the cell periphery, probably the plasma membrane, by presence of MP or AbMV-infection (Zhang et al., 2001; Frischmuth et al., 2007). Bar: 10 μm.

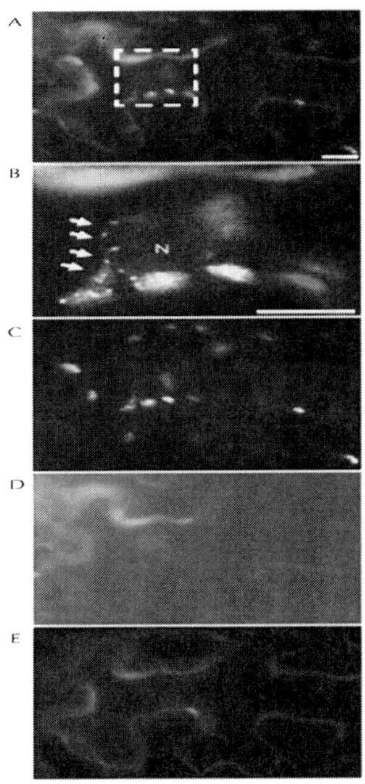

Figure 2: *Abutilon* mosaic virus-induced cpHSC70-1-containing stromules grabbing a nucleus. The experimental set-up is the same as stated in Figure 1. (A) Merged image of cells expressing the three test proteins for 4 dpai

and(B) magnification of the square in (A) which marks cpHSC70-1-oligomers at chloroplasts and stromules (yellow, arrows) which closely associate to a nucleus (blue: NSP:CFP) near to the cell periphery (blue: NSP:CFP, likely plasma membrane and red: PDCB1:mCherry, apoplast), magnified in (B). The separate emissions merged in (A)are shown in (C) YFP, (D) CFP, and (E) mCherry. N, nucleus; bar: 10 μm.

The data obtained by Kwok and Hanson (2004a) suggested that stromules may serve as pathways between nuclei and more distant regions of the cell and possibly even other cells. They observed that clusters of plastids around nuclei are capable of extending stromules both outward, to the cell membrane, as well as inward, through nuclear grooves. Close contact between plastids and the nuclei and the plasma membrane of plant cells suggests that physical interactions may enhance functional interactions between these organelles. Furthermore, Kwok and Hanson (2004a) found that stromules from two adjacent cells appeared to meet at either side of an adjoining cell wall. Consequently, the stromule's structure seems to be suitable for the exchange of molecules between plastids and other organelles or the trafficking of plastidal proteins and metabolites to diverse regions of the plant cell. Remarkably, the cpHSC70-1-containing stromules detected upon an AbMV-infection arose not only on plastids clustered around and in close association with the nucleus, but also appeared to interconnect plastids and extend from plastids outward to the cell periphery (Krenz et al., 2010; Figures 1 and 2).

By contrast, in non-infected tissues only short cpHSC70-1-containing filaments were found which extended solely from cortex positioned plastids to the cell periphery. However, molecular transfer from the plastids to the nucleus or vice versa with the aid of stromules remains to be confirmed. That a retrograde protein exchange between plastids and the nucleus can occur was demonstrated recently. Plastid-encoded HA-tagged Whirly1 protein was translocated to the nucleus in transplastomic tobacco plants, where it stimulated pathogen-related gene expression (Isemer et al., 2012). The chloroplast-localized $NRIP_1$ (N receptor-interacting protein 1) was redirected to the cytoplasm and to the nucleus in presence of the p50 effector, a 50 kDa helicase domain encoded by TMV (Caplan et al., 2008). Upon this recruitment to the nucleus and the cytoplasm $NRIP_1$ binds to the N innate immunity receptor to initiate effector recognition and pathogen defense mechanisms. TMV-infection causes a strong increase in stromule formation, and a localization of fluorescent protein-tagged $NRIP_1$ within stromules was observed (Caplan et al., 2008). Thus, the authors speculated about an involvement of stromules in the nuclear re-localization of $NRIP_1$.

Schattat et al. (2012) do not support a function of stromules in trafficking of macromolecules between plastids. In this thoroughly performed work, interconnectivity of independent plastids was tested with the aid of a photoconvertible stroma-targeted fluorescent protein. Despite the strong microscopic impression of interplastid connectivity via stromules, an exchange of the stroma marker protein could not be visualized by high quality confocal imaging. Various plant materials (e.g., *N. benthamiana* and *Arabidopsis*) were comprehensively analyzed for plastid morphology and marker protein transfer. Although the differently colored plastids and stromules were in very close proximity, the labeled organelles remained separate as indicated by the absence of color mixing. That the method applied in these studies is suitable to detect an exchange of material between organelles upon a fusion was confirmed by analogous experiments using a mitochondria-targeted version of the fluorescent protein. In contrast to our observations from geminivirus-infected plants, the results of Schattat et al. (2012) were obtained working with uninfected plants. Whether these conflicting results are caused by the different experimental set-up, plastid types, and plant material used, or whether indeed a macromolecular trafficking of stroma-proteins through interconnecting stromules is not feasible under any conditions, needs to be elucidated by further experimentation.

In addition to the induction of stromule biogenesis, AbMV-infection influences the localization of a plasmodesmata-associated protein. The plasmodesmata callose binding protein 1 (PDCB1) fused to mCherry was investigated as a marker for plasmodesmata by Simpson et al. (2009). PDCB1 is a glycosylphosphatidylinositol (GPI)-linked protein that exhibits callose binding activity and localizes to the neck region of plasmodesmata in the apoplast. Here it possibly acts as a structural anchor between the plasma membrane component spanning the plasmodesmata and the cell wall. The available data support a function for PDCB1 in plasmodesmata flux control by influencing the callose deposition in the cell wall and as a consequence the aperture of plasmodesmata. Due to its extracellular localization PDCB1 was not expected to interfere with viral proteins like AbMV MP, which is likely to accumulate in the central symplastic cavity region of complex plasmodesmata (Kleinow et al., 2009b; Lee and Lu, 2011). The fluorescence microscopic analyses showed punctate structures in the cell periphery after transient expression of PDCB1:mCherry in epidermal leaf tissues, which are in agreement with the expected plasmodesmata localization (Figure 3).

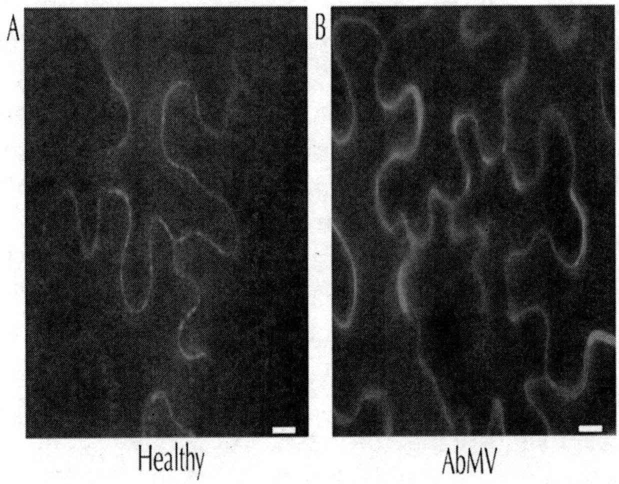

Figure 3: Influence of AbMV-infection on localization of the plasmodesmata marker PDCB1. PDCB1:mCherry (Simpson et al., 2009) was expressed in epidermal tissues of *N. benthamiana* plants either (A) healthy or (B)locally infected with AbMV for 4 days. The experiment was performed as described in Figure 1. (A) Punctate mCherry signals indicate the targeting of PDCB1 to the apoplastic part of the neck region from plasmodesmata according to Simpson et al. (2009). (B) Upon AbMV-infection these plasmodesmata-specific signals disappeared and PDCB1 emerged homogenously in the apoplast. These results showed a virus-induced alteration of the subcellular localization of PDCB1, probably by modifying callose deposition at plasmodesmata. Bar: 10 μm.

Surprisingly, upon AbMV-infection PDCB1:mCherry signals were still distributed at the cell periphery, most likely the cell wall (Figures 1–3), but no punctate structures were detected anymore. Thus, the protein seems to have lost the plasmodesmata localization. We hypothesize this as an AbMV-induced remodeling of the plasmodesmata aperture by callose depletion in the neck region.

ABMV MOVEMENT ALONG STROMULES WITH THE HELP OF A CHAPERONE

A cellular function in plant endogenous macromolecular trafficking is suspected for stromules and chaperones like HSC70s and in addition an involvement for chaperones in viral movement. However, their combined participation in these processes has thus far not been examined. The accumulated data for the geminivirus AbMV indicate that for both factors there is a joint involvement in the viral life cycle, very likely the movement process (Krenz et al., 2010). In the studies of Krenz et al. (2010) stromules were visualized on chlorophyll-containing plastids by BiFC experiments using the chaperone cpHSC70-1. Here, two different types of cpHSC70-1-containing stromules were monitored. Only short stromules extending from plastids to the cell periphery have been found in healthy epidermal tissues, whereas upon AbMV-infection long stromules forming a network between plastids, nucleus, and the cell periphery were detected. For the latter ones, the BiFC signals of cpHSC70-oligomers highlighted mainly stromule structures with elliptical dilations giving them a beaded appearance. This significant difference created by the geminivirus infection might indicate additional functions of cpHSC70 in association with the two stromule types observed. The cpHSC70-containing stromules may function in macromolecule transfer, perhaps just under certain cellular conditions, e.g., a virus infection. This transport may happen intracellulary among plastids and between plastids and other organelles, or even intercellularly through plasmodesmata.

A prerequisite for the traveling of stromal proteins via stromules from an individual plastid to another plastid or organelle (e.g., nucleus) might be the fusion of the outer and inner envelope membrane with the target membrane. AbMV-infection might create a cellular environment that allows such a fusion event of stromules and the consequent transposition of quantities of stromal components. Alternatively, a transfer process might be initiated, which does not comprise a fusion of the inner envelope membrane or the envelope at all. This might consist of transport along the cytoplasmic leaflet of the outer envelope membrane, through the intermembrane space after fusion of the outer envelope with the target membrane or via envelope-coated vesicles released from stromules (Gunning, 2005; Krause and Krupinska, 2009). Irrespective of the underlying mechanism, a movement of a geminiviral nucleoprotein complex in association with interacting plastid stromules and cpHSC70, even with a low efficiency, might be sufficient for intra- and intercellular viral spread. A transport event in close association with membranes or vesicles would be consistent with the geminiviral MP being a membrane-associated protein. AbMV MP was localized to the protoplasmic face of plasma membranes and vesicles, where its C- and N-terminal domains most likely protrude into the cytoplasm (Aberle et al., 2002; Frischmuth et al., 2004, 2007). The central part of MP probably forms an amphipathic helix structure which inserts into one leaflet of the target membrane (Zhang et al., 2002). It has been observed that insertion of amphipathic helices into a monolayer induces bending and generates local curvature (Kozlov et al., 2010; McMahon et al., 2010). Such protein-mediated membrane stresses were found to trigger fusion, fission, and budding events of membranes. Presumably, AbMV MP may be capable of inducing and/or enhancing such membrane remodeling. However, it cannot be excluded that the close association of AbMV MP with plastid structures represents a targeting of the cytoskeletal elements, to which chloroplasts and stromules are usually attached, for cellular transit of viral nucleoprotein complexes. The result of cpHSC70 silencing in *N. benthamiana* revealed that the chaperone/MP interaction is not essential for the systemic spread of AbMV (Krenz et al., 2010), suggesting it plays a role via an alternative path.

In summary, a model (Figure 4) can be proposed in which MP/cpHSC70 interaction and stromule induction facilitate intra- and intercellular macromolecular trafficking along plastids and stromules into the neighboring cell or in the other direction from plastids into the nucleus. Whether this represents an accidental event or is of major significance for AbMV propagation and/or symptom development remains to be investigated.

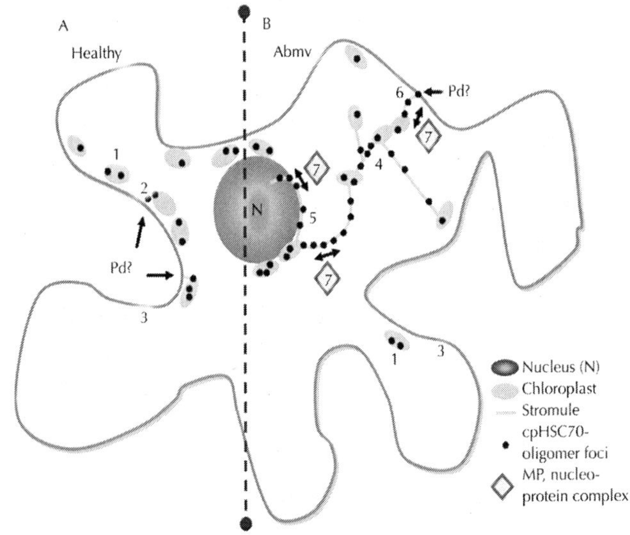

Figure 4: Hypothetical model of AbMV intra- and intercellular trafficking via a plastid network. (A) In healthy plant cells oligomers of the chaperone cpHSC70 locate mainly in small spots at chloroplasts (1), to a lesser amount in small filaments extending from cortical chloroplasts toward the cell periphery (2) and distributed at the cellular margin (3). (B) In AbMV-infected cells homo-oligomers of cpHSC70-1 were found similar to non-infected cells at chloroplasts (1) and to a low extent at the cellular margin (3). However, AbMV-infection establishes the formation of a stromule network interconnecting different chloroplasts (4), but also extending from plastids inward to the nucleus, where they closely attach (5) and outward to the cell periphery/cell wall, assumedly to plasmodesmata (Pd) that transverse the cell wall (6). These stromules exhibit structures where cpHSC70-oligomers appeared mainly in elliptical dilations giving them a "pearls on a string"-appearance. The stromule network might function in intra- and intercellular trafficking of viral nucleoprotein complexes by the interaction of AbMV MP with the chaperone cpHSC70 within stromules (7). The potential underlying transport mechanism is yet unknown, but probably involves membrane fusions or a vesicle formation.

So far geminivirus replication and virion assembly were only detected within the nucleus of infected cells (Rojas et al., 2005; Jeske, 2007, 2009). Thus, it seems to be very unlikely that the observed interaction of AbMV with plastids and stromules plays a role in replication or virion assembly processes. Nevertheless, AbMV-induced stromule formation and/or cpHSC70-1 interaction might be also related to other cellular processes. Plastids were involved in the biosynthesis pathways of many essential compounds (e.g., carbohydrates, fatty acids, purines) and stromules might be important in facilitating metabolic exchanges within the cells (Fester et al., 2001; Hans et al., 2004; Kwok and Hanson, 2004b; Waters et al., 2004; Lohse et al., 2005; Schattat and Klösgen, 2011). Plant cells respond to various abiotic and biotic stress stimuli, causing disturbance in the cellular energy status, by complex changes, which include the carbohydrate metabolism and therefore also plastid activity. It is known that stromules occur more frequently and are longer in plant cells with disturbed metabolism (Hanson and Sattarzadeh, 2008), e.g., cells cultured in liquid medium, callus or suspension culture that shed chlorophyll in their chloroplast. That stromule emergence is triggered by an increased plastid metabolic capacity resulting from biotic stress, is supported by the findings of their strong induction upon symbiotic interaction of root cells with mycorrhiza (Fester et al., 2001;Hans et al., 2004; Lohse et al., 2005). Interestingly, geminiviral proteins other than MP have an impact on a regulatory key component of the stress and glucose signal transduction, the sucrose non-fermenting 1-related (SnRK1) protein kinase (Kong and Hanley-Bowdoin, 2002; Hao et al., 2003;Shen and Hanley-Bowdoin, 2006). Plastids and HSP70s were previously concluded to be involved in the onset of a virus-pathogen response (Noel et al., 2007; Caplan et al., 2008; Nagy et al., 2011); therefore, the interaction of AbMV with the plastidal cpHSC70-1 might depict a counteraction of an antiviral defense mechanism.

Recently, a mutant screen identified a plastid- and a mitochondria-localized RNA helicase to be necessary for plastid development, embryogenesis and unexpectedly for cell-to-cell trafficking (Burch-Smith et al., 2011; Burch-Smith and Zambryski, 2012). For both knock-out plants profound changes in the transcriptome were observed, e.g., a dramatic down-regulation of nucleus-encoded plastid-related genes. Although the proteins are located exclusively in plastids and mitochondria, respectively, loss of one of the RNA helicase functions causes formation of twinned and branched plasmodesmata in *Arabidopsis*. Thus, disruption of the plastid function seems to induce an increased molecule exchange, possibly metabolites or signaling factors, among neighboring cells. The data supports a pathway linking intra- to intercellular communication and therefore a signaling from organelles to the nucleus and plasmodesmata. It can be speculated that viral interactions with plastids also targets this novel regulatory pathway to enhance plasmodesmata trafficking.

CONCLUSIONS AND FUTURE PERSPECTIVES

The AbMV data on movement is consistent with cell-to-cell transport according to the "couple-skating" model. However, the results of Krenz et al. (2010) may demand to widen this concept of geminiviral cellular transfer for AbMV: it may opportunistically hijack different pathways for intracellular transport and plasmodesmata targeting including an alternative route via chloroplasts and stromules with the aid of a plastidal chaperone. It is clear, in any case, that further research is needed to elucidate this hypothesis.

A functional characterization of AbMV MP and the host factor cpHSC70-1 in the assumed transport processes could start with comprehensive analyses of their localization in subcellular compartments and membrane structures. The following points should be considered: (i) To resolve the geometric relation of the different cellular structures high resolution confocal imaging and time lapse imaging is required. (ii) Test protein expression should be done in combination with a set of fluorescent marker proteins, probably photoconvertible, for different compartments to monitor the subcellular distribution, temporal activities, and macromolecular trafficking events. A cytoplasmic marker would be needed to investigate if AbMV-induced stromules actively associate with the nucleus or are just pressed by other organelles, like a large vacuole, toward the nucleus. (iii) Careful controls should be included to account for stresses (including agro-infiltration) that are known to influence stromule formation. (iv) To minimize the influence of agro-infiltration, transgenic plants expressing required fluorescent test proteins and alternate methods to introduce AbMV (biolistic bombardment of vDNA) should be applied. To investigate the precise molecular function of cpHSC70-1 on AbMV infection and presumably macromolecular trafficking, the impact of the following scenarios on AbMV spread could be investigated: (i) transgenic plants overexpressing wild-type cpHSC70-1 or a non-MP-interacting, dominant-negative cpHSC70-1 variant and (ii) infection with AbMV DNA encoding a non-cpHSC70-1-interacting MP mutant. Finally, the relevance of the findings should be tested for additional geminiviruses other than AbMV. This knowledge gained will contribute significantly to the elucidation of the geminiviral intra- and intercellular movement process.

ACKNOWLEDGMENTS

The authors thank Gabi Kepp for excellent technical assistance and Diether Gotthardt and Marvin Müller for taking care of experimental plants. We are grateful to Keith Lloyd Perry (Cornell University, USA) for critical reading of the manuscript, Karin Krupinska (Christian-Albrechts-University of Kiel, Germany) for helpful discussions on stromules and Emmanuelle Bayer and Andrew Maule (John Innes Centre, UK) for providing the expression construct for PDCB1: mCherry. This work was supported by a trilateral ERA-PG/BMBF project (BMBF 0313986) and by Deutsche Forschungsgemeinschaft DFG (KL1366/3-1).

REFERENCES

1. Aberle, H. J., Rütz, M. L., Karayavuz, M., Frischmuth, S., Wege, C., Hülser, D., et al. (2002). Localizing BC1 movement proteins of Abutilon mosaic geminivirus in yeasts by subcellular fractionation and freeze-fracture immunolabelling. *Arch. Virol.* 147, 103–107.
2. Alzhanova, D. V., Prokhnevsky, A. I., Peremyslov, V. V., and Dolja, V. V. (2007). Virion tails of beet yellows virus: coordinated assembly by three structural proteins. *Virology* 359, 220–226.
3. Aoki, K., Kragler, F., Xoconostle-Cazares, B., and Lucas, W. J. (2002). A subclass of plant heat shock cognate 70 chaperones carries a motif that facilitates trafficking through plasmodesmata. *Proc. Natl. Acad. Sci. U.S.A.* 99, 16342–16347.
4. Aparicio, F., Thomas, C. L., Lederer, C., Niu, Y., Wang, D., and Maule, A. J. (2005). Virus induction of heat shock protein 70 reflects a general response to protein accumulation in the plant cytosol. *Plant Physiol.* 138, 529–536.

5. Avisar, D., Prokhnevsky, A. I., and Dolja, V. V. (2008). Class VIII myosins are required for plasmodesmatal localization of a closterovirus HSP70 homolog. *J. Virol.* 82, 2836–2843.

6. Bae, M. S., Cho, E. J., Choi, E. Y., and Park, O. K. (2003). Analysis of the Arabidopsis nuclear proteome and its response to cold stress. *Plant J.* 36, 652–663.

7. Benitez-Alfonso, Y., Faulkner, C., Ritzenthaler, C., and Maule, A. J. (2010). Plasmodesmata: gateways to local and systemic virus infection. *Mol. Plant Microbe Interact.* 23, 1403–1412.

8. Brizard, J. P., Carapito, C., Delalande, F., Van Dorsselaer, A., and Brugidou, C. (2006). Proteome analysis of plant-virus interactome: comprehensive data for virus multiplication inside their hosts. *Mol. Cell. Proteomics* 5, 2279–2297.

9. Bukau, B., Weissman, J., and Horwich, A. (2006). Molecular chaperones and protein quality control. *Cell* 125, 443–451.

10. Burch-Smith, T. M., Brunkard, J. O., Choi, Y. G., and Zambryski, P. C. (2011). Organelle-nucleus cross-talk regulates plant intercellular communication via plasmodesmata. *Proc. Natl. Acad. Sci. U.S.A.* 108, E1451–E1460.

11. Burch-Smith, T. M., and Zambryski, P. C. (2012). Plasmodesmata paradigm shift: regulation from without versus within.*Annu. Rev. Plant Biol.* 63, 239–260.

12. Caplan, J. L., Mamillapalli, P., Burch-Smith, T. M., Czymmek, K., and Dinesh-Kumar, S. P. (2008). Chloroplastic protein NRIP1 mediates innate immune receptor recognition of a viral effector. *Cell* 132, 449–462.

13. Carvalho, C. M., Fontenelle, M. R., Florentino, L. H., Santos, A. A., Zerbini, F. M., and Fontes, E. P. (2008a). A novel nucleocytoplasmic traffic GTPase identified as a functional target of the bipartite geminivirus nuclear shuttle protein. *Plant J.* 55, 869–880.

14. Carvalho, C. M., Machado, J. P., Zerbini, F. M., and Fontes, E. P. (2008b). NSP-interacting GTPase: a cytosolic protein as cofactor for nuclear shuttle proteins. *Plant Signal. Behav.* 3, 752–754.

15. Carvalho, C. M., Santos, A. A., Pires, S. R., Rocha, C. S., Saraiva, D. I., Machado, J. P., et al. (2008c). Regulated nuclear trafficking

of rpL10A mediated by NIK1 represents a defense strategy of plant cells against virus. *PLoS Pathog.* 4:e1000247. doi: 10.1371/journal.ppat.1000247

16. Carvalho, M. F., and Lazarowitz, S. G. (2004). Interaction of the movement protein NSP and the Arabidopsis acetyltransferase AtNSI is necessary for cabbage leaf curl geminivirus infection and pathogenicity. *J. Virol.* 78, 11161–11171.
17. Carvalho, M. F., Turgeon, R., and Lazarowitz, S. G. (2006). The geminivirus nuclear shuttle protein NSP inhibits the activity of AtNSI, a vascular-expressed Arabidopsis acetyltransferase regulated with the sink-to-source transition. *Plant Physiol.*140, 1317–1330.
18. Channarayappa, V. M., Schwegler-Berry, D., and Shivashankar, G. (1992). Ultrastructural changes in tomato infected with tomato leaf curl virus, a whitefly-transmitted geminivirus. *Can. J. Bot.* 70, 1747–1753.
19. Duan, Y. P., Powell, C. A., Purcifull, D. E., Broglio, P., and Hiebert, E. (1997). Phenotypic variation in transgenic tobacco expressing mutated geminivirus movement/pathogenicity (BC1) proteins. *Mol. Plant Microbe Interact.* 10, 1065–1074.
20. Esau, K. (1933). Pathologic changes in the anatomy of leaves of the sugar beet, *Beta vulgaris* L., affected by curly top.*Phytopathology* 23, 679–712.
21. Esau, K. (1944). Anatomical and cytological studies on beet mosaic. *J. Agric. Res.* 69, 95–117.
22. Escaler, M., Aranda, M. A., Roberts, I. M., Thomas, C. L., and Maule, A. J. (2000a). A comparison between virus replication and abiotic stress (heat) as modifiers of host gene expression in pea. *Mol. Plant Pathol.* 1, 159–167.
23. Escaler, M., Aranda, M. A., Thomas, C. L., and Maule, A. J. (2000b). Pea embryonic tissues show common responses to the replication of a wide range of viruses. *Virology* 267, 318–325.
24. Fester, T., Strack, D., and Hause, B. (2001). Reorganization of tobacco root plastids during arbuscule development. *Planta*213, 864–868.
25. Florentino, L. H., Santos, A. A., Fontenelle, M. R., Pinheiro, G. L., Zerbini, F. M., Baracat-Pereira, M. C., et al. (2006). A PERK-

like receptor kinase interacts with the geminivirus nuclear shuttle protein and potentiates viral infection. *J. Virol.* 80, 6648–6656.

26. Flores-Pérez, U., and Jarvis, P. (2012). Molecular chaperone involvement in chloroplast protein import. *Biochim. Biophys. Acta.* doi: 10.1016/j.bbamcr.2012.1003.1019 [Epub ahead of print].
27. Fontes, E. P., Santos, A. A., Luz, D. F., Waclawovsky, A. J., and Chory, J. (2004). The geminivirus nuclear shuttle protein is a virulence factor that suppresses transmembrane receptor kinase activity. *Genes Dev.* 18, 2545–2556.
28. Frischmuth, S., Kleinow, T., Aberle, H.-J., Wege, C., Hülser, D., and Jeske, H. (2004). Yeast two-hybrid systems confirm the membrane-association and oligomerization of BC1 but do not detect an interaction of the movement proteins BC1 and BV1 of Abutilon mosaic geminivirus. *Arch. Virol.* 149, 2349–2364.
29. Frischmuth, S., Wege, C., Hülser, D., and Jeske, H. (2007). The movement protein BC1 promotes redirection of the nuclear shuttle protein BV1 of Abutilon mosaic geminivirus to the plasma membrane in fission yeast. *Protoplasma* 230, 117–123.
30. Gafni, Y., and Epel, B. L. (2002). The role of host and viral proteins in intra- and inter-cellular trafficking of geminiviruses.*Physiol. Mol. Plant Pathol.* 60, 231–241.
31. Gray, J. C., Hansen, M. R., Shaw, D. J., Graham, K., Dale, R., Smallman, P., et al. (2012). Plastid stromules are induced by stress treatments acting through abscisic acid. *Plant J.* 69, 387–398.
32. Gray, J. C., Sullivan, J. A., Hibberd, J. M., and Hansson, M. R. (2001). Stromules: mobile protrusions and interconnections between plastids. *Plant Biol.* 3, 223–233.
33. Gröning, B. R., Abouzid, A., and Jeske, H. (1987). Single-stranded DNA from Abutilon mosaic virus is present in the plastids of infected *Abutilon sellovianum. Proc. Natl. Acad. Sci. U.S.A.* 84, 8996–9000.
34. Gröning, B. R., Frischmuth, T., and Jeske, H. (1990). Replicative form DNA of Abutilon mosaic virus is present in plastids.*Mol. Gen. Genet.* 220, 485–488.
35. Gunning, B. E. (2005). Plastid stromules: video microscopy of their outgrowth, retraction, tensioning, anchoring, branching, bridging, and tip-shedding. *Protoplasma* 225, 33–42.

36. Hans, J., Hause, B., Strack, D., and Walter, M. H. (2004). Cloning, characterization, and immunolocalization of a mycorrhiza-inducible 1-deoxy-d-xylulose 5-phosphate reductoisomerase in arbuscule-containing cells of maize. *Plant Physiol.* 134, 614–624.
37. Hanson, M. R., and Sattarzadeh, A. (2008). Dynamic morphology of plastids and stromules in angiosperm plants. *Plant Cell Environ.* 31, 646–657.
38. Hanson, M. R., and Sattarzadeh, A. (2011). Stromules: recent insights into a long neglected feature of plastid morphology and function. *Plant Physiol.* 155, 1486–1492.
39. Hao, L., Wang, H., Sunter, G., and Bisaro, D. M. (2003). Geminivirus AL2 and L2 proteins interact with and inactivate SNF1 kinase. *Plant Cell* 15, 1034–1048.
40. Harries, P., and Ding, B. (2011). Cellular factors in plant virus movement: at the leading edge of macromolecular trafficking in plants. *Virology* 411, 237–243.
41. Harries, P. A., Park, J. W., Sasaki, N., Ballard, K. D., Maule, A. J., and Nelson, R. S. (2009). Differing requirements for actin and myosin by plant viruses for sustained intercellular movement. *Proc. Natl. Acad. Sci. U.S.A.* 106, 17594–17599.
42. Harries, P. A., Schoelz, J. E., and Nelson, R. S. (2011). Intracellular transport of viruses and their components: utilizing the cytoskeleton and membrane highways. *Mol. Plant Microbe Interact.* 23, 1381–1393.
43. Hehnle, S., Wege, C., and Jeske, H. (2004). Interaction of DNA with the movement proteins of geminiviruses revisited. *J. Virol.* 78, 7698–7706.
44. Holzinger, A., Buchner, O., Lutz, C., and Hanson, M. R. (2007). Temperature-sensitive formation of chloroplast protrusions and stromules in mesophyll cells of *Arabidopsis thaliana*. *Protoplasma* 230, 23–30.
45. Horns, T., and Jeske, H. (1991). Localization of Abutilon mosaic virus DNA within leaf tissue by *in situ* hybridization.*Virology* 181, 580–588.
46. Hou, Y. M., Sanders, R., Ursin, V. M., and Gilbertson, R. L. (2000). Transgenic plants expressing geminivirus movement proteins: abnormal phenotypes and delayed infection by tomato mottle

virus in transgenic tomatoes expressing the bean dwarf mosaic virus BV1 or BC1 proteins. *Mol. Plant Microbe Interact.* 13, 297–308.

47. Ingham, D. J., and Lazarowitz, S. G. (1993). A single missense mutation in the BR1 movement protein alters the host range of the squash leaf curl geminivirus. *Virology* 196, 694–702.
48. Isemer, R., Mulisch, M., Schafer, A., Kirchner, S., Koop, H. U., and Krupinska, K. (2012). Recombinant Whirly1 translocates from transplastomic chloroplasts to the nucleus. *FEBS Lett.* 586, 85–88.
49. Ito, J., Heazlewood, J. L., and Millar, A. H. (2006). Analysis of the soluble ATP-binding proteome of plant mitochondria identifies new proteins and nucleotide triphosphate interactions within the matrix. *J. Proteome Res.* 5, 3459–3469.
50. Itoh, R. D., Yamasaki, H., Septiana, A., Yoshida, S., and Fujiwara, M. T. (2010). Chemical induction of rapid and reversible plastid filamentation in *Arabidopsis thaliana* roots. *Physiol. Plant* 139, 144–158.
51. Jeske, H. (1986). The detection of the Abutilon mosaic virus depends on the experimental daytime. *J. Phytopathol.* 115, 243–256.
52. Jeske, H. (2007). "Replication of geminiviruses and the use of rolling circle amplification for their diagnosis," in *Tomato Yellow Leaf Curl Virus Disease: Management, Molecular Biology, and Breeding for Resistance*, ed. H. Czosnek (Dordrecht: Springer Science Business Media), 141–156.
53. Jeske, H. (2009). Geminiviruses. *Curr. Top. Microbiol. Immunol.* 185–226.
54. Jeske, H., and Schuchalter-Eicke, G. (1984). The Abutilon mosaic virus (AbMV) in different leaf tissue of several host plant species of Malvaceae. *Phytopathol. Z.* 109, 353–362.
55. Jeske, H., and Werz, G. (1978). The influence of light intensity on pigment composition and ultrastructure of plastids in leaves of diseased *Abutilon sellowianum* Reg. *Phytopathol. Z.* 91, 1–13.
56. Jeske, H., and Werz, G. (1980a). Cytochemical characterization of plastidal inclusions in Abutilon mosaic-infected *Malva parviflora* mesophyll cells. *Virology* 106, 155–158.

57. Jeske, H., and Werz, G. (1980b). Ultrastructural and biochemical investigations on the whitefly transmitted Abutilon mosaic virus (AbMV). *Phytopathol. Z.* 97, 43–55.

58. Jimenez, I., Lopez, L., Alamillo, J. M., Valli, A., and Garcia, J. A. (2006). Identification of a plum pox virus CI-interacting protein from chloroplast that has a negative effect in virus infection. *Mol. Plant Microbe Interact.* 19, 350–358.

59. Kampinga, H. H., and Craig, E. A. (2010). The HSP70 chaperone machinery: J proteins as drivers of functional specificity.*Nat. Rev. Mol. Cell Biol.* 11, 579–592.

60. Kanzaki, H., Saitoh, H., Ito, A., Fujisawa, S., Kamoun, S., Katou, S., et al. (2003). Cytosolic HSP90 and HSP70 are essential components of INF1-mediated hypersensitive response and non-host resistance to *Pseudomonas cichorii* in*Nicotiana benthamiana*. *Mol. Plant Pathol.* 4, 383–391.

61. Kleinow, T., Nischang, M., Beck, A., Kratzer, U., Tanwir, F., Preiss, W., et al. (2009a). Three C-terminal phosphorylation sites in the Abutilon mosaic virus movement protein affect symptom development and viral DNA accumulation. *Virology*390, 89–101.

62. Kleinow, T., Tanwir, F., Kocher, C., Krenz, B., Wege, C., and Jeske, H. (2009b). Expression dynamics and ultrastructural localization of epitope-tagged Abutilon mosaic virus nuclear shuttle and movement proteins in *Nicotiana benthamiana* cells.*Virology* 391, 212–220.

63. Köhler, R. H., Cao, J., Zipfel, W. R., Webb, W. W., and Hanson, M. R. (1997). Exchange of protein molecules through connections between higher plant plastids. *Science* 276, 2039–2042.

64. Köhler, R. H., and Hanson, M. R. (2000). Plastid tubules of higher plants are tissue-specific and developmentally regulated.*J. Cell Sci.* 113, 81–89.

65. Köhler, R. H., Schwille, P., Webb, W. W., and Hanson, M. R. (2000). Active protein transport through plastid tubules: velocity quantified by fluorescence correlation spectroscopy. *J. Cell Sci.* 113, 3921–3930.

66. Kong, L. J., and Hanley-Bowdoin, L. (2002). A geminivirus replication protein interacts with a protein kinase and a motor

protein that display different expression patterns during plant development and infection. *Plant Cell* 14, 1817–1832.

67. Kozlov, M. M., McMahon, H. T., and Chernomordik, L. V. (2010). Protein-driven membrane stresses in fusion and fission. *Trends Biochem. Sci.* 35, 699–706.

68. Krause, K., and Krupinska, K. (2009). Nuclear regulators with a second home in organelles. *Trends Plant Sci.* 14, 194–199.

69. Krenz, B., Neugart, F., Kleinow, T., and Jeske, H. (2011). Self-interaction of Abutilon mosaic virus replication initiator protein (Rep) in plant cell nuclei. *Virus Res.* 161, 194–197.

70. Krenz, B., Windeisen, V., Wege, C., Jeske, H., and Kleinow, T. (2010). A plastid-targeted heat shock cognate 70 kDa protein interacts with the Abutilon mosaic virus movement protein. *Virology* 401, 6–17.

71. Krichevsky, A., Kozlovsky, S. V., Gafni, Y., and Citovsky, V. (2006). Nuclear import and export of plant virus proteins and genomes. *Mol. Plant Pathol.* 7, 131–146.

72. Kwok, E. Y., and Hanson, M. R. (2003). Microfilaments and microtubules control the morphology and movement of non-green plastids and stromules in *Nicotiana tabacum*. *Plant J.* 35, 16–26.

73. Kwok, E. Y., and Hanson, M. R. (2004a). Plastids and stromules interact with the nucleus and cell membrane in vascular plants. *Plant Cell Rep.* 23, 188–195.

74. Kwok, E. Y., and Hanson, M. R. (2004b). Stromules and the dynamic nature of plastid morphology. *J. Microsc.* 214, 124–137.

75. Latijnhouwers, M., Xu, X. M., and Moller, S. G. (2010). Arabidopsis stromal 70-kDa heat shock proteins are essential for chloroplast development. *Planta* 232, 567–578.

76. Lee, J. Y., and Lu, H. (2011). Plasmodesmata: the battleground against intruders. *Trends Plant Sci.* 16, 201–210.

77. Lee, J. Y., Yoo, B. C., Rojas, M. R., Gomez-Ospina, N., Staehelin, L. A., and Lucas, W. J. (2003). Selective trafficking of non-cell-autonomous proteins mediated by NtNCAPP1. *Science* 299, 392–396.

78. Levy, A., and Tzfira, T. (2011). Bean dwarf mosaic virus: a model system for the study of viral movement. *Mol. Plant Pathol.*11, 451–461.

79. Lewis, J. D., and Lazarowitz, S. G. (2010). Arabidopsis synaptotagmin SYTA regulates endocytosis and virus movement protein cell-to-cell transport. *Proc. Natl. Acad. Sci. U.S.A.* 107, 2491–2496.

80. Lim, H. S., Vaira, A. M., Bae, H., Bragg, J. N., Ruzin, S. E., Bauchan, G. R., et al. (2010). Mutation of a chloroplast-targeting signal in *Alternanthera* mosaic virus TGB3 impairs cell-to-cell movement and eliminates long-distance virus movement. *J. Gen. Virol.* 91, 2102–2115.

81. Lin, J. W., Ding, M. P., Hsu, Y. H., and Tsai, C. H. (2007). Chloroplast phosphoglycerate kinase, a gluconeogenetic enzyme, is required for efficient accumulation of bamboo mosaic virus. *Nucleic Acids Res.* 35, 424–432.

82. Lohse, S., Schliemann, W., Ammer, C., Kopka, J., Strack, D., and Fester, T. (2005). Organization and metabolism of plastids and mitochondria in arbuscular mycorrhizal roots of *Medicago truncatula*. *Plant Physiol.* 139, 329–340.

83. Lozano-Duran, R., Rosas-Diaz, T., Luna, A. P., and Bejarano, E. R. (2011). Identification of host genes involved in geminivirus infection using a reverse genetics approach. *PLoS ONE* 6:e22383. doi: 10.1371/journal.pone.0022383

84. Lucas, W. J. (2006). Plant viral movement proteins: agents for cell-to-cell trafficking of viral genomes. *Virology* 344, 169–184.

85. Mariano, A. C., Andrade, M. O., Santos, A. A., Carolino, S. M., Oliveira, M. L., Baracat-Pereira, M. C., et al. (2004). Identification of a novel receptor-like protein kinase that interacts with a geminivirus nuclear shuttle protein. *Virology* 318, 24–31.

86. Maule, A. J., Benitez-Alfonso, Y., and Faulkner, C. (2011). Plasmodesmata – membrane tunnels with attitude. *Curr. Opin. Plant Biol.* 14, 683–690.

87. Maule, A. J., Escaler, M., and Aranda, M. A. (2000). Programmed responses to virus replication in plants. *Mol. Plant Pathol.* 1, 9–15.

88. Mayer, M. P. (2010). Gymnastics of molecular chaperones. *Mol. Cell* 39, 321–331.

89. Mayer, M. P., and Bukau, B. (2005). HSP70 chaperones: cellular functions and molecular mechanism. *Cell. Mol. Life Sci.*62, 670–684.

90. McGarry, R. C., Barron, Y. D., Carvalho, M. F., Hill, J. E., Gold, D., Cheung, E., et al. (2003). A novel Arabidopsis acetyltransferase interacts with the geminivirus movement protein NSP. *Plant Cell* 15, 1605–1618.

91. McMahon, H. T., Kozlov, M. M., and Martens, S. (2010). Membrane curvature in synaptic vesicle fusion and beyond. *Cell*140, 601–605.

92. Nagy, P. D., Wang, R. Y., Pogany, J., Hafren, A., and Makinen, K. (2011). Emerging picture of host chaperone and cyclophilin roles in RNA virus replication. *Virology* 411, 374–382.

93. Natesan, S. K., Sullivan, J. A., and Gray, J. C. (2009). Myosin XI is required for actin-associated movement of plastid stromules. *Mol. Plant* 2, 1262–1272.

94. Natesan, S. K. A., Sullivan, J. A., and Gray, J. C. (2005). Stromules: a characteristic cell-specific feature of plastid morphology. *J. Exp. Bot.* 56, 787–797.

95. Newell, C. A., Natesan, S. K. A., Sullivan, J. A., Jouhet, J., Kavanagh, T. A., and Gray, J. C. (2012). Exclusion of plastid nucleoids and ribosomes from stromules in tobacco and Arabidopsis. *Plant J.* 69, 399–410.

96. Niehl, A., and Heinlein, M. (2011). Cellular pathways for viral transport through plasmodesmata. *Protoplasma* 248, 75–99.

97. Noel, L. D., Cagna, G., Stuttmann, J., Wirthmuller, L., Betsuyaku, S., Witte, C. P., et al. (2007). Interaction between SGT1 and cytosolic/nuclear HSC70 chaperones regulates Arabidopsis immune responses. *Plant Cell* 19, 4061–4076.

98. Noueiry, A. O., Lucas, W. J., and Gilbertson, R. L. (1994). Two proteins of a plant DNA virus coordinate nuclear and plasmodesmal transport. *Cell* 76, 925–932.

99. Oparka, K. J. (2004). Getting the message across: how do plant cells exchange macromolecular complexes? *Trends Plant Sci.* 9, 33–41.

100. Pascal, E., Goodlove, P. E., Wu, L. C., and Lazarowitz, S. G. (1993). Transgenic tobacco plants expressing the geminivirus BL1 protein exhibit symptoms of viral disease. *Plant Cell* 5, 795–807.

101. Pascal, E., Sanderfoot, A. A., Ward, B. M., Medville, R., Turgeon, R., and Lazarowitz, S. G. (1994). The geminivirus BR1 movement protein binds single-stranded DNA and localizes to the cell nucleus. *Plant Cell* 6, 995–1006.

102. Peltier, J. B., Cai, Y., Sun, Q., Zabrouskov, V., Giacomelli, L., Rudella, A., et al. (2006). The oligomeric stromal proteome of*Arabidopsis thaliana* chloroplasts. *Mol. Cell. Proteomics* 5, 114–133.

103. Peltier, J. B., Emanuelsson, O., Kalume, D. E., Ytterberg, J., Friso, G., Rudella, A., et al. (2002). Central functions of the lumenal and peripheral thylakoid proteome of Arabidopsis determined by experimentation and genome-wide prediction.*Plant Cell* 14, 211–236.

104. Prod'homme, D., Jakubiec, A., Tournier, V., Drugeon, G., and Jupin, I. (2003). Targeting of the turnip yellow mosaic virus 66K replication protein to the chloroplast envelope is mediated by the 140K protein. *J. Virol.* 77, 9124–9135.

105. Pyke, K. A., and Howells, C. A. (2002). Plastid and stromule morphogenesis in tomato. *Ann. Bot.* 90, 559–566.

106. Qin, S. W., Ward, B. M., and Lazarowitz, S. G. (1998). The bipartite geminivirus coat protein aids BR1 function in viral movement by affecting the accumulation of viral single-stranded DNA. *J. Virol.* 72, 9247–9256.

107. Reinero, A., and Beachy, R. N. (1986). Association of TMV coat protein with chloroplast membranes in virus-infected leaves. *Plant Mol. Biol.* 291–301.

108. Rocha, C. S., Santos, A. A., Machado, J. P., and Fontes, E. P. (2008). The ribosomal protein L10/QM-like protein is a component of the NIK-mediated antiviral signaling. *Virology* 380, 165–169.

109. Rojas, M. R., Hagen, C., Lucas, W. J., and Gilbertson, R. L. (2005). Exploiting chinks in the plant's armor: evolution and emergence of geminiviruses. *Annu. Rev. Phytopathol.* 43, 361–394.

110. Rojas, M. R., Noueiry, A. O., Lucas, W. J., and Gilbertson, R. L. (1998). Bean dwarf mosaic geminivirus movement proteins

recognize DNA in a form- and size-specific manner. *Cell* 95, 105–113.

111. Rushing, A. E., Sunter, G., Gardiner, W. E., Dute, R. R., and Bisaro, D. M. (1987). Ultrastructural aspects of tomato golden mosaic virus infection in tobacco. *Phytopathology* 77, 1231–1236.
112. Sanderfoot, A. A., Ingham, D. J., and Lazarowitz, S. G. (1996). A viral movement protein as a nuclear shuttle. The geminivirus BR1 movement protein contains domains essential for interaction with BL1 and nuclear localization. *Plant Physiol.* 110, 23–33.
113. Sanderfoot, A. A., and Lazarowitz, S. G. (1995). Cooperation in viral movement: the geminivirus BL1 movement protein interacts with BR1 and redirects it from the nucleus to the cell periphery. *Plant Cell* 7, 1185–1194.
114. Santos, A. A., Lopes, K. V., Apfata, J. A., and Fontes, E. P. (2010). NSP-interacting kinase, NIK: a transducer of plant defence signalling. *J. Exp. Bot.* 61, 3839–3845.
115. Sattarzadeh, A., Krahmer, J., Germain, A. D., and Hanson, M. R. (2009). A myosin XI tail domain homologous to the yeast myosin vacuole-binding domain interacts with plastids and stromules in *Nicotiana benthamiana*. *Mol. Plant* 2, 1351–1358.
116. Saunders, K., Wege, C., Veluthambi, K., Jeske, H., and Stanley, J. (2001). The distinct disease phenotypes of the common and yellow vein strains of tomato golden mosaic virus are determined by nucleotide differences in the 3′-terminal region of the gene encoding the movement protein. *J. Gen. Virol.* 82, 45–51.
117. Schattat, M., Barton, K., Baudisch, B., Klosgen, R. B., and Mathur, J. (2011). Plastid stromule branching coincides with contiguous endoplasmic reticulum dynamics. *Plant Physiol.* 155, 1667–1677.
118. Schattat, M. H., Griffiths, S., Mathur, N., Barton, K., Wozny, M. R., Dunn, N., et al. (2012). Differential coloring reveals that plastids do not form networks for exchanging macromolecules. *Plant Cell* 24, 1465–1477.
119. Schattat, M. H., and Klösgen, R. B. (2011). Induction of stromule formation by extracellular sucrose and glucose in epidermal leaf tissue of *Arabidopsis thaliana*. *BMC Plant Biol.* 11:115. doi: 10.1186/1471-2229-11-115

120. Schoelz, J. E., Harries, P. A., and Nelson, R. S. (2011). Intracellular transport of plant viruses: finding the door out of the cell. *Mol. Plant* 4, 813–831.

121. Schoelz, J. E., and Zaitlin, M. (1989). Tobacco mosaic virus RNA enters chloroplasts in vivo. *Proc. Natl. Acad. Sci. U.S.A.*86, 4496–4500.

122. Schuchalter-Eicke, G., and Jeske, H. (1983). Seasonal changes in the chloroplast ultrastructure in Abutilon mosaic virus (AbMV) infected *Abutilon* spec. (Malvaceae). *Phytopathol. Z.* 108, 172–184.

123. Shalla, T. A. (1964). Assembly and aggregation of tobacco mosaic virus in tomato leaflets. *J. Cell Biol.* 21, 253–264.

124. Shen, W., and Hanley-Bowdoin, L. (2006). Geminivirus infection up-regulates the expression of two Arabidopsis protein kinases related to yeast SNF1 and mammalian AMPK activating kinases. *Plant Physiol.* 142, 1642–1655.

125. Simpson, C., Thomas, C., Findlay, K., Bayer, E., and Maule, A. J. (2009). An Arabidopsis GPI-anchor plasmodesmal neck protein with callose binding activity and potential to regulate cell-to-cell trafficking. *Plant Cell* 21, 581–594.

126. Su, P.-H., and Li, H.-M. (2008). Arabidopsis stromal 70-kD heat shock proteins are essential for plant development and important for thermotolerance of germinating seeds. *Plant Physiol.* 146, 1231–1241.

127. Su, P. H., and Li, H. M. (2010). Stromal Hsp70 is important for protein translocation into pea and Arabidopsis chloroplasts.*Plant Cell* 22, 1516–1531.

128. Sung, D. Y., Vierling, E., and Guy, C. L. (2001). Comprehensive expression profile analysis of the Arabidopsis HSP70 gene family. *Plant Physiol.* 126, 789–800.

129. Swindell, W. R., Huebner, M., and Weber, A. P. (2007). Transcriptional profiling of Arabidopsis heat shock proteins and transcription factors reveals extensive overlap between heat and non-heat stress response pathways. *BMC Genomics* 8:125. doi: 10.1186/1471-2164-8-125

130. Tirlapur, U. K., Dahse, I., Reiss, B., Meurer, J., and Oelmuller, R. (1999). Characterization of the activity of a plastid-targeted green fluorescent protein in Arabidopsis. *Eur. J. Cell Biol.* 78, 233–240.

131. Torrance, L., Cowan, G. H., Gillespie, T., Ziegler, A., and Lacomme, C. (2006). Barley stripe mosaic virus-encoded proteins triple-gene block 2 and γb localize to chloroplasts in virus-infected monocot and dicot plants, revealing hitherto-unknown roles in virus replication. *J. Gen. Virol.* 87, 2403–2411.

132. Trejo-Saavedra, D. L., Vielle-Calzada, J. P., and Rivera-Bustamante, R. F. (2009). The infective cycle of cabbage leaf curl virus (CaLCuV) is affected by CRUMPLED LEAF (CRL) gene in *Arabidopsis thaliana*. *Virol. J.* 6, 169.

133. Ueki, S., and Citovsky, V. (2011). To Gate, or not to gate: regulatory mechanisms for intercellular protein transport and virus movement in plants. *Mol. Plant* 4, 782–793.

134. von Arnim, A., and Stanley, J. (1992). Determinants of tomato golden mosaic virus symptom development located on DNA B.*Virology* 186, 286–293.

135. Waigmann, E., Ueki, S., Trutnyeva, K., and Citovsky, V. (2004). The ins and outs of nondestructive cell-to-cell and systemic movement of plant viruses. *Crit. Rev. Plant Sci.* 23, 195–250.

136. Waters, M. T., Fray, R. G., and Pyke, K. A. (2004). Stromule formation is dependent upon plastid size, plastid differentiation status and the density of plastids within the cell. *Plant J.* 39, 655–667.

137. Wege, C. (2007). "Movement and localization of tomato yellow leaf curl viruses in the infected plant," in *Tomato Yellow Leaf Curl Virus Disease: Management, Molecular Biology, and Breeding for Resistance*, ed. H. Czosnek (Dordrecht: Springer Science Business Media), 185–206.

138. Weibezahn, J., Schlieker, C., Tessarz, P., Mogk, A., and Bukau, B. (2005). Novel insights into the mechanism of chaperone-assisted protein disaggregation. *Biol. Chem.* 386, 739–744.

139. Xiang, Y., Kakani, K., Reade, R., Hui, E., and Rochon, D. (2006). A 38-amino-acid sequence encompassing the arm domain of the cucumber necrosis virus coat protein functions as a chloroplast transit peptide in infected plants. *J. Virol.* 80, 7952–7964.

140. Zavaliev, R., Ueki, S., Epel, B. L., and Citovsky, V. (2011). Biology of callose (beta-1,3-glucan) turnover at plasmodesmata. *Protoplasma* 248, 117–130.

141. Zhang, S. C., Ghosh, R., and Jeske, H. (2002). Subcellular targeting domains of Abutilon mosaic geminivirus movement protein BC1. *Arch. Virol.* 147, 2349–2363.

142. Zhang, S. C., Wege, C., and Jeske, H. (2001). Movement proteins (BC1 and BV1) of Abutilon mosaic geminivirus are cotransported in and between cells of sink but not of source leaves as detected by green fluorescent protein tagging. *Virology*290, 249–260.

143. Zhou, Y., Rojas, M. R., Park, M.-R., Seo, Y.-S., Lucas, W. J., and Gilbertson, R. L. (2011). Histone H3 interacts and co-localizes with the nuclear shuttle protein and movement protein of a geminivirus. *J. Virol.* 85, 11821–11832.

144. Zhou, Y. C., Garrido-Ramirez, E. R., Sudarshana, M. R., Yendluri, S., and Gilbertson, R. L. (2007). The N-terminus of the begomovirus nuclear shuttle protein (BV1) determines virulence or avirulence in *Phaseolus vulgaris*. *Mol. Plant Microbe Interact.* 20, 1523–1534.

Chapter 7

Real-Time PCR in Virology

Ian M. Mackay[1, 2, 3,] Katherine E. Arden[4,] and Andreas Nitsche5

[1]Clinical Virology Research Unit, Sir Albert Sakzewski Virus Research Centre, Royal Children's Hospital, Brisbane, Australia

[2]Department of Paediatrics and

[3]Clinical Medical Virology Centre, University of Queensland, Queensland, Australia

[4]Membrane Transport Laboratory, Division of Cancer and Cell Biology, Queensland Institute of Medical Research, Queensland, Australia and 5TIB MOLBIOL, Tempelhofer Weg 11–12, D-10829 Berlin, Germany

ABSTRACT

The use of the polymerase chain reaction (PCR) in molecular diagnostics has increased to the point where it is now accepted as the gold standard for detecting nucleic acids from a number of origins and it has become an essential tool in the research laboratory. Real-time PCR has engendered wider acceptance of the PCR due to its improved rapidity, sensitivity, reproducibility and the reduced risk of carry-over contamination. There are currently five main chemistries used for the detection of PCR product during real-time PCR. These are the DNA binding fluorophores, the 5′ endonuclease, adjacent linear and hairpin oligoprobes and the self-fluorescing amplicons, which are described in detail. We also discuss factors that have restricted the development of multiplex real-time PCR as well as the role of real-time PCR in quantitating nucleic acids. Both amplification hardware and the fluorogenic detection chemistries have evolved rapidly as the understanding of real-time PCR has developed and this review aims to update the scientist on the current state of the art. We describe the background, advantages and limitations of real-time PCR and we review the literature as it applies to virus detection in the routine and research laboratory in order to focus on one of the many areas in which the application of real-time PCR has provided significant methodological benefits and improved patient outcomes. However, the technology discussed has been applied to other areas of microbiology as well as studies of gene expression and genetic disease.

BACKGROUND

The polymerase chain reaction (PCR) (1, 2) has been used as the new gold standard for detecting a wide variety of templates across a range of scientific specialties, including virology. The method utilises a pair of synthetic oligonucleotides or primers, each hybridising to one strand of a double-stranded DNA (dsDNA) target, with the pair spanning a region that will be exponentially reproduced. The hybridised primer acts as a substrate for a DNA polymerase (most commonly derived from the thermophilic bacterium Thermus aquaticus and called Taq), which creates a complementary strand via sequential addition of deoxynucleotides. The process can be summarised in three steps: (i) dsDNA separation at temperatures >90°C, (ii) primer annealing at 50–75°C, and (iii) optimal extension at 72–78°C (Fig. 1A). The rate of temperature change or ramp rate, the length of the incubation at each temperature and the number of times each set of temperatures (or cycle) is repeated are controlled by a programmable thermal cycler. Current technologies have significantly shortened the ramp times using electronically controlled heating blocks or fan-forced heated air flows to moderate the reaction temperature. Consequently, PCR is displacing some of the gold standard cell culture, antigenaemia and serological assays (3). Existing combinations of PCR and detection assays (called 'conventional PCR' here) have been used to obtain quantitative data with promising results. However, these approaches have suffered from the laborious post-PCR handling steps required to evaluate the amplicon (4).

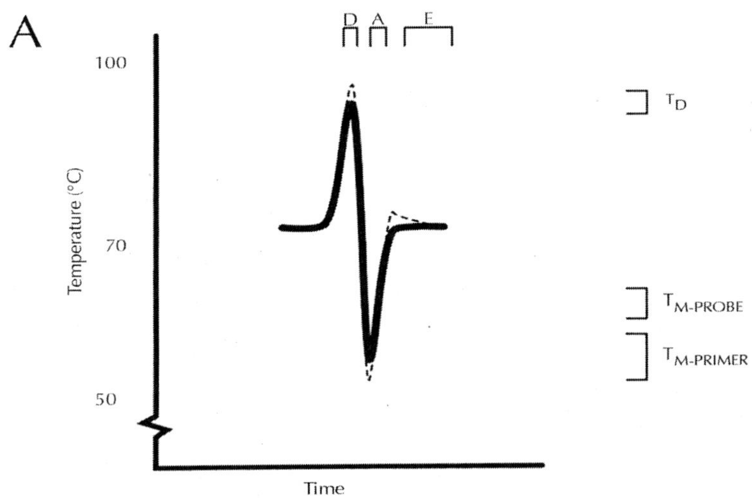

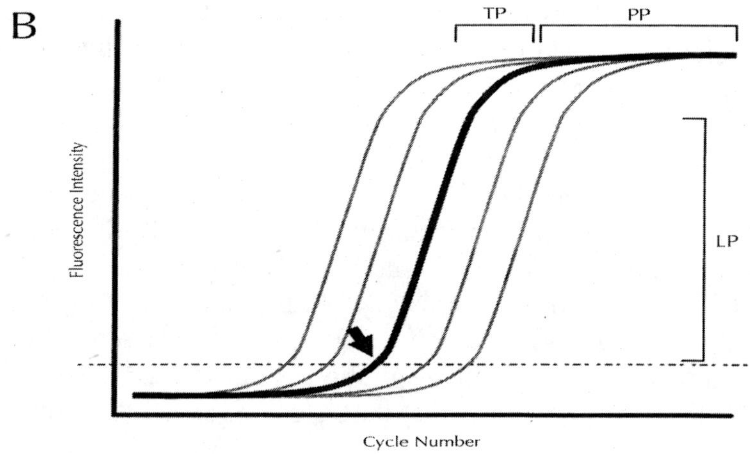

Figure 1: Kinetic amplification. (A) An idealised plot of temperature versus time during a single PCR cycle comprised of the denaturation (D), primer and probe annealing (A) and primer extension (E) steps. At the indicated optimal temperature ranges, dsDNA is denatured (T_D), oligoprobes anneal ($T_{M\text{-PROBE}}$) and finally the primers anneal as a precursor to their extension ($T_{M\text{-PRIMER}}$). The actual temperature, shown as a dashed line, may overshoot the desired temperature to varying degrees, depending on the quality of the thermocycler employed. (B) The ideal amplification curve of a real-time PCR (bold), when plotted as fluorescence intensity against the cycle number, is a typical sigmoidal growth curve. Early amplification cannot be viewed because the detection signal is indistinguishable from the background. However, when enough amplicon is present, the assay's exponential progress can be monitored as the rate of amplification enters a log-linear phase (LP). Under ideal conditions, the amount of amplicon increases at a rate of approximately one $\log_{10}$ every three cycles. As primers and enzyme become limiting and products inhibitory to the PCR accumulate, the reaction slows, entering a transition phase (TP), eventually reaching the plateau phase (PP) where there is little or no further increase in product yield. The point at which the fluorescence passes from insignificant levels to clearly detectable is called the threshold cycle (C_T; indicated by an arrow), and this value is used in the calculation of template quantity during quantitative real-time PCR. Also shown are curves representing an optimal titration of template (grey), consisting of higher and lower starting template concentrations, which produce lower or higher C_T values, respectively. Data for the construction of a standard curve are taken from the LP, which subsequently allows the concentration of unknown samples to be determined.

Traditional detection of amplified DNA relies upon electrophoresis of the nucleic acids in the presence of ethidium bromide and visual or densitometric analysis of the resulting bands after irradiation by ultraviolet light (5). Southern blot detection of amplicon using hybridisation with a labelled oligonucleotide probe is also time consuming and requires multiple PCR product handling steps, further risking a spread of amplicon throughout the laboratory (6). Alternatively, PCR–ELISA may be used to capture amplicon onto a solid phase using biotin or digoxigenin-labelled primers, oligonucleotide probes (oligoprobes) or directly after incorporation of the digoxigenin into the amplicon (7–12). Once captured, the amplicon can be detected using an enzyme-labelled avidin or anti-digoxigenin reporter molecule similar to a standard ELISA format.

The possibility that, in contrast to conventional assays, the detection of amplicon could be visualised as the amplification progressed was a welcome one (13). This approach has provided a great deal of insight into the kinetics of the reaction and it is the foundation of kinetic or 'real-time' PCR (Fig. 1B) (6, 14–17). Real-time PCR has already proven itself valuable in laboratories around the globe, building on the enormous amount of data generated by conventional PCR assays.

The monitoring of accumulating amplicon in real time has been made possible by the labelling of primers, probes or amplicon with fluorogenic molecules, this chemistry has clear benefits over radiogenic oligoprobes that include an avoidance of radioactive emissions, ease of disposal and an extended shelf life (18).

The increased speed of real-time PCR is largely due to reduced cycle times, removal of post-PCR detection procedures and the use of fluorogenic labels and sensitive methods of detecting their emissions (19, 20). The reduction in amplicon size generally recommended by the creators of commercial real-time assays may also play a role in this speed, however we have shown that decreased product size does not necessarily improve PCR efficiency (21).

The disadvantages of using real-time PCR in comparison with conventional PCR include the inability to monitor amplicon size without opening the system, the incompatibility of some platforms with some fluorogenic chemistries, and the relatively restricted multiplex capabilities of current applications. Also, the start-up expense of real-time PCR may be prohibitive when used in low-throughput laboratories.

These shortcomings are mostly due to limitations in the system hardware or the available fluorogenic dyes or 'fluorophores', both of which will be discussed in more detail. Because most of the popular real-time PCR chemistries depend upon the hybridisation of an oligoprobe to its complementary sequence on one of the strands of the amplicon, the use of more of the primer that creates this strand is beneficial to the generation of an increased fluorescent signal (22). Asymmetric PCR, as this is known, has been shown to produce improved fluorescence from a hairpin oligoprobe PCR (23) and we have found it directly applicable to other oligoprobe-hybridisation assays. The most commonly used fluorogenic oligoprobes rely upon fluorescence resonance energy transfer (FRET; Fig. 2) between fluorogenic labels or between one fluorophore and a dark or 'black-hole' non-fluorescent quencher (NFQ), which disperses energy as heat rather than fluorescence. FRET is a spectroscopic process by which energy is passed between molecules separated by 10–100 Å that have overlapping emission and absorption spectra (24, 25). Förster primarily developed the theory behind this process: the mechanism is a non-radiative induced-dipole interaction (26). The efficiency of energy transfer is proportional to the inverse sixth power of the distance (R) between the donor and acceptor ($1/R^6$) fluorophores (27, 28).

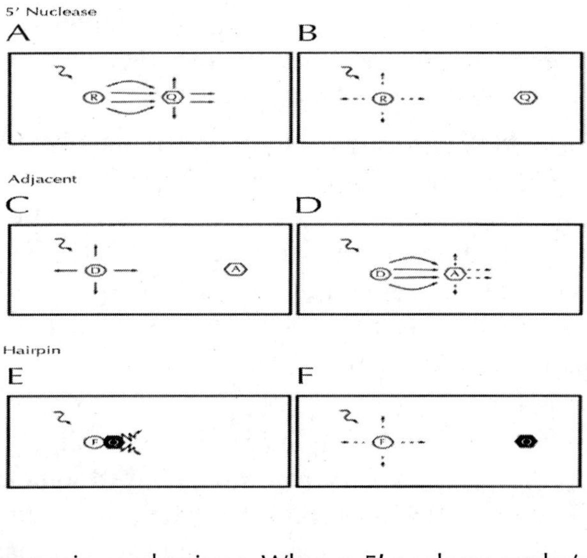

Figure 2: Fluorogenic mechanisms. When a 5′ nuclease probe's reporter (R) and quencher (Q, open) are in close proximity as in (A), the quencher 'hi-

jacks' the emissions that have resulted from excitation of the reporter by the instrument's light source. The quencher then emits this energy (solid arrows). When the fluorophores are separated beyond a certain distance, as occurs upon hydrolysis as depicted in (B), the quencher no longer exerts any influence and the reporter emits at a distinctive wavelength (dashed arrows) which is recorded by the instrument. In the reverse process as depicted in (C) using adjacent oligoprobes, the fluorophores begin as separated entities. A signal from the acceptor (A) can only be generated when the donor (D) comes into close proximity as shown in (D). This occurs as the result of adjacent hybridisation of the oligoprobes to the target amplicon. In (E) another form of quenching is shown, caused by the intimate contact of labels attached to hairpin oligoprobes (molecular beacon, sunrise or scorpion). The fluorophore (F) and a NFQ (Q, closed) interact more by collision than FRET, disrupting each other's electronic structure and directly passing on the excitation energy which is dissipated as heat (jagged, dashed arrows). When the labels are separated by disruption of the hairpin, as is the case in (F), the fluorophore is free to fluoresce (dashed arrows).

Post-amplification manipulation of the amplicon is not required for real-time PCR, therefore these assays are described as 'closed' or homogeneous systems. The advantages of homogeneous systems include a reduced result turnaround, minimisation of the potential for carry-over contamination and the ability to closely scrutinise the assay's performance (29). In addition, real-time PCR has proven cost effective when implemented in a high throughput laboratory (30), particularly when replacing conventional, culture-based approaches to virus detection.

In the remainder of this article, the theory behind real-time PCR will be reviewed and its rapidly increasing use in the fields of human virology will be used as an illustration.

AMPLICON DETECTION

In the following section we will focus on the detection processes that discriminate real-time PCR from conventional PCR assays. There are five major chemistries currently in use, and they can be classified into amplicon sequence specific or non-specific methods of real-time PCR detection (31). Each of the chemistries has an associated nomenclature to describe the fluorescent labels; however, for general discussion, fluorophore will continue to be used to describe these moieties. Although this review focuses on the use of these chemistries in real-time applications, they can also be used as a label for end-point amplicon detection.

DNA-Binding Fluorophores

The basis of the sequence non-specific detection methods is the DNA-binding fluorogenic molecule. Included in this group are the earliest and simplest approaches to real-time PCR. Ethidium bromide (32), YO-PRO-1 (33, 34) and SYBR® green 1 (35) all fluoresce when associated with dsDNA which is exposed to a suitable wavelength of light. This approach requires less specialist knowledge than the design of fluorogenic oligoprobes, is less expensive and does not suffer when the template sequence varies, which may abrogate hybridisation of an oligoprobe (36). Formation of primer-dimer (37) is common and, together with the formation of specific products, is strongly associated with entry of the PCR into the plateau phase (Fig. 1B) (38, 39). Association of a DNA-binding fluorophore with primer-dimer or other non-specific amplification products can confuse interpretation of the results. Adding a short, higher temperature incubation after the extension step in which fluorescence data are acquired minimises the contribution of these products to the fluorescence signal (40). The problem of primer-dimer can also be addressed using software capable of fluorescent melting curve analysis. This method makes use of the temperature at which the dsDNA amplicon is denatured (T_D; Fig. 1A). The shorter primer-dimer can be discriminated by its reduced T_D compared with the full-length amplicon. Analysis of the melting curves of amplicon in the presence of SYBR green 1 has demonstrated that the practical sensitivity of DNA-binding fluorophores is limited by non-specific amplification at low initial template concentrations.

DNA binding fluorophores also increase the T_D and broaden the melting transition, requiring substantial sequence change to produce a shift in theT_D. Oligoprobes are able to discriminate single point mutations using the temperature at which 50% of oligoprobe-target duplexes separate (41). This temperature is called the melting temperature (T_M) and it is dependent upon the concentration of the dsDNA, its length, nucleotide sequence and the solvent composition, and is often confused with T_D (Fig.1A) (42).

Linear Oligoprobes

The use of a pair of adjacent, fluorogenic hybridisation oligoprobes was first described in the late 1980s (43, 44) and, now known as 'HybProbes', they have become the method of choice for the LightCycler™ (Roche Molecular Biochemicals, Germany), a capillary-based, microvolume fluorimeter and thermocycler with rapid temperature control (20,45). The upstream oligoprobe is labelled with a 3′ donor fluorophore (FITC) and the downstream probe is commonly labelled with either a LightCycler Red 640 or Red 705 acceptor fluorophore at the 5′ terminus so that when both oligoprobes are hybridised, the two fluorophores are located within 10 nt of each other, sometimes attracting the name 'kissing' probes (Figs 2C, D and 3C). The plastic and glass composite capillaries are optically clear and act as cuvettes for fluorescence analysis, as well as facilitating rapid heat transfer. Capillaries are rotated past a blue light-emitting diode and fluorescence is monitored by three photodetection diodes with different wavelength filters. The temperature is varied by rapidly heating and cooling air using a heating element and fan which produce ramp rates of 20°C/s, prolonging polymerase survival (46). Additionally, because the oligoprobes are not significantly hydrolysed during amplification (47) and the LightCycler is able to monitor the changes in fluorescence emission during denaturation of the adjacent oligoprobes from their amplicon, this system can perform single tube genotyping. This capability, which makes use of fluorescent melting curve analysis, provides significant information about the sequence to which the oligoprobes are binding. Mutation(s) under one or both oligoprobes can be determined by the decrease in melting temperature that they incur due to destabilisation of the oligoprobe/target duplex (Fig. 4). This has imparted significant improvements in speed upon the diagnosis of genetic disease as well as a growing number of multiplex PCR approaches for the detection of related viral pathogens. Despite the fact that the hybridisation does not reach equilibrium using these ramp rates, the apparent T_M values are both reproducible and characteristic of a given probe/target duplex (48). However, the capillaries are fragile and require some experience to handle (49).

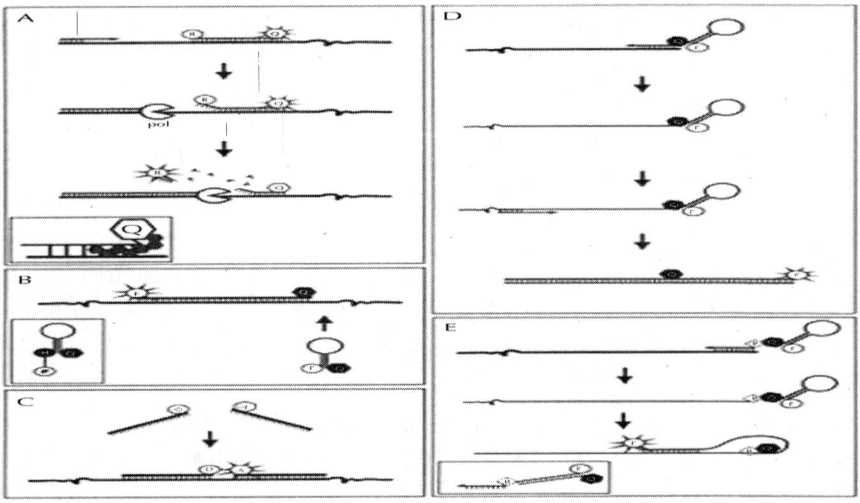

Figure 3: Oligoprobe chemistries. (A) 5′ Nuclease oligoprobes. As the DNA polymerase (pol) progresses along the relevant strand, it displaces and then hydrolyses the oligoprobe via its 5′→3′ endonuclease activity. Once the reporter (R) is removed from the extinguishing influence of the quencher (Q, open), it is able to release excitation energy at a wavelength that is monitored by the instrument and different from the emissions of the quencher. Inset shows the NFQ and MGB molecule that make up the improved MGB nuclease-oligoprobes. (B) Hairpin oligoprobes. Hybridisation of the oligoprobe to the target separates the fluorophore (F) and non-fluorescent quencher (Q, closed) sufficiently to allow emission from the excited fluorophore, which is monitored. Inset shows a wavelength-shifting hairpin oligoprobe incorporating a harvester molecule. (C) Adjacent oligoprobes. Adjacent hybridisation results in a FRET signal due to interaction between the donor (D) and acceptor (A) fluorophores. This bimolecular system acquires its data from the acceptor's emissions in an opposite manner to the function of nuclease oligoprobe chemistry. (D) Sunrise primers. The opposite strand is duplicated so that the primer's hairpin structure can be disrupted. This separates the labels, eliminating the quenching in a similar manner to the hairpin oligoprobe. (E) Scorpion primers. The primer does not require extension of the complementary strand; in fact it blocks extension to ensure that the hairpin in the probe is only disrupted by specific hybridisation with a complementary sequence designed to occur downstream of its own, nascent strand. Inset shows a duplex scorpion that exchanges the stem–loop structure for a primer element terminally labelled with the fluorophore and a separate complementary oligonucleotide labelled with a quencher at the 5′ terminus.

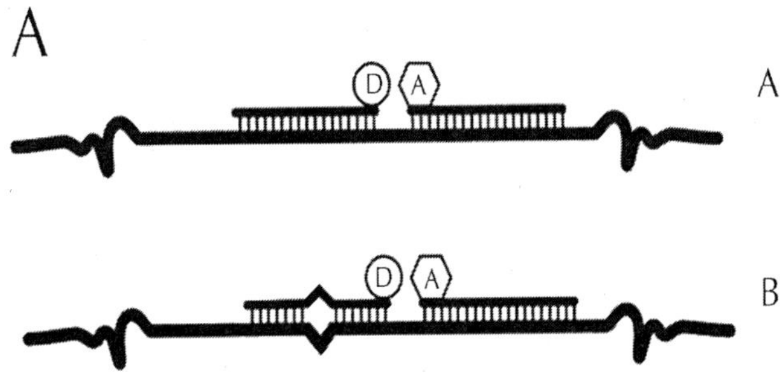

Figure 4: Fluorescent melting curve analysis. At the completion of a real-time PCR using a pair of adjacent oligoprobes, the reaction can be cooled to a temperature below the expected T_M of the oligoprobes then heated above 85°C at a fraction of a degree per second (B). During heating, the emissions of the acceptor can be constantly acquired (C). Software calculates the negative derivative of the fluorescence over time, producing clear peaks that indicate theT_M of the oligoprobe-target melting transition (D). When one or more mutations are present under one or both oligoprobes (A), this T_M is shifted and this shift can be used diagnostically to discriminate single nucleotide polymorphisms in the template. Ideally, one of the oligoprobes, the anchor, is designed to bind to a stable sequence region, whereas the other, sensor, will span the mismatch. Mismatches near the centre of the probe and flanked by G: C pairs are more destabilising than mismatches near the ends of the oligoprobe flanked by A: T pairs.

When comparing signals from the different chemistries, the destruction of nuclease oligoprobes continues despite a plateau in product accumulation whereas SYBR green 1 fluorescence in the no template control generally increases non-specifically during later cycles. Adjacent oligoprobe fluorescence begins to decrease as the rate of collision between the growing numbers of complementary amplicon strands increases favouring the formation of dsDNA over the hybridisation of oligoprobe to its target DNA strand. Additionally, there is the possibility that some oligoprobe is consumed by sequence-related endonuclease activity (50, 51). All three oligoprobe chemistries (SYBR Green I, nuclease and adjacent oligoprobes) seem capable of detecting amplified product with approximately the same sensitivity (20).

Combinations of the above approaches are now appearing as more users of the instrumentation become familiar with the concepts behind real-time PCR and contribute to the literature. If a sequence-specific, fluorophore-labelled linear oligoprobe is added to a SYBR green 1 mix, currently called the Bi-probe system, FRET will occur and an additional layer of specificity can be obtained (44, 52, 53). An assay using a BODIPY®FL-labelled oligoprobe was adapted to run in the Light Cycler using a -globin target sequence (54). The probe was designed so that the fluorophore was located on a terminal cytosine and was quenched by proximity with a complementary guanine. The assay demonstrated that quenching varies linearly with the concentration of template across a defined concentration range. The commonly used fluorophore FITC is inherently quenched by deoxyguanosine nucleotides. The level of quenching can be increased if more guanines are present or a single guanine is located in the first overhang position, 1 nt beyond the fluorophore-labelled terminus of the probe. This approach to amplicon detection is easier to design than fluorogenic oligoprobes, simpler to synthesise and use in real-time PCR and does not require a DNA polymerase with nuclease activity (55).

The light-up probe is a peptide nucleic acid to which the asymmetric cyanine fluorophore thiazole orange is attached (56). When hybridised with a nucleic acid target, either as a duplex or triplex, depending on the oligoprobe's sequence, the fluorophore becomes strongly fluorescent. These probes do not interfere with the PCR, do not require conformational change, are sensitive to single nucleotide mismatches allowing fluorescence melting analysis, and because a single reporter is used, a direct measurement of fluorescence can be made instead of the measurement of a change in fluorescence between two fluorophores (56,57). However, non-specific fluorescence has been reported during later cycles using these probes (58).

5′ Nuclease Oligoprobes

In the late 1980s homogeneous assays were few and far between, but rapid advances in thermocycler instrumentation and the chemistry of nucleic acid manipulation have since made these assays commonplace. The success of these assays revolves around a signal changing in some rapid and measurable way upon hybridisation of a probe to its target (59). By using an excess, the time required for hybridisation of an oligoprobe to its target, especially when the amount of that target has been increased by PCR or some other amplifying process, is significantly reduced (41,59).

In 1991, Holland et al. (6) described a technique that was to form the foundation for homogeneous PCR using fluorogenic oligoprobes. Amplicon was detected by monitoring the effect of Taq DNA polymerase's 5′→3′ endonuclease activity on specific oligoprobe/ target DNA duplexes. The radiolabelled products were examined using thin layer chromatography and the presence or absence of hydrolysis was used as an indicator of duplex formation. These oligoprobes contained a 3′ phosphate moiety, which blocked their extension by the polymerase, but otherwise had no effect on the amplicon's yield.

The desirable criteria for an oligoprobe label are (i) easy attachment of the label to DNA, (ii) detectability at low concentrations, (iii) detectability using simple instrumentation, (iv) production of an altered signal upon specific hybridisation, (v) biological safety, (vi) stability at elevated temperatures and (vii) an absence of interference with the activity of the polymerase (6, 18).

An innovative approach used nick-translation PCR in combination with dual-fluorophore labelled oligoprobes (14). In the first truly homogenous assay of its kind, one fluorophore was added to the 5′ terminus and one to the middle of a sequence specific oligonucleotide probe. When in such close proximity, the 5′ reporter fluorophore (6-carboxy-fluoroscein) transferred laser-induced excitation energy by FRET to the 3′ quencher fluorophore (6-carboxy-tetramethyl-rhodamine; TAMRA), which reduced the lifetime of the reporter's excited state by taking its excess energy and emitting it as a fluorescent signal of its own (Fig. 2A and B). TAMRA emitted the new energy at a wavelength that was monitored but not utilised in the presentation of data. However, when the oligoprobe hybridised to its template, the fluorophores were released due to hydrolysis of the oligoprobe component of the probe/target duplex. Once the labels were separated, the reporter's emissions were no longer quenched and the instrument monitored the resulting fluorescence. These oligoprobes have been called 5′ nuclease, hydrolysis or TaqMan®oligoprobes (Fig. 3A). Nuclease oligoprobes have design requirements that are applicable to the other linear oligoprobe chemistries, including (i) a length of 20–40 nt, (ii) a GC content of 40–60%, (iii) no runs of a single nucleotide, particularly G, (iv) no repeated sequence motifs, (v) an absence of hybridisation or overlap with the forward or reverse primers and (vi) a T_M at least 5°C higher than that of the primers, to ensure the oligoprobe has bound to the template before extension of the primers can occur (60).

This technology, however, required the development of a platform to excite and detect fluorescence as well as perform thermal cycling. A charge-coupled device had been described in 1992 for the quantification of conventional reverse transcription (RT)–PCR products (61). In 1993 this approach was combined with a thermal cycler resulting in the first real-time PCR fluorescence excitation and detection platform (29). To date, the ABI Prism® 7700 sequence detection system (Perkin Elmer Corporation/Applied Biosystems, USA) has been the main instrument used for 5′ nuclease oligoprobes. Non-PCR related fluorescence fluctuations have been normalised using a non-participating or 'passive' internal reference fluorophore (6-carboxy-N,N,N′,N′-tetramethylrhodamine; ROX). The corrected values, obtained from a ratio of the emission intensity of the reporter signal and ROX, are called RQ^+. To further control amplification fluctuations, the fluorescence from a 'no-template' control reaction (RQ^-) is subtracted from RQ^+ resulting in the RQ value that indicates the magnitude of the signal generated for the given PCR (62).

The fractional cycle number at which the real-time fluorescence signal mirrors progression of the reaction above the background noise was used as an indicator of successful target amplification (63). This threshold cycle (C_T) is defined as the PCR cycle in which the gain in fluorescence generated by the accumulating amplicon exceeds 10 standard deviations of the mean baseline fluorescence, using data taken from cycles 3 to 15 (Fig. 1B) (64). The C_T is proportional to the number of target copies present in the sample (17).

A recent improvement to the nuclease oligoprobe has resulted in the minor groove binding (MGB) oligoprobes (Fig. 3A, inset). This chemistry replaces the standard TAMRA quencher with an NFQ and incorporates a molecule that stabilises the oligoprobe-target duplex by folding into the minor groove of the dsDNA (65). This allows the use of very short (14 nt) oligoprobes, which are ideal for detecting single nucleotide polymorphisms (SNPs). A related use of dual labelled oligonucleotide sequences has been to provide the signal-generating portion of the DzyNA–PCR system (66). Here, the reporter and quencher are separated after cleavage of the probe by a DNAzyme, which is created during PCR as the complement of an antisense DNAzyme sequence included in the 5′ tail of one of the primers. Upon cleavage, the dual labelled substrate releases the fluorophores and generates a signal in an analogous manner to the 5′ nuclease probe.

HAIRPIN OLIGOPROBES

Molecular beacons were the first hairpin oligoprobes and are a variation of the dual-labelled nuclease oligoprobe (Fig. 3B). The hairpin oligoprobe's fluorogenic labels are called fluorophore and quencher, and they are positioned at the termini of the oligoprobe. The labels are held in close proximity by distal stem regions of homologous base pairing deliberately designed to create a hairpin structure which results in quenching either by FRET or a direct energy transfer by a collisional mechanism due to the intimate proximity of the labels (Fig. 2E and F) (67). In the presence of a complementary sequence, designed to occur within the bounds of the primer binding sites, the oligoprobe will hybridise, shifting into an open configuration. The fluorophore is now spatially removed from the quencher's influence and fluorescent emissions are monitored during each cycle (68). The occurrence of a mismatch between a hairpin oligoprobe and its target has a greater destabilising effect on the duplex than the introduction of an equivalent mismatch between the target and a linear oligoprobe. This is because the hairpin structure provides a highly stable alternate conformation. Therefore, hairpin oligoprobes have been shown to be more specific than the more common linear oligoprobes making them ideal candidates for detecting SNPs (67). The quencher, 4-(4′-dimethylamino-phenylazo)-benzene (DABCYL), differs from that described for the nuclease oligoprobes because it is an NFQ.

The wavelength-shifting hairpin probe is a recent improvement to this chemistry which makes use of a second, harvesting fluorophore. The harvester passes excitation energy acquired from a blue light source and releases it as fluorescent energy in the far-red wavelengths. The energy can then be used by a receptive 'emitter' fluorophore that produces light at characteristic wavelengths (Fig. 3B, inset). This offers the potential for improved multiplex real-time PCR and SNP analysis (Fig. 4), using currently available instruments (69). Because the function of these oligoprobes depends upon correct hybridisation of the stem, accurate design is crucial to their function (47).

Self-fluorescing Amplicon

The self-priming amplicon is similar in concept to the hairpin oligoprobe, except that the label becomes irreversibly incorporated into the PCR product (Fig. 3D and E). Two approaches have been described: sunrise primers (now commercially called Amplifluor™ hairpin primers) and scorpion primers (31, 70). The sunrise primer consists of a 5′ fluorophore and a DABCYL NFQ. The labels are separated by complementary stretches of sequence that create a stem when the sunrise primer is closed. At the 3′ terminus is a target-specific primer sequence. The sunrise primer's sequence is intended to be duplicated by the nascent complementary strand and, in this way, the stem is destabilised, the two fluorophores are held ~20 nt (70 Å) apart and the fluorophore is free to emit its excitation energy for monitoring (70). This system could suffer from non-specific fluorescence due to duplication of the sunrise primer sequence during the formation of primer-dimer.

The scorpion primer is almost identical in design except for an adjacent hexethylene glycol molecule that blocks duplication of the signalling portion of the scorpion. In addition to the difference in structure, the function of scorpion primers differs slightly in that the 5′ region of the oligonucleotide is designed to hybridise to a complementary region within the amplicon. This hybridisation forces the labels apart disrupting the hairpin and permitting emission in the same way as hairpin probes (31).

VIRAL QUANTITATION

The majority of diagnostic PCR assays reported to date have been used in a qualitative, or 'yes/no' format. The development of real-time PCR has brought true quantitation of target nucleic acids out of the pure research laboratory and into the diagnostic laboratory.

Determining the amount of template by PCR can be performed in two ways: as relative quantitation and as absolute quantitation. Relative quantitation describes changes in the amount of a target sequence compared with its level in a related matrix. Absolute quantitation states the exact number of nucleic acid targets present in the sample in relation to a specific unit (71). Generally, relative quantitation provides sufficient information and is simpler to develop. However, when monitoring the progress of an infection, absolute quantitation is useful in order to express the results in units that are common to both scientists and clinicians and across different platforms. Absolute quantitation may also be necessary when there is a lack of sequential specimens to demonstrate changes in virus levels, no suitably standardised reference reagent or when the viral load is used to differentiate active versus persistent infection.

A very accurate approach to absolute quantitation by PCR is the use of competitive co-amplification of an internal control nucleic acid of known concentration and a wild-type target nucleic acid of unknown concentration, with the former designed or chosen to amplify with an equal efficiency to the latter (72–76). However, while conventional competitive PCR is relatively inexpensive, real-time PCR is far more convenient, reliable and better suited to quick decision making in a clinical situation (77, 78). This is because conventional, quantitative, competitive PCR (qcPCR) requires significant development and optimisation to ensure reproducible performance and a predetermined dynamic range for both the amplification and detection components (79).

Although a comparison of absolute standard curves, relative standard curves and C_T values produces similar final values (80), the general belief remains that an internal control in combination with replicates of each sample are essential for reliable quantitation by PCR (38,39). Unfortunately, real-time PCR software with the ability to calculate the concentration of an unknown by comparing signals generated by an amplified target and internal control is only beginning to emerge. This issue will hopefully be addressed in upcoming commercial releases (81). Therefore, the next best approach to quantitation by PCR is the use of an external standard curve. This approach relies upon titration of an identically amplified template, in a related sample matrix, within the same experimental run. While the external standard curve is the more commonly described approach, it suffers from uncontrolled and unmonitored inter-tube variations. Because of this omission, such experiments should be described as semi-quantitative. Despite this suboptimal approach, fluorescence data is generally collected from PCR cycles that span the linear amplification portion of the reaction where the fluorescent signal and the accumulating DNA are proportional. Because the emissions from fluorescent chemistries are temperature dependent, data is generally acquired only once per cycle at the same temperature in order to monitor amplicon yield (45). The C_T of the sample at a specific fluorescence value can then be compared with similar data collected from a series of standards by the calculation of a standard curve.The determination of the C_T depends upon the sensitivity and ability of the instrument to discriminate specific fluorescence from background noise, the concentration and nature of the fluorescence-generating component and the amount of template initially present.

Real-time PCR offers significant improvements to the quantitation of viral load because of its enormous dynamic range that can accommodate at least eight $\log_{10}$ copies of nucleic acid template (33, 52, 77, 82–89). This is made possible because the data are chosen from the linear phase (LP; Fig.1B) of amplification where conditions are optimal, rather than the end-point where the final amount of amplicon present may have been affected by inhibitors, poorly optimised reaction conditions or saturation by inhibitory PCR by-products and double-stranded amplicon. The result of taking data from the end-point is that there may not be a relationship between the initial template and final amplicon concentrations.

Real-time PCR is also an attractive alternative to conventional PCR for the study of viral load because of its low inter-assay and intra-assay variability (77,87,90) and its equivalent or greater analytical sensitivity in comparison with traditional viral culture, or conventional single-round, and nested PCR (77,85,91–96). Real-time PCR has been reported to be at least as sensitive as Southern blot (92). However, these reports could be an over-estimate due to the choice of smaller targets, which amplify more efficiently, or due to the use of different or improved primers for the real-time assays because the use of software to design optimised primers and oligoprobes is more common.

When this increased sensitivity and broad dynamic range are combined, it is possible to quantitate template from samples containing a large range of concentrations, as is often the case in patient samples. This avoids the need for dilution of the amplicon prior to conventional detection or repeat of the assay using a diluted sample because the first test result falls outside the limits of the assay. These are problems encountered when using some conventional qcPCR assay kits, which cannot encompass high viral loads whilst maintaining suitable sensitivity (52, 97–99). The flexibility of real-time PCR is also demonstrated by its ability to detect one target in the presence of a vast excess of another target in duplexed assays (84).

Viral load is also a useful indicator of the extent of active infection, virus–host interactions and the response to antiviral therapy, all of which can play a role in the treatment regimen selected (100,101). Conventional quantitative PCR has already proven the benefits of applying nucleic acid amplification to the monitoring of viral load as a useful marker of disease progression and as a component of studies into the efficacy of antiviral compounds (74,100,102–104). The severity of some diseases has been shown to correlate with the viral load making real-time PCR quantitation useful to study not simply the presence of a virus but the role of viral reactivation or persistence in the progression of disease (78,82,91,105–112).

An example of the benefits which real-time PCR has brought to the quantitative detection of human cytomegalovirus (CMV) is seen in patients who are immunosuppressed following solid organ or bone marrow transplantation. Although qualitative detection of CMV DNA by PCR has been used as an indicator for the success of antiviral therapy, quantitative assays are preferred in order to monitor patient's therapeutic responses. Moreover, since it has been postulated that the monitoring of viral replication over time is a more reliable indicator of a developing viral disease than the determination of absolute viral amounts at a single point of time, several quantitative assays have been established and evaluated to increase diagnostic accuracy. Quantitative competitive assays based on end-point analysis have displayed detection limits of 5×10^1 genome equivalents (ge) per assay and a dynamic range of 5×10^1–5×10^4ge/assay (113). Hybridisation-based assays covered approximately the same dynamic range of four orders of magnitude with detection limits of 20 ge/assay (114,115). Although these assays possess dynamic ranges that may be sufficient for most clinical applications, they display a high inter- and intra-assay variability, up to 40% (115).

In contrast, one of the first published real-time PCR assays for the detection of CMV DNA could be performed in <90 min, spanned a dynamic range of six to seven orders of magnitude with a detection limit of at least 10 ge/assay and an inter-assay and intra-assay variability of <10% and <5%, respectively, using plasma samples from bone marrow transplant patients (116).

MULTIPLEX REAL-TIME PCR

Multiplexing (using multiple primers to allow amplification of multiple templates within a single reaction) is a useful application of conventional PCR (117). However, its transfer to real-time PCR has confused its traditional terminology. The term multiplex real-time PCR is more commonly used to describe the use of multiple fluorogenic oligoprobes for the discrimination of multiple amplicons. The transfer of this technique has proven problematic because of the limited number of fluorophores available (14) and the common use of a monochromatic energizing light source. Although excitation by a single wavelength produces bright emissions from a suitably selected fluorophore, this restricts the number of fluorophores that can be included (69). Recent improvements to the design of the hairpin primers, and hairpin and nuclease oligoprobes as well as novel combinations of fluorophores such as in the bi-probe and light-up probe systems, have promised the ability to discriminate an increasing number of targets.

The discovery and application of the non-fluorescent quenchers has liberated some wavelengths that were previously occupied by the emissions from the early quenchers themselves. This breakthrough has permitted the inclusion of a greater number of spectrally discernable oligoprobes per reaction, and highlighted the need for a single non-fluorescent quencher, which can quench a broad range of emission wavelengths (e.g. 400–600 nm). Early real-time PCR systems contained optimised filters to minimise overlap of the emission spectra from the fluorophores. Despite this, the number of fluorophores that could be combined and clearly distinguished was limited when compared with the discriminatory abilities of conventional multiplex PCR. More recent real-time PCR platforms have incorporated either multiple light-emitting diodes to span the entire visible spectrum, or a tungsten light source, which emits light over a broad range of wavelengths. When these platforms incorporate high quality optical filters it is possible to apply any current real-time PCR detection chemistries on the one machine. Nonetheless, these improvements generally allow only four-colour oligoprobe multiplexing, of which one colour is ideally set aside for an internal control to monitor inhibition and perhaps even act as a co-amplified competitor. Some real-time PCR designs have made use of single or multiple nucleotide changes between similar templates to allow their differentiation by T_M (Fig. 4) thus avoiding the need for multiple fluorophores (49,91,118–122). This approach has been used far more commonly in the detection of human genetic diseases where as many as 27 possible nucleotide substitutions have been detected using only one or two fluorophores (48,123–128).

To date, there have only been a handful of truly multiplexed real-time PCR assays described in the literature and few of these have been applied to the diagnosis of infectious disease. Some of these approaches cannot, technically, be considered real-time, homogenous assays because they require interruption of the procedure to transfer template, their fluorescence is detected by end-point analysis or the assays are not performed within the same tube. One of the best viral, multiplex, real-time PCR protocols can discriminate between four retroviral target sequences (129), however, conventional multiplex PCR using end-point detection can easily discriminate more than five different amplified sequences, indicating a greater flexibility when compared with real-time PCR (130–133).

Despite these limitations, a number of assays have been applied to the detection of several viral genomes at once, including the use of non-specific label, SYBR green 1 to detect herpes simplex viruses (HSV), varicella zoster virus (VZV) or CMV, in separate tubes (134) or, by adaptation of a conventional multiplex PCR, to identify HSV-1 and HSV-2, VZV and enteroviruses within a single capillary by applying fluorescent melting curve analysis (121). Another single-tube multiplexed nuclease oligoprobe RT–PCR was capable of simultaneously detecting Influenza A and B in patient's respiratory samples with improved sensitivity compared with conventional or shell-viral culture (95).

Future developments of novel chemistries such as combinatorial fluorescence energy transfer tags (135), and improvements to the design of real-time instrumentation and software will greatly enhance the future of multiplex real-time PCR.

APPLICATIONS TO VIROLOGY

Real-time PCR has been extremely useful for studying viral agents of infectious disease and helping to clarify disputed infectious disease processes. Most of the assays presented in the literature allow an increased frequency of virus detection compared with conventional techniques, which makes the implementation of real-time PCR attractive to many areas of virology.

Of course, real-time PCR has proven increasingly valuable for general virological studies, although increasingly, these applications are difficult to review due to the nature of their use as a tool rather than the focus of published study. Such studies have investigated the role of viruses in a range of diseases by simply confirming the presence or absence of the virus (136,137) or, in the future, by monitoring the levels of specific gene activity (138) as a result of growth under manipulated conditions. Altered viral entry or replication, caused by the modification of target tissues, can also be followed using real-time PCR as can links between virus replication and the expression of cellular genes (139–141).

Real-time PCR has enhanced the speed and scope of measuring viral strain and titre differences in patients displaying different syndromes due to varieties of the same virus (106). Also, epidemiological studies have been improved in speed and scope through the use of real-time PCR because it can reliably measure the amount of two nucleic acid targets within a single reaction (91,142). New chemistries have allowed better discrimination of multiple viral genotypes within a single reaction vessel (143) and provided an alternative to methods of virus detection based on morbidity and mortality assays.

The use of real-time PCR has provided insight into the role(s) some compounds have on PCR inhibition as well as shedding light on the efficiency of different nucleic acid extraction methods, from a diverse range of sample types (144–146). This ability to utilise template from a range of sample types fulfils a requirement for an ideal detection system, which is the ability to apply a single technology across many fields. This flexibility is highlighted by the detection of viral nucleic acids derived, in different ways, from plants (147–149), animals (86,89,94,150–152), urban sludge (85,153), tissue culture (23,77,96,154–160), various solid tissues (108,118,161–166), cerebrospinal fluid (49,167,168), peripheral blood mononuclear cells (82,93,105,112,169–171), plasma (81,88,90,172–176), serum (30,33,36,87,97,99,109,177–181), swabs (182,183), saliva (106) and urine (78,122,184,185). Also, chronic conditions such as sarcoma (186–189), carcinoma (92,111), cervical intraepithelial neoplasia (190–192), and lymphoproliferative disorders (193,194) can be relatively easily studied to investigate direct or indirect links with viral infection. These studies have targeted viruses which include the flaviviruses (33,36,96,109,195), hepadnaviridae (52,97), herpesviruses (21,77,78,82–84,91,92,98,105,106,108, 111,112,144,167,171), orthomyxoviruses (95), parvoviruses (88), papovaviruses (122,143,145), paramyxoviruses (94), picornaviruses (85,86), retroviruses (90,93,156,196,197) and TT virus (137). Viral load monitoring by real-time PCR has also proven beneficial for studying patients following organ transplant (21,83,107,198–201).

This technology has now become an essential tool in the thorough assessment of viral gene therapy vectors prior to their use in clinical trials. Nuclease oligoprobes have been most commonly reported for these studies, which assess the biodistribution, function and purity of these drug preparations (164,197,202–205).

Additionally, the study of emerging viruses has been complimented by the use of real-time PCR as a tool to demonstrate links between unique viral sequences and patient clinical signs and symptoms (94, 96, 150, 160, 206, and 207).

The speed and flexibility of real-time PCR has also proven useful to commercial interests for the screening of microbial contamination of large-scale reagent preparations produced from eukaryotic expression systems (208,209).

CONCLUSIONS AND SUMMARY

Advances in the development of fluorophores, nucleotide labelling chemistries, and the novel applications of oligoprobe hybridisation have provided real-time PCR technology with a broad enough base to ensure its acceptance. Recently, instrumentation has appeared that is capable of incredibly short cycling times combined with the ability to detect and differentiate multiple amplicons. New instruments are also flexible enough to allow the use of any of the chemistries described in this review making real-time nucleic acid amplification an increasingly attractive and viable proposition for the routine diagnostic laboratory. In many cases these laboratories perform tissue culture to isolate virus and serological methods to confirm the identity of the isolate, which may take a considerable, and clinically relevant, amount of time.

The familiarity that leads to comfortable routine use of a technology is now apparent in the inclusion of the fluorogenic oligoprobe chemistries in many laboratories. According to the literature, the most widely used format is the 5′ endonuclease oligoprobe although that is most likely due to its commercial maturity. The more recently developed oligoprobe chemistries have been used in a number of innovative applications and it is apparent from the rate and content of real-time PCR-related publications that they are becoming more widely accepted.

Recent developments in multiplex real-time PCR have suggested a future in which easy identification, genotyping and quantitation of viral targets in single, rapid reactions will be commonplace. Of course, this technology is by no means restricted to virology, as significant achievements have appeared in the area of mutation detection, applying all the benefits described above to enhance the detection of genetic disease and, where applicable, allow quantitation of the extent of such genetic changes.

Micro- and macro-array technology will surely play some chimeric role in the real-time PCR of future research, but for now there is significant potential for routine, research and commercial interests to redesign existing systems for greater sophistication, flexibility and the ability to generate high quality quantitative data. The development of assays capable of real-time PCR that can discriminate as many targets as conventional multiplex PCR assays, whilst producing quantitative data at a greatly increased speed, will consolidate fluorogenic nucleic acid amplification as a routine tool for the laboratory of tomorrow.

ACKNOWLEDGEMENTS

The Royal Children's Hospital Foundation Grant number 922-034, which was sponsored by the Woolworth's 'Care for Kids' campaign, supported this work. This is publication number 139 from the Sir Albert Sakzewski Virus Research Centre.

REFERENCES

1. Freymuth,F., Eugene,G., Vabret,A., Petitjean,J., Gennetay,E., Brouard,J., Duhamel,J.F. and Guillois,B. (1995) Detection of respiratory syncytial virus by reverse transcription–PCR and hybridization with a DNA enzyme immunoassay. J. Clin. Microbial. 33, 3352–3355.
2. Mullis,K.B. and Faloona,F. (1987) Specific synthesis of DNA in vitro via a polymerase-catalysed chain reaction. Methods Enzymol., 155, 335–350.
3. Niubo,J., Perez,J.L., Carvajal,A., Ardanuy,C. and Martin,R. (1994) Effect of delayed processing of blood samples on performance of cytomegalovirus antigenemia assay. J. Clin. Microbiol. , 32, 1119–1120.
4. Guatelli,J.C., Gingeras,T.R. and Richman,D.D. (1989) Nucleic acid amplificationin vitro: detection of sequences with low copy numbersandapplicationtodiagnosisofhumanimmunodeficiency virus type 1 infection. Clin. Microbiol. Rev., 2, 217–226.
5. Kidd,I.M., Clark,D.A. and Emery,V.C. (2000) A non-radioisotopic quantitative competitive polymerase chain reaction method: application in measurement of human herpesvirus 7 load. J. Virol. Methods, 87, 177–181.
6. Holland,P.M., Abramson,R.D., Watson,R. and Gelfand,D.H. (1991) Detection of specific polymerase chain reaction product by utilizing the 5'-3' exonuclease activity of Thermus aquaticus. Proc. Natl Acad. Sci. USA, 88, 7276–7280.
7. van der Vliet,G.M.E., Hermans,C.J. and Klatser,P.R. (1993) Simple colorimetric microtiter plate hybridization assay for detection of amplified Mycobacterium lepraeDNA. J. Clin. Microbiol., 31, 665–670.
8. Keller,G.H., Huang,D.-P., Shih,J.W.-K. and Manak,M.M. (1990) Detection of hepatitis B virus DNA in serum by polymerase chain reaction amplification and microtiter sandwich hybridization. J. Clin. Microbiol., 28, 1411–1416.

9. Kemp,D.J., Churchill,M.J., Smith,D.B., Biggs,B.A., Foote,S.J., Peterson,M.G., Samaras,N., Deacon,N.J. and Doherty,R. (1990) Simplified colorimetric analysis of polymerase chain reactions: detection of HIV sequences in AIDS patients. Gene, 94,223–228.

10. Kox,L.F.F., Noordhoek,G.T., Kunakorn,M., Mulder,S., Sterrenburg,M. and Kolk,A.H.J. (1996) Microwell hybridization assay for detection of PCR products fromMycobacterium tuberculosis complex and the recombinant Mycobacterium smegmatis strain 1008 used as an internal control. J. Clin. Microbiol., 34, 2117–2120.

11. Dekonenko,A., Ibrahim,M.S. and Schmaljohn,C.S. (1997) A colorimetric PCR-enzyme immunoassay to identify hantaviruses. Clin. Diag. Virol., 8, 113–121.

12. Watzinger,F., Hörth,E. and Lion,T. (2001) Quantitation of mRNA expression by competititive PCR using non-homologous competitors containing a shifted restriction site. Nucleic Acids Res., 29, e52.

13. Lomeli,H., Tyagi,S., Pritchard,C.G., Lizardi,P.M. and Kramer,F.R. (1989) Quantitative assays based on the use of replicatable hybridization probes. Clin. Chem., 35, 1826–1831.

14. Lee,L.G., Connell,C.R. and Bloch,W. (1993) Allelic discrimination by nick-translation PCR with fluorogenic probes. Nucleic Acids Res., 21, 3761–3766.

15. Livak,K.J., Flood,S.J.A., Marmaro,J., Giusti,W. and Deetz,K. (1995) Oligonucleotides with fluorescent dyes at opposite ends provide a quenched probe system useful for detecting PCR product and nucleic acid hybridization. PCR Methods Appl., 4, 357–362.

16. Heid,C.A., Stevens,J., Livak,K.J. and Williams,P.M. (1996) Real time quantitative PCR. Genome Res., 6, 986–994.

17. Gibson,U.E.M., Heid,C.A. and Williams,P.M. (1996) A novel method for real time quantitative RT–PCR. Genome Res., 6, 995–1001.

18. Matthews,J.A. and Kricka,L.J. (1988) Analytical strategies for the use of DNA probes Anal. Biochem., 169, 1–25.

19. Wittwer,C.T., Fillmore,G.C. and Garling,D.J. (1990) Minimizing the time required for DNA amplification by efficient heat transfer to small samples. Anal. Biochem., 186, 328–331.

20. Wittwer,C.T., Ririe,K.M., Andrew,R.V., David,D.A., Gundry,R.A. and Balis,U.J. (1997) The LightCycler™: a microvolume multisample fluorimeter with rapid temperature control. Biotechniques, 22, 176–181.

21. Nitsche,A., Steuer,N., Schmidt,C.A., Landt,O., Ellerbrok,H., Pauli,G. and Siegert,W. (2000) Detection of human cytomegalovirus DNA by real-time quantitative PCR. J. Clin. Microbiol., 38, 2734–2737.

22. Gyllensten,U.B. and Erlich,H.A. (1988) Generation of single-stranded DNA by the polymerase chain reaction and its application to direct sequencing of the HLA-DQA locus. Proc. Natl Acad. Sci. USA, 85, 7652–7656.

23. Poddar,S.K. (2000) Symmetric vs asymmetric PCR and molecular beacon probe in the detection of a target gene of adenovirus. Mol. Cell. Probes, 14, 25–32.

24. Stryer,L. and Haugland,R.P. (1967) Energy transfer: a spectroscopic ruler.Proc. Natl Acad. Sci. USA, 58, 719–726.

25. Clegg,R.M. (1992) Fluorescence resonance energy transfer and nucleic acids.Methods Enzymol., 211, 353–388.

26. Förster,T. (1948) Zwischenmolekulare energiewanderung und fluoreszenz.Ann. Phys., 6, 55–75.

27. Selvin,P. (1995) Fluorescence resonance energy transfer. Methods Enzymol.,246, 300–334.

28. Didenko,V.V. (2001) DNA probes using fluorescence resonance energy transfer (FRET): designs and applications. Biotechniques, 31, 1106–1121.

29. Higuchi,R., Fockler,C., Dollinger,G. and Watson,R. (1993) Kinetic PCR analysis: real-time monitoring of DNA amplification reactions. Biotechnology (NY), 11, 1026–1030.

30. Martell,M., Gómez,J., Esteban,J.I., Sauleda,S., Quer,J., Cabot,B., Esteban,R. and Guardia,J. (1999) High-throughput real-time reverse transcription–PCR quantitation of hepatitis C virus RNA. J. Clin. Microbiol., 37, 327–332.

31. Whitcombe,D., Theaker,J., Guy,S.P., Brown,T. and Little,S. (1999) Detection of PCR products using self-probing amplicons and fluorescence. Nat. Biotechnol., 17,804–807.
32. Higuchi,R., Dollinger,G., Walsh,P.S. and Griffith,R. (1992) Simulutaneous amplification and detection of specific DNA sequences. Biotechnology (NY), 10, 413–417.
33. Ishiguro,T., Saitoh,J., Yawata,H., Yamagishi,H., Iwasaki,S. and Mitoma,Y. (1995) Homogeneous quantitative assay of hepatitis C virus RNA by polymerase chain reaction in the presence of a fluorescent intercalater. Anal. Biochem., 229,207–213.
34. Tseng,S.Y., Macool,D., Elliott,V., Tice,G., Jackson,R., Barbour,M. and Amorese,D. (1997) An homogeneous fluorescence polymerase chain reaction assay to identify Salmonella. Anal. Biochem., 245, 207–212.
35. Morrison,T.M., Weis,J.J. and Wittwer,C.T. (1998) Quantification of low-copy transcripts by continuous SYBR green I monitoring during amplification.Biotechniques, 24, 954–962.
36. Komurian-Pradel,F., Paranhos-Baccalà,G., Sodoyer,M., Chevallier,P., Mandrand,B., Lotteau,V. and André,P. (2001) Quantitation of HCV RNA using real-time PCR and fluorimetry. J. Virol. Methods, 95, 111–119.
37. Chou,Q., Russell,M., Birch,D.E., Raymond,J. and Bloch,W. (1992) Prevention of pre-PCR mis-priming and primer dimerization improves low-copy-number amplifications. Nucleic Acids Res., 20, 1717–1723.
38. Halford,W.P. (1999) The essential prerequisites for quantitative RT–PCR. Nat. Biotechnol., 17, 835.
39. Halford,W.P., Falco,V.C., Gebhardt,B.M. and Carr,D.J.J. (1999) The inherent quantitative capacity of the reverse transcription–polymerase chain reaction. Anal. Biochem., 266, 181–191.
40. Pfaffl,M. (2001) In Meuer,S., Wittwer,C. and Nakagawara,K.-I. (eds), Rapid Cycle Real-Time PCR: Methods and Applications. Springer, Berlin, pp. 281–291.
41. Wetmur,J.G. (1991) DNA probes: applications of the principles of nucleic acid hybridization. Crit. Rev. Biochem. Mol. Biol., 26, 227–259.

42. Ririe,K.M., Rasmussen,R.P. and Wittwer,C.T. (1997) Product differentiation by analysis of DNA melting curves during the polymerase chain reaction. Anal. Biochem., 245, 154–160.

43. Heller,M.J. and Morrison,L.E. (1985) In Kingsbury,D.T. and Falkow,S. (eds),Rapid Detection and Identification of Infectious Agents. Academic Press, New York, pp. 245–256.

44. Cardullo,R.A., Agrawai,S., Flores,C., Zamecnik,P.C. and Wolf,D.E. (1988) Detection of nucleic acid hybridization by nonradiative fluoresence resonance energy transfer. Proc. Natl Acad. Sci. USA, 85, 8790–8794.

45. Wittwer,C.T., Herrmann,M.G., Moss,A.A. and Rasmussen,R.P. (1997) Continuous fluorescence monitoring of rapid cycle DNA amplification.Biotechniques, 22, 130–138.

46. Weis,J.H., Tan,S.S., Martin,B.K. and Wittwer,C.T. (1992) Detection of rare mRNAs via quantitative RT–PCR. Trends Genet., 8, 263–264.

47. Bustin,S.A. (2000) Absolute quantification of mRNA using real-time reverse transcription polymerase chain reaction assays. J. Mol. Endocrinol., 25, 169–193.

48. Gundry,C.N., Bernard,P.S., Herrmann,M.G., Reed,G.H. and Wittwer,C.T. (1999) Rapid F508del and F508C assay using fluorescent hybridization probes. Genet. Test,3, 365–370.

49. Schalasta,G., Arents,A., Schmid,M., Braun,R.W. and Enders,G. (2000) Fast and type-specific analysis of herpes simplex virus types 1 and 2 by rapid PCR and fluorescence melting-curve-analysis. Infection, 28, 85–91.

50. Wilhelm,J., Pingoud,A. and Hahn,M. (2001) Comparison between Taq DNA polymerase and its Stoffel fragment for quantitative real-time PCR with hybridization probes. Biotechniques, 30, 1052–1062.

51. Lyamichev,V., Brow,M.A.D. and Dahlberg,J.E. (1993) Structure-specific endonucleolytic cleavage of nucleic acids by eubacterial DNA polymerases. Science,260, 778–783.

52. Brechtbuehl,K., Whalley,S.A., Dusheiko,G.M. and Saunders,N.A. (2001) A rapid real-time quantitative polymerase chain reaction for hepatitis B virus. J. Virol.Methods, 93, 105–113.

53. Walker,R.A., Saunders,N., Lawson,A.J., Lindsay,E.A., Dassama,M., Ward,L.R., Woodward,M.J., Davies,R.H., Liebana,E. and Threlfall,E.J. (2001) Use of a LightCyclergyrA mutation assay for rapid identification of mutations conferring decreased susceptibility to ciprofloxacin in multiresistant Salmonella enterica serotype typhimurium DT104 isolates. J. Clin. Microbiol., 39, 1443–1448.
54. Kurata,S., Kanagawa,T., Yamada,K., Torimura,M., Yokomaku,T., Kamagata,Y. and Kurane,R. (2001) Fluorescent quenching-based quantitative detection of specific DNA/RNA using a BODIPY®FL-labeled probe or primer. Nucleic Acids Res., 29, e34.
55. Crockett,A.O. and Wittwer,C.T. (2001) Fluorescein-labelled oligonucleotides for real-time PCR: using the inherent quenching of deoxyguanosine nucleotides.Anal. Biochem., 290, 89–97.
56. Svanvik,N., Westman,G., Wang,D. and Kubista,M. (2001) Light-up probes: thiazole orange-conjugated peptide nucleic acid for the detection of target nucleic acid in homogensous solution. Anal. Biochem., 281, 26–35.
57. Isacsson,J., Cao,H., Ohlsson,L., Nordgren,S., Svanvik,N., Westman,G., Kubista,M., Sjöback,R. and Sehlstedt,U. (2000) Rapid and specific detection of PCR products using light-up probes. Mol. Cell. Probes, 14, 321–328.
58. Svanvik,N., Sehlstedt,U., Sjöback,R. and Kubista,M. (2000) Detection of PCR products in real time using light-up probes. Anal. Biochem., 287, 179–182.
59. Morrison,L.E., Halder,T.C. and Stols,L.M. (1989) Solution-phase detection of polynucleotides using interacting fluorescent labels and competitive hybridization.Anal. Biochem., 183, 231–244.
60. Landt,O. (2001) In Meuer,S., Wittwer,C. and Nakagawara,K. (eds), Rapid Cycle Real-time PCR: Methods and Applications. Springer Verlag, Germany, pp. 35–41.
61. Nakayama,H., Yokoi,H. and Fujita,J. (1992) Quantification of mRNA by non-radioactive RT–PCR and CCD imaging system. Nucleic Acids Res., 20, 4939.
62. Gelmini,S., Orlando,C., Sestini,R., Vona,G., Pinzani,P., Ruocco,L. and Pazzagli,M. (1997) Quantitative polymerase chain reaction-based homogeneous assay with fluorogenic probes to measure c-cerB-2 oncogene amplification. Clin. Chem., 43, 752–758.

63. Wilhelm,J., Hahn,M. and Pingoud,A. (2001) Influence of DNA target melting behavior on real-time PCR quantification. Clin. Chem., 46, 1738–1743.

64. Jung,R., Soondrum,K. and Neumaier,M. (2000) Quantitative PCR. Clin. Chem. Lab. Med., 38, 833–836.

65. Kutyavin,I.V., Afonina,I.A., Mills,A., Gorn,V.V., Lukhtanov,E.A., Belousov,E.S., Singer,M.J., Walburger,D.K., Lokhov,S.G., Gall,A.A. et al. (2000) 3′-Minor groove binder-DNA probes increase sequence specificity at PCR extension temperatures. Nucleic Acids Res., 28, 655–661.

66. Todd,A.V., Fuery,C.J., Impey,H.L., Applegate,T.L. and Haughton,M.A. (2000) DzyNA–PCR: use of DNAzymes to detect and quantify nucleic acid sequences in a real-time fluorescent format. Clin. Chem., 46, 625–630.

67. Tyagi,S., Bratu,D.P. and Kramer,F.R. (1998) Multicolor molecular beacons for allele discrimination. Nat. Biotechnol., 16, 49–53.

68. Tyagi,S. and Kramer,F.R. (1996) Molecular beacons: probes that fluoresce upon hybridization. Nat. Biotechnol., 14, 303–308.

69. Tyagi,S., Marras,S.A.E. and Kramer,F.R. (2000) Wavelength-shifting molecular beacons. Nat. Biotechnol., 18, 1191–1196.

70. Nazarenko,I.A., Bhatnager,S.K. and Hohman,R.J. (1997) A closed tube format for amplification and detection of DNA based on energy transfer. Nucleic Acids Res.,25, 2516–2521.

71. Freeman,W.M., Walker,S.J. and Vrana,K.E. (1999) Quantitative RT–PCR: pitfalls and potential. Biotechniques, 26, 112–125.

72. Orlando,C., Pinzani,P. and Pazzagli,M. (1998) Developments in quantitative PCR. Clin. Chem. Lab. Med., 36, 255–269.

73. Becker-Andre,M. and Hahlbrock,K. (1989) Absolute mRNA quantification using the polymerase chain reaction (PCR). A novel approach by a PCR aided transcript titration assay (PATTY). Nucleic Acids Res., 17, 9437–9447.

74. Clementi,M., Menzo,S., Manzin,A. and Bagnarelli,P. (1995) Quantitative molecular methods in virology. Arch. Virol., 140, 1523–1539.

75. Gilliland,G., Perrin,S. and Bunn,H.F. (1990) PCR Protocols: A Guide to Methods and Applications. Academic Press, San Diego, CA, pp. 60–69.

76. Siebert,P.D. and Larrick,J.W. (1992) Competitive PCR. Nature, 359, 557–558.
77. Locatelli,G., Santoro,F., Veglia,F., Gobbi,A., Lusso,P. and Malnati,M.S. (2000) Real-time quantitative PCR for human herpesvirus 6 DNA. J. Clin. Microbiol., 38,4042–4048.
78. Tanaka,N., Kimura,H., Iida,K., Saito,Y., Tsuge,I., Yoshimi,A., Matsuyama,T. and Morishima,T. (2000) Quantitative analysis of cytomegalovirus load using a real-time PCR assay. J. Med. Virol., 60, 455–462.
79. Ferré,F. (1992) Quantitative or semi-quantitative PCR: reality versus myth.PCR Methods Appl., 2, 1–9.
80. Johnson,M.R., Wang,K., Smith,J.B., Heslin,M.J. and Diasio,R.B. (2000) Quantitation of dihydropyrimidine dehydrogenase expression by real-time reverse transcription polymerase chain reaction. Anal. Biochem., 278, 175–184.
81. Kleiber,J., Walter,T., Haberhausen,G., Tsang,S., Babiel,R. and Rosenstraus,M. (2000) Performance characteristics of a quantitative, homogeneous TaqMan RT–PCR test for HCV RNA. J. Mol. Diagn., 2, 158–166.
82. Kimura,H., Morita,M., Yabuta,Y., Kuzushima,K., Kato,K., Kojima,S., Matsuyama,T. and Morishima,T. (1999) Quantitative analysis of Epstein–Barr virus load by using a real-time PCR assay. J. Clin. Microbiol., 37, 132–136.
83. Najioullah,F., Thouvenot,D. and Lina,B. (2001) Development of a real-time PCR procedure including an internal control for the measurement of HCMV viral load.J. Virol. Methods, 92, 55–64.
84. Ryncarz,A.J., Goddard,J., Wald,A., Huang,M.-L., Roizman,B. and Corey,L. (1999) Development of a high-throughput quantitative assay for detecting herpes simplex virus DNA in clinical samples. J. Clin. Microbiol., 37, 1941–1947.
85. Monopoeho,S., Mignotte,B., Schwartzbrod,L., Marechal,V., Nicolas,J.-C., Billaudel,S. and Férré,V. (2000) Quantification of enterovirus RNA in sludge samples using single tube real-time RT–PCR. Biotechniques, 29, 88–93.
86. Alexandersen,S., Oleksiewicz,M.B. and Donaldson,A.I. (2001) The early pathogenesis of foot-and-mouth disease in pigs infected by contact: a quantitative time-course study using TaqMan RT–PCR. J. Gen. Virol., 82, 747–755.

87. Abe,A., Inoue,K., Tanaka,T., Kato,J., Kajiyama,N., Kawaguchi,R., Tanaka,S., Yoshiba,M. and Kohara,M. (1999) Quantitation of hepatitis B virus genomic DNA by real-time detection PCR. J. Clin. Microbiol., 37, 2899–2903.

88. Gruber,F., Falkner,F.G., Dorner,F. and Hämmerle,T. (2001) Quantitation of viral DNA by real-time PCR applying duplex amplification, internal standardization and two-colour fluorescence detection. Appl. Environ. Microbiol., 67, 2837–2839.

89. Moody,A., Sellers,S. and Bumstead,N. (2000) Measuring infectious bursal disease virus RNA in blood by multiplex real-time quantitative RT–PCR. J. Virol.Methods, 85, 55–64.

90. Schutten,M., van den Hoogen,B., van der Ende,M.E., Gruters,R.A., Osterhaus,A.D.M.E. and Niesters,H.G.M. (2000) Development of a real-time quantitative RT–PCR for the detection of HIV-2 RNA in plasma. J. Virol. Methods, 88,81–87.

91. Kearns,A.M., Turner,A.J.L., Taylor,C.E., George,P.W., Freeman,R. and Gennery,A.R. (2001) LightCycler-based quantitative PCR for rapid detection of human herpesvirus 6 DNA in clinical material. J. Clin. Microbiol., 39, 3020–3021.

92. Capone,R.B., Pai,S.I., Koch,W.M. and Gillison,M.L. (2001) Detection and quantitation of human papillomavirus (HPV) DNA in the sera of patients with HPV-associated head and neck squamous cell carcinoma. Clin. Cancer Res., 6, 4171–4175.

93. Leutenegger,C.M., Klein,D., Hofmann-Lehmann,R., Mislin,C., Hummel,U., Boni,J., Boretti,F., Guenzburg,W.H. and Lutz,H. (1999) Rapid feline immunodeficiency virus provirus quantitation by polymerase chain reaction using the TaqMan fluorogenic real-time detection system. J. Virol. Methods, 78, 105–116.

94. Smith,I.L., Halpin,K., Warrilow,D. and Smith,G.A. (2001) Development of a fluorogenic RT–PCR assay (TaqMan) for the detection of Hendra virus. J. Virol.Methods, 98, 33–40.

95. van Elden,L.J.R., Nijhuis,M., Schipper,P., Schuurman,R. and van Loon,A.M. (2001) Simultaneous detection of influenza viruses A and B using real-time quantitative PCR. J. Clin. Microbiol., 39, 196–200.

96. Lanciotti,R.S., Kerst,A.J., Nasci,R.S., Godsey,M.S., Mitchell,C.J., Savage,H.M., Komar,N., Panella,N.A., Allen,B.C., Volpe,K.E. et al. (2000) Rapid detection of west nile virus from human clinical specimens, field-collected mosquitoes and Avian samples by a TaqMan reverse transcriptase–PCR assay. J. Clin. Microbiol., 38, 4066–4071.

97. Weinberger,K.M., Wiedenmann,E., Böhm,S. and Jilg,W. (2000) Sensitive and accurate detection of hepatitis B virus DNA using a kinetic fluorescence detection system (TaqMan PCR). J. Virol. Methods, 85, 75–82.

98. Schaade,L., Kockelkorn,P., Ritter,K. and Kleines,M. (2000) Detection of cytomegalovirus DNA in human specimens by LightCycler PCR. J. Clin. Microbiol., 38,4006–4009.

99. Kawai,S., Yokosuka,O., Kanda,T., Imazeki,F., Maru,Y. and Saisho,H. (1999) Quantification of hepatitis C virus by TaqMan PCR: comparison with HCV amplicor monitor assay. J. Med. Virol., 58, 121–126.

100. Clementi,M. (2000) Quantitative molecular analysis of virus expression and replication. J. Clin. Microbiol., 38, 2030–2036.

101. Limaye,A.P., Jerome,K.R., Kuhr,C.S., Ferrenberg,J., Huang,M.-L., Davis,C.L., Corey,L. and Marsh,C.L. (2000) Quantitation of BK virus load in serum for the diagnosis of BK virus-associated nephropathy in renal transplant recipients. J. Infect. Dis., 183, 1669–1672.

102. Holodniy,M., Katzenstein,D., Sengupta,S., Wang,A.M., Casipit,C., Schwartz,D.H., Konrad,M., Groves,E. and Merigan,T.C. (1991) Detection and quantification of human immunodeficiency virus RNA in patient serum by use of the polymerase chain reaction. J. Infect. Dis., 163, 862–866.

103. Kaneko,S., Murakami,S., Unoura,M. and Kobayashi,K. (1992) Quantitation of hepatitis C virus RNA by competitive polymerase chain reaction. J. Med. Virol., 37,278–282.

104. Menzo,S., Bagnarelli,P., Giacca,M., Manzin,A., Varaldo,P.E. and Clementi,M. (1992) Absolute quantitation of viremia in human immunodeficiency virus infection by competitive reverse transcription and polymerase chain reaction. J. Clin. Microbiol., 30, 1752–1757.

105. Ohyashiki,J.K., Suzuki,A., Aritaki,K., Nagate,A., Shoji,N., Ohyashiki,K., Ojima,T., Abe,K. and Yamamoto,K. (2000) Use of real-time PCR to monitor human herpesvirus 6 reactivation after allogeneic bone marrow transplantation. Int. J. Mol. Med., 6, 427–432.

106. Furuta,Y., Ohtani,F., Sawa,H., Fukuda,S. and Inuyama,Y. (2001) Quantitation of varicella-zoster virus DNA in patients with ramsay hunt syndrome and zoster sine herpete. J. Clin. Microbiol., 39, 2856–2859.

107. Limaye,A.P., Huang,M.-L., Leisenring,W., Stensland,L., Corey,L. and Boeckh,M. (2001) Cytomegalovirus (CMV) DNA load in plasma for the diagnosis of CMV disease before engraftment in hematopoietic stem-cell transplant recipients. J. Infect. Dis., 183, 377–382.

108. Lallemand,F., Desire,N., Rozenbaum,W., Nicolas,J.-C. and Marechal,V. (2000) Quantitative analysis of human herpesvirus 8 viral load using a real-time PCR assay.J. Clin. Microbiol., 38, 1404–1408.

109. Laue,T., Emmerich,P. and Schmitz,H. (1999) Detection of dengue virus RNA inpatients after primary or secondary dengue infection by using the TaqMan automated amplification system. J. Clin. Microbiol., 37, 2543–2547.

110. Chang,L.-J., Urlacher,V., Iwakuma,T., Cui,Y. and Zucali,J. (1999) Efficacy and safety anlyses of a recombinant human immunodeficieny virus type 1 derived vector system. Gene Ther., 6, 715–728.

111. Lo,Y.M.D., Chan,L.Y.S., Lo,K.-W., Leung,S.-F., Zhang,J., Chan,A.T.C., Lee,J.C.K., Hjelm,N.M., Johnson,P.J. and Huang,D.P. (1999) Quantitative analysis of cell-free Epstein–Barr virus DNA in plasma of patients with nasopharyngeal carcinoma. Cancer Res., 59, 1188–1191.

112. Hawrami,K. and Breur,J. (1999) Development of a flurogenic polymerase chain reaction assay (TaqMan) for the detection and quantitation of varicella zoster virus. J. Virol. Methods, 79, 33–40.

113. Roberts,T.C., Brennan,D.C., Buller,R.S., Gaudreault-Keener,M., Schnitzler,M.A., Sternhell,K.E., Garlock,K.A., Singer,G.G. and Storch,G.A. (1998) Quantitative polymerase chain reaction to

predict occurrence of symptomatic cytomegalovirus infection and assess response to ganciclovir therapy in renal transplant recipients. J. Infect. Dis., 178, 626–636.

114. Rollag,H., Sagedal,S., Holter,E., Degre,M., Ariansen,S. and Nordal,K.P. (1998) Diagnosis of cytomegalovirus infection in kidney transplant recipients by a quantitative RNA–DNA hybrid capture assay for cytomegalovirus DNA in leukocytes.Eur. J. Clin. Microbiol. Infect. Dis., 17, 124–127.

115. Boivin,G., Handfield,J., Murray,G., Toma,E., Lalonde,R., Lazar,J.G. and Bergeron,M.G. (1997) Quantitation of cytomegalovirus (CMV) DNA in leukocytes of human immunodeficiency virus-infected subjects with and without CMV disease by using PCR and the SHARP Signal Detection System. J. Clin. Microbiol., 35, 525–526.

116. Nitsche,A., Steuer,N., Schmidt,C.A., Landt,O. and Siegert,W. (1999) Different real-time PCR formats compared for the quantitative detection of human cytomegalovirus DNA. Clin. Chem., 45, 1932–1937.

117. Chamberlain,J.S., Gibbs,R.A., Ranier,J.E., Nguyen,P.N. and Caskey,C.T. (1988) Deletion screening of the Duchenne muscular dystrophy locis via multiplex DNA amplification. Nucleic Acids Res., 16, 11141–11156.

118. Espy,M.J., Uhl,J.R., Mitchell,P.S., Thorvilson,J.N., Svien,K.A., Wold,A.D. and Smith,T.F. (2000) Diagnosis of herpes simplex virus infections in the clinical laboratory by LightCycler PCR. J. Clin. Microbiol., 38, 795–799.

119. Espy,M.J., Ross,T.K., Teo,R., Svien,K.A., Wold,A.D., Uhl,J.R. and Smith,T.F. (2000) Evaluation of lightcycler PCR for implementation of laboratory diagnosis of herpes simplex virus infections. J. Clin. Microbiol., 38, 3116–3118.

120. Loparev,V.N., McCaustland,K., Holloway,B.P., Krause,P.R., Takayama,M. and Schmid,D.S. (2000) Rapid genotyping of varicella-zoster virus vaccine and wild-type strains with fluorophore-labeled hybridization probes. J. Clin. Microbiol., 38, 4315–4319.

121. Read,S.J., Mitchell,J.L. and Fink,C.G. (2001) LightCycler multiplex PCR for the laboratory diagnosis of common viral infections of the central nervous system. J. Clin. Microbiol., 39, 3056–3059.

122. Whiley,D.M., Mackay,I.M. and Sloots,T.P. (2001) Detection and differentiation of human polyomaviruses JC and BK by LightCycler PCR. J. Clin. Microbiol., 39,4357–4361.

123. Schütz,E., von Ashen,N. and Oellerich,M. (2000) Genotyping of eight thiopurine methyltransferase mutations: Three-color multiplexing, 'two-color/shared' anchor and fluorescence-quenching hybridization probe assays based on thermodynamic nearest-neighbour probe design. Clin. Chem., 46, 1728–1737.

124. Lay,M.J. and Wittwer,C.T. (1997) Real-time fluorescence genotyping of factor V Leiden during rapid-cycle PCR. Clin. Chem., 43, 2262–2267.

125. Herrmann,M.G., Dobrowolski,S.F. and Wittwer,C.T. (2000) Rapid β-globin genotyping by multiplexing probe melting temperature and color. Clin. Chem., 46,425–429.

126. Lee,L.G., Livak,K.J., Mullah,B., Graham,R.J., Vinayak,R.S. and Woudenberg,T.M. (1999) Seven-color, homogeneous detection of six PCR products. Biotechniques, 27,342–349.

127. Bernard,P.S. and Wittwer,C.T. (2000) Homogeneous amplification and variant detection by fluorescent hybridization probes. Clin. Chem., 46, 147–148.

128. Elenitoba,K.S.J., Bohling,S.D., Wittwer,C.T. and King,T.C. (2001) Multiplex PCR by multicolor fluorimetry and fluorescence melting curve analysis. Nature Med.,7, 249–253.

129. Vet,J.A.M., Majithia,A.R., Marras,S.A.E., Tyagi,S., Dube,S., Poiesz,B.J. and Kramer,F.R. (1999) Multiplex detection of four pathogenic retroviruses using molecular beacons. Proc. Natl Acad. Sci. USA, 96, 6394–6399.

130. Quereda,C., Corral,I., Laguna,F., Valencia,M.E., Tenorio,A., Echevarria,J.E., Navas,E., Martín-Davilla,P., Moreno,A., Moreno,V. et al. (2000) Diagnostic utility of a multiplex herpesvirus PCR assay performed with cerebrospinal fluid from human immunodeficiency virus-infected patients with neurological disorders. J. Clin. Microbiol., 38, 3061–3067.

131. Kehl,S.C., Henrickson,K.J., Hua,W. and Fan,J. (2001) Evaluation of the hexaplex assay for detection of respiratory viruses in children. J. Clin. Microbiol., 39, 1696–1701.

132. Echevarria,J.E., Erdman,D.D., Swierkosz,E.M., Holloway,B.P. and Anderson,L.J. (1998) Simultaneous detection and identification of human parainfluenza viruses 1, 2 and 3 from clinical samples by multiplex PCR. J. Clin. Microbiol., 36, 1388–1391.

133. Henegariu,O., Heerema,N.A., Dloughy,S.R., Vance,G.H. and Vogt,P.H. (1997) Multiplex PCR: critical parameters and step-by-step protocol. Biotechniques, 23,504–511.

134. Nicoll,S., Brass,A. and Cubie,H.A. (2001) Detection of herpes viruses in clinical samples using real-time PCR. J. Virol. Methods, 96, 25–31.

135. Tong,A.K., Li,Z., Jones,G.S., Russo,J.J. and Ju,J. (2001) Combinatorial fluorescence energy transfer tags for multiplex biological assays. Nat. Biotechnol.,19, 756–759.

136. Kato,T., Mizokami,M., Mukaide,M., Orito,E., Ohno,T., Nakano,T., Tanaka,Y., Kato,H., Sugauchi,F., Ueda,R. et al. (2000) Development of a TT virus DNA quantification system using real-time detection PCR. J. Clin. Microbiol., 38, 94–98.

137. Iriyama,M., Kimura,H., Nishikawa,K., Yoshioka,K., Wakita,T., Nishimura,N., Shibata,M., Ozaki,T. and Morishima,T. (1999) The prevalence of TT virus (TTV) infection and its relationship to hepatitis in children. Med. Microbiol. Immunol. (Berl.), 188, 83–89.

138. Liu,J., Feldman,P. and Chung,T.D.Y. (2002) Real-time monitoring in vitrotranscription using molecular beacons. Anal. Biochem., 300, 40–45.

139. Kennedy,M.M., O'Leary,J.J., Oates,J.L., Lucas,S.B., Howells,D.D., Picton,S. and McGee,J.O. (1998) Human herpes virus 8 (HHV-8) in Kaposi's sarcoma: lack of association with Bcl-2 and p53 protein expression. J. Clin. Pathol., 51, 155–159.

140. Kennedy,M.M., Biddolph,S., Lucas,S.B., Howells,D.D., Picton,S., McGee,J.O. and O'Leary,J.J. (1999) CD40 upregulation is independent of HHV-8 in the pathogenesis of Kaposi's sarcoma. J. Clin. Pathol., 52, 32–36.

141. Kennedy,M.M., Biddolph,S., Lucas,S.B., Howells,D.D., Picton,S., McGee,J.O., Silva,I., Uhlmann,V., Luttich,K. and O'Leary,J. (1999) Cyclin D1 expression and HHV-8 in Kaposi sarcoma. J. Clin. Pathol., 52, 569–573.

142. Zerr,D.M., Huang,M.-L., Corey,L., Erickson,M., Parker,H.L. and Frenkel,L.M. (2000) Sensitive method for detection of human herpesvirus 6 and 7 in saliva collected in field studies. J. Clin. Microbiol., 38, 1981–1983.

143. Jordens,J.Z., Lanham,S., Pickett,M.A., Amarasekara,S., Aberywickrema,I. and Watt,P.J. (2000) Amplification with molecular beacon primers and reverse line blotting for the detection and typing of human papillomaviruses. J. Virol. Methods,89, 29–37.

144. Niesters,H.G., van Esser,J., Fries,E., Wolthers,K.C., Cornelissen,J. and Osterhaus,A.D. (2000) Development of a real-time quantitative assay for detection of Epstein–Barr virus. J. Clin. Microbiol., 38, 712–715.

145. Biel,S.S., Held,T.K., Landt,O., Niedrig,M., Gelderblom,H.R., Siegert,W. and Nitsche,A. (2000) Rapid quantification and differentiation of human polyomavirus DNA in undiluted urine from patients after bone marrow transplantation. J. Clin. Microbiol., 38, 3689–3695.

146. Petrik,J., Pearson,G.J.M. and Allain,J.-P. (1997) High throughput PCR detection of HCV based on semiautomated multisample RNA capture. J. Virol.Methods, 64, 147–159.

147. Skaf,J.S., Schultz,M.H., Hirata,H. and de Zoeten,G.A. (2000) Mutational evidence that VPg is involved in the replication and not the movement of Pea enation mosaic virus-1. J. Gen. Virol., 81, 1103–1109.

148. Roberts,C.A., Dietzgen,R.G., Heelan,L.A. and Maclean,D.J. (2000) Real-time RT–PCR fluorescent detection of tomato spotted wilt virus. J. Virol. Methods, 88, 1–8.

149. Eun,A.J.-C., Seoh,M.-L. and Wong,S.W. (2000) Simultaneous quantitation of two orchid viruses by the TaqMan® real-time RT–PCR. J. Virol. Methods, 87, 151–160.

150. Halpin,K., Young,P.L., Field,H.E. and Mackenzie,J.S. (2000) Isolation of Hendra virus from pteropid bats: a natural reservoir of Hendra virus. J. Gen. Virol.,81, 1927–1932.

151. Egli,C., Thür,B., Liu,L. and Hofmann,M.A. (2001) Quantitative TaqMan® RT–PCR for the detection and differentiation of European and North American strains of porcine reproductive and respiratory syndrome virus. J. Virol. Methods, 98, 63–75.

152. Gut,M., Leutenegger,C.M., Huder,J.B., Pedersen,N.C. and Lutz,H. (1999) One-tube fluorogenic reverse transcription–polymerase chain reaction for the quantitation of feline coronaviruses. J. Virol. Methods, 77, 37–46.
153. Monopoeho,S., Maul,A., Mignotte-Cadiergues,B., Schwartzbrod,L., Billaudel,S. and Ferré,F. (2001) Best viral elution method available for quantification of enteroviruses in sludge by both cell culture and reverse transcription. Appl. Environ. Microbiol., 67, 2484–2488.
154. Garin,D., Peyrefitte,C., Crance,J.-M., Le Faou,A., Jouan,A. and Bouloy,M. (2001) Highly sensitive Taqman® PCR detection of Puumala hantavirus. Microbes Infect., 3, 739–745.
155. O'Doherty,U., Swiggard,W.J. and Malim,M.H. (2000) Human immunodeficiency virus type 1 spinoculation enhances infection through virus binding. J. Virol., 74,10074–10080.
156. Klein,D., Janda,P., Steinborn,R., Salmons,B. and Günzburg,W.H. (1999) Proviral load determination of different feline immunodeficiency virus isolates using real-time polymerase chain reaction: influence of mismatches on quantification. Electrophoresis, 20, 291–299.
157. Hadfield,T.L., Turell,M., Dempsey,M.P., David,J. and Park,E.J. (2001) Detection of West Nile virus in mosquitoes by RT–PCR. Mol. Cell. Probes, 15, 147–150.
158. Saldanha,C.E., Lubinski,J., Martin,C., Nagashunmugam,T., Wang,L., van der Keyl,H., Tal-Singer,R. and Friedman,H.M. (2000) Herpes simplex virus type 1 glycoprotein E domains involved in virus spread and disease. J. Virol., 74, 6712–6719.
159. Poddar,S.K. (1999) Detection of adenovirus using PCR and molecular beacon. J. Virol. Methods, 82, 19–26.
160. Gibb,T.R., Norwood,D.A.,Jr, Woollen,N. and Henchal,E.A. (2001) Development and evaluation of a fluorogenic 5'-nuclease assay to identify Marburg virus. Mol. Cell. Probes, 15, 259–266.
161. Pevenstein,S.R., Williams,R.K., McChesney,D., Mont,E.K., Smialek,J.E. and Straus,S.E. (1999) Quantitation of latent varicella-zoster virus and herpes simplex virus genomes in human trigeminal ganglia. J. Virol., 73, 10514–10518.

162. Espy,M.J., Rys,P.N., Wold,A.D., Uhl,J.R., Sloan,L.M., Jenkins,G.D., Ilstrup,D.M., Cockerill,F.R.,III, Patel,R., Rosenblatt,J.E. et al. (2001) Detection of herpes simplex virus DNA in genital and dermal specimens by LightCycler PCR after extraction using the IsoQuick, MagNA Pure and BioRobot 9604 methods. J. Clin. Microbiol., 39, 2233–2236.

163. Espy,M.J., Teo,R., Ross,T.K., Svien,K.A., Wold,A.D., Uhl,J.R. and Smith,T.F. (2000) Diagnosis of Varicella-Zoster Virus infections in the clinical laboratory by LightCycler PCR. J. Clin. Microbiol., 38, 3187–3189.

164. Josefsson,A.M., Magnusson,P.K.E., Ylitalo,N., Sørensen,P., Qwarforth-Tubbin,P., Andersen,P.K., Melbye,M., Adami,H.-O. and Gyllensten,U. (2000) Viral load of human papilloma virus 16 as a determinant for development of cervical carcinoma in situ: a nested case-control study. Lancet, 355, 2189–2193.

165. Johnston,C. (2000) Quantitative tests for human papillomavirus. Lancet, 355,2179–2180.

166. Watanabe,M., Lee,B.-J., Kamitani,W., Kobayashi,T., Taniyama,H., Tomonaga,K. and Ikuta,K. (2001) Neurological diseases and viral dynamics in the brains of neonatally borna disease virus-infected gerbils. Virology, 282, 65–76.

167. Peter,J.B. and Sevall,J.S. (2001) Review of 3200 serially received CSF samples submitted for type-specific HSV detection by PCR in the reference laboratory setting.Mol. Cell. Probes, 15, 177–182.

168. Kessler,H.H., Mühlbauer,G., Rinner,B., Stelzl,E., Berger,A., Dörr,H.-W., Santner,B., Marth,E. and Rabenau,H. (2000) Detection of herpes simplex virus DNA by real-time PCR. J. Clin. Microbiol. , 38, 2638–2642.

169. Kimura,H., Kido,S., Ozaki,T., Tanaka,N., Ito,Y., Williams,R.K. and Morishima,T. (2000) Comparison of quantitations of viral load in varicella and zoster.J. Clin. Microbiol. , 38, 2447–2449.

170. Ohga,S., Nomura,A., Takada,H., Thara,K., Kawakami,K., Yanai,F., Takahata,Y., Tanaka,T., Kasuga,N. and Hara,T. (2001) Epstein–Barr virus (EBV) load and cytokine gene expression in activated T cells of chronic active EBV infection. J. Infect. Dis , 183, 1–7.

171. White,I. and Campbell,T.C. (2000) Quantitation of cell-free cell-associated Kaposi's sarcoma associated herpesvirus DNA by real-time PCR. J. Clin. Microbiol.,38, 1992–1995.

172. Mercier,B., Burlot,L. and Ferec,C. (1999) Simultaneous screening for HBV DNA and HCV RNA genomes in blood donations using a novel TaqMan PCR assay. J. Virol. Methods, 77, 1–9.

173. Nitsche,A., Müller,C.W., Radonic,A., Landt,O., Ellerbrok,H., Pauli,G. and Siegert,W. (2001) Human herpesvirus 6A DNA is detected frequently in plasma but rarely in peripheral blood leukocytes of patients after bone marrow transplantation.J. Infect. Dis., 183, 130–133.

174. Lei,K.I.K., Chan,L.Y.S., Chan,W.Y., Johnson,P.J. and Lo,Y.M.D. (2000) Quantitative analysis of circulating cell-free Epstein–Barr virus (EBV) DNA levels in patients with EBV-associated lymphoid malignancies. Brit. J. Haematol., 111, 239–246.

175. Quackenbush,S.L., Casey,R.N., Murcek,R.J., Paul,T.A., Work,T.M., Limpus,C.J., Chaves,A., duToit,L., Perez,J.V., Aguirre,A.A. et al. (2001) Quantitative analysis of herpesvirus sequences from normal tissue and fibropapillomas of marine turtles with real-time PCR. Virology, 187, 105–111.

176. Aberham,C., Pendl,C., Gross,P., Zerlauth,G. and Gessner,M. (2001) A quantitative, internally controlled real-time PCR assay for the detection of parvovirus B19 DNA. J. Virol. Methods, 92, 183–191.

177. Sauleda,S., Reesink,H.J., Esteban,J.I., Hess,G., Esteban,R. and Guardia,J. (1999) Profiles of GBV-C/hepatitis G virus markers in patients coinfected with hepatitis C virus. J. Med. Virol., 59, 45–51.

178. Fujisawa,T., Horiike,N., Michitaka,K. and Onji,M. (2000) Influence of RNA titre and amino acid changes in the NS5A region of GB virus C/hepatitis virus G virus on the effectiveness of interferon therapy. J. Gastroenterol. Hepatol., 15, 632–639.

179. Houng,H.-S.H., Chen,R.C.-M., Vaughn,D.W. and Kanesa-thasan,N. (2001) Development of a fluorogenic RT–PCR system for quantitative identification of dengue virus serotypes 1–4 using conserved and serotype-specific 3′ noncoding sequences. J. Virol. Methods, 95, 19–32.

180. Saito,T., Shinzawa,H., Uchida,T., Kawamata,O., Honma,S., Watanabe,H., Shao,L., Saito,K., Togashi,H. and Takahashi,T. (1999) Quantitative DNA analysis of low-level hepatitis B viremia

in two patients with serologically negative chronic hepatitis B. J. Med. Virol., 58, 325–331.

181. Pas,S.D., Fries,E., De Man,R.A., Osterhaus,A.D.M.E. and Niesters,H.G. (2000) Development of a quantitative real-time detection assay for hepatitis B virus DNA and comparison with two commercial assays. J. Clin. Microbiol., 38, 2897–2901.

182. Schweiger,B., Zadow,I., Heckler,R., Timm,H. and Pauli,G. (2000) Application of a fluorogenic PCR assay for typing and subtyping influenza viruses in respiratory samples. J. Clin. Microbiol., 38, 1552–1558.

183. Koenig,M., Reynolds,K.S., Aldous,W. and Hickman,M. (2001) Comparison of Light-Cycler PCR, enzyme immunoassay and tissue culture for detection of Herpes Simplex Virus. Diagn. Microbiol. Infect. Dis., 40, 107–110.

184. Schalasta,G., Eggers,M., Schmid,M. and Enders,G. (2000) Analysis of human cytomegalovirus DNA in urines of newborns and infants by means of a new ultrarapid real-time PCR-system. J. Clin. Virol., 19, 175–185.

185. Kearns,A.M., Draper,B., Wipat,W., Turner,A.J.L., Wheeler,J., Freeman,R., Harwood,J., Gould,F.K. and Dark,J.H. (2001) LightCycler-based quantitative PCR for detection of cytomegalovirus in blood, urine and respiratory samples. J. Clin. Microbiol., 39, 2364–2365.

186. Kennedy,M.M., Lucas,S.B., Jones,R.R., Howells,D.D., Picton,S.J., Hanks,E.E., McGee,J.O. and O'Leary,J. (1997) HHV8 and Kaposi's sarcoma: a time cohort study. J. Clin. Pathol., 50, 96–100.

187. Kennedy,M.M., Cooper,K., Howells,D.D., Picton,S., Biddolph,S., Lucas,S.B., McGee,J.O. and O'Leary,J.J. (1998) Identification of HHV8 in early Kaposi's sarcoma: implications for Kaposi's sarcoma pathogenesis. J. Clin. Pathol., 51, 14–20.

188. O'Leary,J.J., Kennedy,M., Luttich,K., Uhlmann,V., Silva,I., Russell,J., Sheils,O., Ring,M., Sweeney,M., Kenny,C. et al. (2000) Localisation of HHV-8 in AIDS related lymphadenopathy. J. Clin. Pathol., 53, 43–47.

189. O'Leary,J., Kennedy,M., Howells,D., Silva,I., Uhlmann,V., Luttich,K., Biddolph,S., Lucas,S., Russell,J., Bermingham,N. et al. (2000) Cellular localisation of HHV-8 in Castleman's disease:

is there a link with lymphnode vascularity? J. Clin. Pathol., 53, 69–76.

190. Josefsson,A., Livak,K. and Gyllensten,U. (1999) Detection and quantitation of human papillomavirus by using the fluorescent 5′ exonuclease assay. J. Clin. Microbiol., 37, 490–496.

191. Swan,D.C., Tucker,R.A., Holloway,B.P. and Icenogle,J.P. (1997) A sensitive, type-specific, fluorogenic probe assay for detection of human papillomavirus DNA. J. Clin. Microbiol., 35, 886–891.

192. Lanham,S., Herbert,A. and Watt,P. (2001) HPV detection and measurement of HPV-16, telomerase and survivin transcripts in colposcopy clinic patients. J. Clin. Pathol., 54, 304–308.

193. MacKenzie,J., Gallagher,A., Clayton,R.A., Perry,J., Eden,O.B., Ford,A.M., Greaves,M.F. and Jarrett,R.F. (2001) Screening for herpesvirus genomes in common acute lymphoblastic leukemia. Leukemia, 15, 415–421.

194. Jabs,W.J., Hennig,H., Kittel,M., Pethig,K., Smets,F., Bucsky,P., Kirchner,H. and Wagner,H.J. (2001) Normalized quantification by real-time PCR of Epstein–Barr virus load in patients at risk for posttransplant lymphoproliferative disorders. J. Clin. Microbiol., 39, 564–569.

195. Beames,B., Chavez,D., Guerra,B., Notvall,L., Brasky,K.M. and Lanford,R.E. (2000) Development of a primary tamarin hepatocyte culture system for GB virus-B: a surrogate model for hepatitis C virus. J. Virol., 74, 11764–11772.

196. Lewin,S.R., Vesanen,M., Kostrikis,L., Hurley,A., Duran,M., Zhang,L., Ho,D.D. and Markowitz,M. (1999) Use of real-time PCR and molecular beacons to detect virus replication in human immunodeficiency virus type 1-infected individuals on prolonged effective antiretroviral therapy. J. Virol., 73, 6099–6103.

197. Choo,C.K., Ling,M.T., Suen,C.K.M., Chan,K.W. and Kwong,Y.L. (2000) Retrovirus-mediated delivery of HPV16 E7 antisense RNA inhibited tumorigenicity of CaSki cells. Gynecol. Oncol., 78, 293–301.

198. Hoshino,Y., Kimura,H., Kuzushima,K., Tsurumi,T., Nemoto,K., Kikuta,A., Nishiyama,Y., Kojima,S., Matsuyama,T. and Morishima,T. (2000) Early intervention in post-transplant lymphoproliferative disorders based on Epstein–Barr viral load. Bone Marrow Transplant, 26, 199–201.

199. Machida,U., Kami,M., Fukui,T., Kazuyama,Y., Kinoshita,M., Tanaka,Y., Kanda,Y., Ogawa,S., Honda,H., Chiba,S. et al. (2000) Real-time automated PCR for early diagnosis and monitoring of cytomegalovirus infection after bone marrow transplantation. J. Clin. Microbiol., 38, 1536–1542.

200. Gault,E., Michel,Y., Nicolas,J.-C., Belabani,C., Nicolas,J.-C. and Garbarg-Chenon,A. (2001) Quantification of human cytomegalovirus DNA by real-time PCR. J. Clin. Microbiol., 39, 772–775.

201. Tanaka,N., Kimura,H., Hoshino,Y., Kato,K., Yoshikawa,T., Asano,Y., Horibe,K., Kojima,S. and Morishima,T. (2000) Monitoring four herpesviruses in unrelated cord blood transplantation. Bone Marrow Transplant, 26, 1193–1197.

202. Gerard,C.J., Arboleda,M.J., Solar,G., Mule,J.J. and Kerr,W.G. (1996) A rapid and quantitative assay to estimate gene transfer into retrovirally transduced hematopoietic stem/progenitor cells using a 96-well format PCR and fluorescent detection system universal for MMLV-based proviruses. Hum. Gene Ther., 7, 343–354.

203. Hackett,N.R., El Sawy,T., Lee,L.Y., Silva,I., O'Leary,J., Rosengart,T.K. and Crystal,R.G. (2000) Use of quantitative TaqMan real-time PCR to track the time-dependent distribution of gene transfer vectors in vivo. Mol. Ther., 2, 649–656.

204. Sanburn,N. and Cornetta,K. (1999) Rapid titer determination using quantitative real-time PCR. Gene Ther., 6, 1340–1345.

205. Scherr,M., Battmer,K., Blömer,U., Ganser,A. and Grez,M. (2001) Quantitative determination of lentiviral vector particle numbers by real-time PCR. Biotechniques,31, 520–526.

206. Lanciotti,R.S. and Kerst,A.J. (2001) Nucleic acid sequence-based amplification assays for rapid detection of West Nile and St. Louis Encephalitis. J. Clin. Microbiol., 39, 4506–4513.

207. Gibb,T.R., Norwood,D.A.,Jr, Woollen,N. and Henchal,E.A. (2001) Development and evaluation of a fluorgenic 5' nuclease assay to detect and differentiate between Ebola virus subtypes Zaire and Sudan. J. Clin. Microbiol., 39,4125–4130.

208. Brorson,K., Swann,P.G., Lizzio,E., Maudru,T., Peden,K. and Stein,K.E. (2001) Use of a quantitative product-enhanced reverse

transcriptase assay to monitor retrovirus levels in mAb cell-culture and downstream processing. Biotechnol. Prog.,17, 188–196.

209. de Wit,C., Fautz,C. and Xu,Y. (2000) Real-time quantitative PCR for retrovirus-like particle formation in CHO cell culture. Biologicals, 28, 137–148.

Chapter 8

Sequence Exploration Reveals Information bias Among Molecular Markers Used in Phylogenetic Reconstruction for Colletotrichum Species

Sephra n rampersad[1], fazeeda n hosein[1], and christine VF carrington[2]

[1]Department of Life Sciences, Faculty of Science and Technology, The University of the West Indies, St. Augustine, Trinidad and Tobago - West Indies

[2]Department of Pre-clinical Science, Faculty of Medical Sciences, The University of the West Indies, St. Augustine, Trinidad and Tobago - West Indies

ABSTRACT

The *Colletotrichum gloeosporioides* species complex is among the most destructive fungal plant pathogens in the world, however, identification of isolates of quarantine importance to the intra-specific level is confounded by a number of factors that affect phylogenetic reconstruction. Information bias and quality parameters were investigated to determine whether nucleotide sequence alignments and phylogenetic trees accurately reflect the genetic diversity and phylogenetic relatedness of individuals. Sequence exploration of GAPDH, ACT, TUB2 and ITS markers indicated that the query sequences had different patterns of nucleotide substitution but were without evidence of base substitution saturation. Regions of high entropy were much more dispersed in the ACT and GAPDH marker alignments than for the ITS and TUB2 markers. A discernible bimodal gap in the genetic distance frequency histograms was produced for the ACT and GAPDH markers which indicated successful separation of intra- and inter-specific sequences in the data set. Overall, analyses indicated clear differences in the ability of these markers to phylogenetically separate individuals to the intra-specific level which coincided with information bias.

INTRODUCTION

Colletotrichum gloeosporioides is among the most pervasive and destructive fungal plant pathogens in the world (Sutton 1992; Cannon et al. 2008). *C. gloeosporioides* exists as a species complex (*Colletotrichum gloeosporioides sensu lato* (Weir et al. 2012) whose segregate taxa cannot be easily separated morphologically or phylogenetically and the approaches currently used to assign intra-specific ranking to these segregate taxa are still to be resolved and universally applied. Consequently, there is a preference among many for using the broad, group-species nomenclature rather than using names at the intra-specific level. However, this can cause confusion concerning the segregate taxa because there is overlap in morphological, biological and genetic variation at the intra-specific level. Correct identification of isolates of the *C. gloeosporioides sensu lato* species complex is important as some species pose quarantine risks some *Colletotrichum* species can infect multiple hosts.

Phylogenetic analysis of nucleotide sequences is one approach to achieving this level of taxonomic resolution. To date, low-level phylogenetic relationships in fungi have been inferred based on the sequence data for a number of nuclear and mitochondrial markers

(Bridge et al., 2005; Vialle et al. 2009). However, the rationale behind the selection of these genes is not always apparent or reported – an important consideration since accurate tree reconstruction is critically dependent on the availability of suitably informative characters for phylogenetic analyses (Brito and Edwards2009). Very few studies have been carried out to understand why some nuclear genes are better suited for phylogenetic reconstruction than others. Meaningful recovery of the phylogenetic hypothesis from a specific genetic marker may be complicated by factors such as a heterogeneous base composition (Lockhart et al. 1994), codon position saturation and transition/transversion rate bias (Phillips et al. 2004). Additionally, since some of these genes are duplicated in some fungal taxa there is the potential to infer phylogenetically inaccurate relationships (Ayliffe et al. 2001; Landvik et al. 2001; Tanabe et al. 2002; Tanabe et al. 2004 ; Fitzpatrick et al. 2006).

Multi-gene phylogenies are now commonly used to accurately define species in fungi and can be used to understand population history, demography and speciation (Taylor et al. 2000). However, different genes can evolve along different evolutionary trajectories which may result in tree inconsistencies (Brito and Edwards 2009; Aguileta et al. 2008; Townsend et al. 2008). In cases of very closely or distantly related taxa, certain aspects of the sequence data may contradict each other or carry insufficient or ambiguous information. Some researchers seek to address these irregularities and increase the robustness of phylogenetic inferences by increasing the sequence length of the marker used, thus increasing the amount of available data for inference (Bremer et al. 1999; van Oppen et al. 2001). This can be achieved artificially by concatenating the sequences of a number of different markers of distinct loci (Weir et al. 2012; Bremer et al. 1999; van Oppen et al. 2001; Slowinski and Lawson 2002). Consequently, before concatenation, it is important to screen sequences for homogeneity of base substitution and other congruence parameters. Ideally, the phylogenetic signal of each marker should be determined. Where there is incongruence, partitioning the data based on marker-specific substitution models is one approach to using concatenated sequences, however, combining sequences that possess a high phylogenetic signal with sequences that possess a low phylogenetic signal will not improve tree accuracy (Whelan et al. 2001; Posada and Crandall 2002; Egger et al. 2007).

Posada and Crandall (2002), therefore, recommended the use of single gene trees for phylogenetic tree construction.

In an effort to regularize any reference to and application of the name *C. gloeosporioides*, Cannon et al. (2008) carried out epitypification work and presented epitype sequences for *C. gloeosporioides sensu stricto*. The identification of many species within the *C. gloeosporioides*species complex based on multi-gene analyses has indicated that these new species are phylogenetically distinct from the epitype strain of *C. gloeosporioides sensu stricto* (Rojas et al.2010). Among the genetic markers used for species assignations were partial Actin (ACT), -Tubulin (TUB2), Calmodulin (CAL), Glutamine synthetase (GS) and Glyceraldehyde 3-phosphate dehydrogenase (GAPDH) and the nuclear rDNA internally transcribed spacer (ITS) region (Weir et al. 2012; Cai et al. 2009). However, even based on an 8-gene multilocus data set, accurate assignment of isolates at the intra-specific level was not achieved in some cases and in others, there was considerable overlap (Weir et al. 2012). Weir et al. (2012) and Cai et al. (2009) concluded that within the *C. gloeosporioides* species complex GAPDH, CAL, and ACT genetic markers can be used as DNA barcodes but, ITS sequences do not facilitate intraspecies discrimination. Doyle et al. (2013) separated *C. gloeosporioides sensu lato* isolates infecting cranberry in the United States, however, when the Apn2/Mat genetic marker was compared with TUB2 and ITS markers (Silva et al. 2012) some isolates still had ambiguous phylogenetic placement based on separate gene tree assessment or as part of a concatenated data set.

Studies on information bias and quality parameters are especially important when controversial and poorly supported relationships are to be investigated as in the case of the *C. gloeosporioides*species complex. However, there is a consensus that (i) the use of multilocus genetic data sets is required to better resolve phylogenetic relationships within the species complex, (ii) there is a low level of phylogenetic resolution for some intra-specific members of the complex, and (iii) current markers may not allow phylogenetic assignment of all identified intra-specific members of this species complex (Cannon et al. 2008; Weir et al. 2012; Rojas et al. 2010; Cai et al. 2009; Silva et al. 2012).

It is hypothesized that information bias among currently used molecular markers is one reason why it is difficult to generate phylogenetic trees that accurately reflect the genetic diversity and phylogenetic relatedness of individuals of the *C. gloeosporioides* species complex and which will allow intra-specific demarcation among such individuals. The main objectives of this study were (i) to conduct pre-phylogenetic sequence data exploration of parameters that are known to affect phylogenetic signal and resolution; these included determining indices of disparity, base substitution bias, entropy variability and sequence heterogeneity among a select sequence data set: *C. gloeosporioides sensu lato* isolates associated with anthracnose of papaya in Trinidad were selected to be the test population, and (ii) to compare three algorithms with gene-specific models of evolution in creating gene trees for the identification and assignment of query isolates and epitypes to the intra-specific level based on multi-locus nucleotide sequence data. It is not the objective of this study to present new or novel isolates or to give definitive species assignment or phylogeny if the analyses applied do not allow it.

MATERIALS AND METHODS

DNA Extraction, PCR, and Sequencing

DNA was extracted using the E.Z.N.A. fungal DNA extraction kit® according to manufacturer's instructions (Omega bio-tek Ltd., USA). Four markers were selected to generate sequences for quality analysis: partial Actin (ACT), partial -Tubulin (TUB2), Glyceraldehyde-3-phosphate dehydrogenase (GAPDH) and internally transcribed spacer regions I and II (ITS) of the rDNA gene regions. The universal primer pair ITS4/5 was used in PCR to amplify the ITS region (496 bp) of the nuclear ITS1-5.8S-ITS2 rDNA (White et al. 1990); Bt2a/b primers allowed amplification of a 560 bp fragment of the TUB2 gene region (Glass and Donaldson 1995); GDF/GDR (Templeton et al.1992) and ACT-512 F/ACT-783R (Carbone and Kohn 1999) primers were used to amplify 300 bp and 290 bp fragments, respectively. The standard calmodulin (CL1/CL2A) primers (O'Donnell et al.2000) were not successful in generating any amplions for most isolates and this gene region was not subsequently used in the analysis. Each 25 μL PCR reaction contained 1 × PCR buffer; 1.5 mM $MgCl_2$, 0.2 mM dNTP, 2.5 U Taq DNA Polymerase (Invitrogen by Life Technologies Co., USA) and 50 pmoles of each primer (Integrated DNA Technologies, USA). Thermal cycling conditions applied were: an initial denaturation of 5 min at 94°C followed by 35 cycles of 1 min at 94°C, 1 min at 55°C, 1 min at 72°C with a final extension of 5 min at 72°C. PCR products were sequenced directly (Amplicon Express, WA, USA).

Data sets

Only holotype and epitype sequences of *C. gloeosporioides sensu stricto*, *C. asianum*, *C. fructicola*,*C. siamense* and *C. tropicale* were used in the study (Phoulivong et al., 2010; Cannon et al. 2012) (Table 1). These selected species have been analyzed in previous independent phylogenetic studies, and thus provides an objective reference in phylogenetic relationships.

Table 1: Authentic sequences for accepted *Colletotrichum* species (Cannon et al.2008)

Authentic sequences for accepted *Colletotrichum* species (Cannon et al.2008)				
Marker	**GenBank Accession No.**	**Strain1**	**Source**	**Species**
GAPDH	FJ972582	CBS 95397	Culture from epitype	*Colletotrichum gloeosporioides sensu stricto*
GAPDH	FJ972577	MFU090228, ICMP 18581, CBS 130416	Culture from holotype	*Colletotrichum fructicola*
GAPDH	FJ972576	MFU 090233, ICMP 18580, CBS 130418	Culture from holotype	*Colletotrichum asianum*
GAPDH	FJ972575	MFU 090230, ICMP 18578, CBS 130417	Culture from holotype	*Colletotrichum siamense*
GAPDH	FJ972572	MFU 090230, ICMP 18578, CBS 130417	Culture from holotype	*Colletotrichum siamense*
GAPDH	FJ972571	MFU 090233, ICMP 18580, CBS 130418	Culture from holotype	*Colletotrichum asianum*
GAPDH	JX010078	ICMP 17673, ATCC 201874	Culture from holotype	*Colletotrichum tropicale*
GAPDH	GU888192	CBS:151.28	Culture from holotype	*Colletotrichum lindemuthianum*
ACT	FJ907430	CBS 95397	Culture from epitype	*Colletotrichum gloeosporioides sensu stricto*
ACT	FJ907426	MFU090228, ICMP 18581, CBS 130416	Culture from holotype	*Colletotrichum fructicola*

ACT	FJ907423	MFU 090230, ICMP 18578, CBS 130417	Culture from holotype	*Colletotrichum siamense*
ACT	FJ907420	MFU 090230, ICMP 18578, CBS 130417	Culture from holotype	*Colletotrichum siamense*
ACT	FJ903188	MFU 090233, ICMP 18580, CBS 130418	Culture from holotype	*Colletotrichum asianum*
ACT	JX009489	CBS 124949, ICMP18653	Culture from holotype	*Colletotrichum tropicale*
ACT	GU227898	CBS:151.28	Culture from holotype	*Colletotrichum lindemuthianum*
TUB	FJ907445	CBS 95397	Culture from epitype	*Colletotrichum gloeosporioides sensu stricto*
TUB	FJ907439	MFU 090233, ICMP 18580, CBS 130418	Culture from holotype	*Colletotrichum asianum*
TUB	FJ907435	MFU 090230, ICMP 18578, CBS 130417	Culture from holotype	*Colletotrichum siamense*
TUB	FJ907434	MFU 090233, ICMP 18580, CBS 130418	Culture from holotype	*Colletotrichum asianum*
TUB	JX010407	CBS 124949, ICMP18653	Culture from holotype	*Colletotrichum tropicale*
TUB	GU228094	CBS:151.28	Culture from holotype	*Colletotrichum lindemuthianum*
ITS	FJ972612	MFU 090233, ICMP 18580, CBS 130418	Culture from holotype	*Colletotrichum asianum*
ITS	FJ972611	MFU090228, ICMP 18581, CBS 130416	Culture from holotype	*Colletotrichum fructicola*
ITS	FJ972609	CBS 95397	Culture from epitype	*Colletotrichum gloeosporioides sensu stricto*

ITS	FJ972605	MFU 090233, ICMP 18580, CBS 130418	Culture from holotype	*Colletotrichum asianum*
ITS	FJ972604	MFU 090230, ICMP 18578, CBS 130417	Culture from holotype	*Colletotrichum siamense*
ITS	FJ972603	MFU090228, ICMP 18581, CBS 130416	Culture from holotype	*Colletotrichum fructicola*
ITS	JX010264	CBS 124949, ICMP18653	Culture from holotype	*Colletotrichum tropicale*
ITS	AY376534	STE-U 5297	Culture from ex-epitype	*Colletotrichum gloeosporioides sensu stricto*
ITS	EU371022	IMI 356878	Culture from epitype	*Colletotrichum gloeosporioides sensu stricto*
ITS	GU227800	CBS:151	Culture from holotype	*Colletotrichum lindemuthianum*

[1]Strain:

ATCC: American type culture collection, 10801 University Boulevard, Manassas, Virginia, USA.

CBS: Culture collection of the Centraalbureau voor Schimmelcultures, Fungal Biodiversity Centre, Utrecht, The Netherlands.

ICMP: International Collection of Microorganisms from Plants, Landcare Research, Auckland, New Zealand.

IMI: Culture collection of CABI Europe UK Centre, Egham, UK.

MFU: fungarium of Mae Fah Luang University, Thailand (cultures in BCC (BIOTEC Culture Collection, Thailand).

STE-U: Culture collection of the Department of Plant Pathology, University of Stellenbosch, South Africa.

Rampersad *et al.*

In addition to the sequences derived in this study (Table 2), query sequences were also extracted from a previous study by Rampersad et al. (2013) and used in the final data set. Representative sequences were submitted to GenBank (GenBank: JQ218143, JQ218144, JQ218145, JQ218146). A total of 143 sequences for all markers was used in the final data set - 35 taxa for GAPDH; 34 taxa for ACT; 37 taxa for TUB2; 37 taxa for ITS. Alignments were carried out using the MAFFT server (http://mafft.cbrc.jp/alignment/server/ *webcite*) under default parameters. BioEdit software (http://www.mbio.ncsu.edu/BioEdit/bioedit.html *webcite*). was used to edit the alignments of each marker data set. A schematic outlining the approach used to explore sequences is illustrated in Figure 1.

Table 2: Isolate data

Isolate	Field	Location	Cultivar	Year
PAW-Cg-110	Tableland	South Trinidad	Tainung No. 2-F1 hybrid	2011
PAW-Cg-111	Tableland	South Trinidad	Tainung No. 2-F1 hybrid	2011
PAW-Cg-112	Tableland	South Trinidad	Tainung No. 2-F1 hybrid	2011
PAW-Cg-113	Tableland	South Trinidad	Tainung No. 2-F1 hybrid	2011
PAW-Cg-114	Tableland	South Trinidad	Tainung No. 2-F1 hybrid	2011
PAW-Cg-115	Tableland	South Trinidad	Tainung No. 2-F1 hybrid	2011
PAW-Cg-116	Tableland	South Trinidad	Tainung No. 2-F1 hybrid	2011
PAW-Cg-118	Tableland	South Trinidad	Red Lady	2011
PAW-Cg-119	Tableland	South Trinidad	Red Lady	2011
PAW-Cg-120	Tableland	South Trinidad	Tainung No. 2-F1 hybrid	2011
PAW-Cg-121	Tableland	South Trinidad	Tainung No. 2-F1 hybrid	2011
PAW-Cg-122	Tableland	South Trinidad	Tainung No. 2-F1 hybrid	2011

P A W - Cg-5	Tableland	South Trinidad	Tainung No. 2-F1 hybrid	2011
P A W - Cg-6	Tableland	South Trinidad	Tainung No. 2-F1 hybrid	2011
P A W - Cg-7	Tableland	South Trinidad	Red Lady	2011
P A W - Cg-8	Barrackpore	South Trinidad	Red Lady	2011
P A W - Cg-102	Barrackpore	South Trinidad	Tainung No. 2-F1 hybrid	2011
P A W - Cg-103	Barrackpore	South Trinidad	Tainung No. 2-F1 hybrid	2011
P A W - Cg-104	Barrackpore	South Trinidad	Tainung No. 2-F1 hybrid	2011
P A W - Cg-105	Barrackpore	South Trinidad	Red Lady	2011
P A W - Cg-106	Santa Cruz	North Trinidad	Tainung No. 2-F1 hybrid	2011
P A W - Cg-107	Santa Cruz	North Trinidad	Tainung No. 2-F1 hybrid	2011
P A W - Cg-108	Santa Cruz	North Trinidad	Tainung No. 2-F1 hybrid	2011
P A W - Cg-109	Santa Cruz	North Trinidad	Red Lady	2011
P A W - Cg-9	Paramin	North Trinidad	Red Lady	2011
P A W - Cg-10	Paramin	North Trinidad	Red Lady	2011
P A W - Cg-41	Paramin	North Trinidad	Tainung No. 2-F1 hybrid	2011

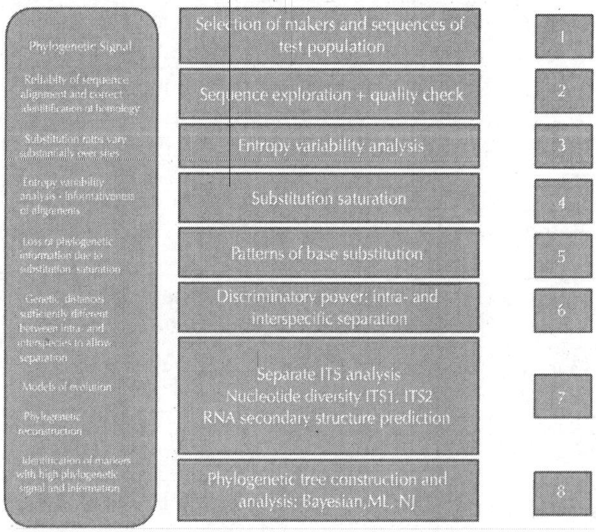

Figure 1: Schematic of methods used in sequence exploration prior to phylogenetic reconstruction.

Net Base Composition Bias Disparity between Sequences

Homogeneity tests can indicate different rates of substitution characteristic of certain lineages across different genes (Hedges and Kumar 2003). Several factors can influence substitution rates including the functionality of proteins and RNA, generation time, metabolic rate, population size, and life histories (Thomas et al. 2006; Smith and Donoghue 2008). Base composition for each gene according to test species was calculated in MEGA 5 (Tamura et al. 2011). The disparity index (ID) corresponds to the observed difference in base composition between two sequences compared to the compositional difference that would be expected under homogeneity of the evolutionary process. I_D equals 0 when the homogeneity assumption is satisfied. Values greater than 0 indicate that the larger differences in base composition bias than expected under homogeneity. Disparity Index (I_D) per site was calculated for all sequence pairs. All positions containing gaps were omitted from the data set (complete deletion option). The significance of a given I_D value was then assessed using a Monte Carlo approach.

Substitution Saturation

Base substitution saturation decreases the amount of phylogenetic information contained in a sequence data set and can disrupt analysis involving deep phylogeny (Xia and Xie 2001). Homoplasy due to multiple substitutions was tested with the index of substitution saturation (ISS) (Xia et al. 2003a), which assumes a critical index of substitution saturation (ISSc) that defines a threshold for significant saturation in the data. The level of substitution saturation was assessed by using the substitution saturation test of the program package DAMBE v. 5.3.46 software package (Xia et al. 2003b) which determines an "index of substitution saturation", based on the notion of entropy in information theory.

Entropy Variability Analyses

Entropy is a measure of sequence variation and can be used to quantify the level of phylogenetic information available in sequence alignments (Shannon 1948; Krüger et al. 2012). A position has zero entropy (invariant alignment column) if it has the same character state in every sequence in the alignment, thereby making this nucleotide position phylogenetically uninformative. The maximum Shannon entropy is dependent on the number of discrete variable in the data set. We investigated the entropy landscape of each alignment by constructing entropy plots for each marker in BioEdit.

Recombination Detection

Multiple recombination detection methods were applied as implemented in RDP3 to detect recombination events in the aligned sequences for each gene. Using RDP3, four methods were applied: RDP (Martin and Rybicki 2000), GENECONV (Padidam et al. 1999), MaxChi (Maynard 1992) and 3Seq (Boni et al. 2007). Sequences were treated as linear and the threshold *P*-value was set at 0.05, using Bonferroni correction. All other settings were set to default. Methods like GENECONV and MAXCHI do not include phylogenetic information in their recombination detection algorithms and phyogenetic trees generated did not include any recombination event information.

Differentiation between Intra- and Interspecific Sequences

The genetic distances were calculated under the Kimura 2 parameter (K2P) model using MEGA 5 for all sequences of the data set for each marker. Frequency histograms were generated to examine the intra- and possible interspecific variation among genetic distances. Using the genetic distances from pair-wise comparisons, the percentage comparisons with genetic distances greater than 5% are reported. These cut-off values are arbitrary but may indicate less inclusive clades. The distribution and size of the modes depend on the degree and range of sequence divergence and the number of intra-specific and interspecific taxa in the data set.

IT'S Sequence Analysis

Errors can occur during amplification when two different sequences are used as template DNA. The PCR product may include a combination of these original sequences (Jumpponen 2011). Such chimeric sequences may be misinterpreted as novel which can artificially inflate estimates of diversity and interfere with phylogenetic inference and species discrimination if undetected (Hugenholtz and Huber 2003). ITS sequences were checked for putative chimeras using the UNITE PlutoF Chimera checker (Edgar et al. 2011) and the Chimera Test developed in the Fungal Metagenomics Project at the University of Alaska (https://biotech.inbre.alaska.edu/fungal_portal/?program=chimera_test *webcite*).

Verifying ITS Sequence Validity

Alignments were carried out using the online version of the sequence alignment program MAFFT version 6 ((http://mafft.cbrc.jp/alignment/server/ *webcite*) to ensure that the 5.8S region of the ITS array forms an anchor region in the middle of the alignment. This check will determine if the ITS sequences were composed of stochastic, artifactual nucleotide data.

Phylogenetic Hypotheses

Multiple sequence alignments of each gene were made and manually adjusted where necessary with BioEdit. Individual gene trees were reconstructed using the Bayesian Markov chain Monte Carlo (MCMC) approach as implemented in Mr. Bayes 3.2.1 (Huelsenbeck and Ronquist 2001), and Neighbour joining (NJ) and Maximum Likelihood (ML) algorithms implemented in PAUP* version 4.0b8 (D. L. Swofford, Sinauer Associates, Sunderland, MA, Swofford 2003). For each of the markers, tree topologies were compared and evaluated for disagreement.

The Mr. Bayes analyses were performed under the GTR + Γ + I model. To ensure accuracy, four parallel runs were carried out with one cold and three heated MCMC chains per run for 1,100,000 generations; trees were sampled every 200 generations and the first 100,000 trees were discarded as these represented the trees of the burn-in phase of the analysis. Convergence was assessed by examining the stationarity of the ln-likelihood and effective sample size which was >500 for all analyses. The remaining trees were summarized as a Maximum Clade Credibility (MCC) tree (i.e. tree topology with the highest clade posterior probabilities across all nodes).

Bootstrap-supported (1,000 replicates) maximum likelihood (ML) and neighbour-joining (NJ) trees were estimated under their respective best-fit models of nucleotide substitution as determined using ModelTest v.3.7 (Posada 2008) with corrected Akaike information criteria (AICc). The models used were TUB2: HKY85; ACT: F81; ITS: GTR and GAPDH: GTR. Separate sequences of the ITS1 and ITS2 loci were used in a second and third data set to generate Bayesian trees under the GTR + Γ + I model, and ML and NJ phylogenetic trees based on the best-fit evolutionary model which were HKY85 for ITS1 and GTR for ITS2.

We included sequence data from representative epitype isolates which were selected as authentic cultures by the CBS (Centraalbureau voor Schimmelcultures institute, Utretcht, Netherlands), IMI (CABI Europe – UK, Bakeham Lane, Egham, Surrey, UK) and/or as cultures listed as authentic by Cannon et al. (2012) (Table 1). We also sought to reduce the level of incongruence between individual gene trees by using the approach of conditional editing of sequences (Salichos and Rokas 2013) where (i) sites containing gaps and any aberrant sequences that produced a "bad" alignment were removed from the data set, (ii) quickly evolving taxa that are attracted by the distant outgroup or taxa that resulted in long branch attraction were removed, (iii) a limited number of taxa in each gene tree (query and epitype), (iv) only established or known ex-epitype and epitype sequences and (v) genes that recover specific internodes and where species placement has proven consistent among independent studies were used. FigTree was used to annotate trees and export graphics.

RESULTS

MAFFT analysis revealed no evidence of chimeric sequences present in the ITS data set and indicated good quality ITS sequences with no artifactual nucleotide data or pseudogenes. When the separate ITS loci were analyzed, it was found that the ITS1 region was more variable than the ITS2 region ($Pi_{ITS1}=0.01068$; $Pi_{ITS2}=0.00390$).

In this study, tests of the homogeneity of substitution patterns in query sequences indicated that isolates had different patterns of nucleotide substitution for ACT and TUB2 markers (Table 3). For the ITS sequence data, all isolates had *P*-values lower than 0.05; for the TUB2 sequence data set (*N* =29) 12 isolates had *P*-values lower than 0.05; for the GAPDH sequence data set (*N* =27) 9 isolates had *P*-values lower than 0.05. None of the isolates had a *P*-value lower than 0.05 for the ACT data set (*N* =27).

Table 3: Tests of homogeneity of base substitution patterns

Tests of homogeneity of base substitution patterns			
Marker	**Number of positions in final data set**	**Number of isolates1**	**Isolate ID**
ACT	158	3	PAW-Cg-118, PAW-Cg-107 and PAW-Cg-108
GAPDH	164	0	...
ITS	481	0	...

TUB2	287	11	PAW-Cg-9, PAW-Cg-111, PAW-Cg-112, PAW-Cg-114, PAW-Cg-116, PAW-Cg-119, PAW-Cg-121, PAW-Cg-8, PAW-Cg-9 and PAW-Cg-41

[1]A Monte Carlo test (1000 replicates) was used to estimate the *P*-values ($P \leq 0.05$); the number and identity of isolates with a *P*-value smaller than 0.05. All positions containing gaps and missing data were eliminated from the data set (complete deletion option).

Rampersad *et al.*

Base Substitution Saturation

Tests to determine base substitution saturation were conducted for all markers separately, for combined ITS array data set, and for the stem and loop regions of ITS1 and ITS2 regions separately (Table 4). Results indicated that for each data set, ISS values were consistently lower than the ISSc values which indicated little saturation in base substitution. The test also revealed little difference between paired (stem) and unpaired positions (loops). In the stem and loop regions of all relevant ITS data sets, the ISS value was always lower significantly ($P < 0.0001$) than the observed ISSc. The complete results of the saturation tests are summarised in Table 3. The findings of the saturation test suggested that there was no need to partition the stem and loop regions of the ITS1 and ITS2 data sets for phylogenetic analyses as these regions had little saturation.Estimation of saturation can also be determined based on the slope of the regression line in the saturation plot; data sets without any saturation have a slope =1, data sets with high saturation have a slope =0. Generally, base saturation by transitions was higher than base saturation by transversions except for the ITS markers which had near equivalent saturation for both transition- and transversion-type base substitution (Figure 2). Slopes for transversion-type and transition-type base substitutions were closer to 1 for ACT, GAPDH, TUB2 and ITS, in the order of least to most saturated.

Table 4: Comparison of the index of substitution saturation (ISS) with the critical index of substitution saturation (ISSc) that defines a threshold for significant saturation in the data

Comparison of the index of substitution saturation (ISS) with the critical index of substitution saturation (ISSc) that defines a threshold for significant saturation in the data					
Marker	**Iss**	**Issc[1]**	**T**	**DF**	***P***
ACT	0.0609	0.6383	37.0781	155.0000	0.0000
GAPDH	0.0454	0.6162	36.7266	163.0000	0.0000
ITS	0.0167	0.7042	146.3186	477.0000	0.0000
TUB2	0.0089	0.6636	87.1685	292.0000	0.0000
ITS1-Loops	0.1665	0.9113	13.6810	50.0000	0.0000
ITS2-Loops	0.0425	0.8548	34.1213	59.0000	0.0000
ITS1-Stem	0.0390	0.4671	19.9356	102.0000	0.0000
ITS2-Stem	0.0481	0.4980	16.5953	96.0000	0.0000

[1]Only the symmetrical tree topology is presented.

Rampersad *et al.*

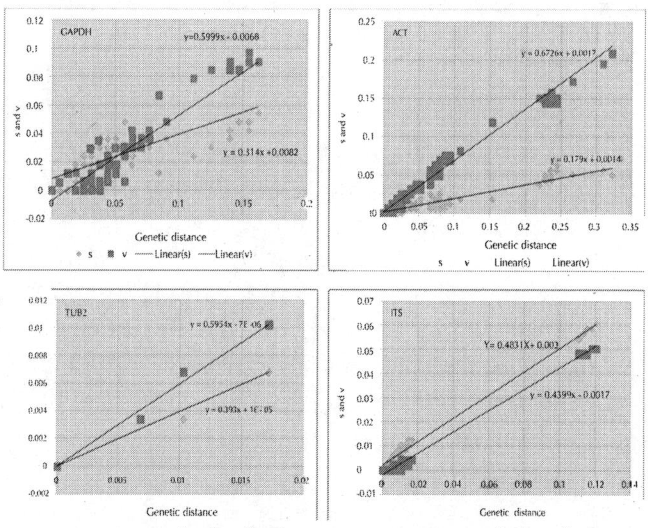

Figure 2: DAMBE base substitution saturation plots for GAPDH, ACT, TUB2 and ITS sequence data sets.

Entropy Variability Analysis

Shannon entropy distributions within the GAPDH, ACT, TUB2 and ITS DNA sequence alignments were examined. Regions of high entropy were the most dispersed in the GAPDH marker alignment. Entropy plots (Figure 3) revealed only two high entropy clusters extending greater than eight sites. GAPDH had the highest number of peaks over 0.25 (18 peaks with three over 0.5 maximum), followed by ACT with eight peaks over 0.25 (only one over 0.5 maximum), ITS had seven peaks (three over 0.5 maximum) and TUB2 had five peaks (all of which were over a 0.5 maximum. Similar results were obtained for the translated amino acid alignment for reading frames 1, 2 and 3 of the TUB2 gene, data not shown). With respect to the ACT marker, the plot indicated a distinct cluster of sites towards the end of the alignment (alignment position 110 to 141) with entropy values greater than 0.25. For the ITS alignments, most of the entropy values occurred in two distinct clusters and which corresponded to the more variable ITS1 and ITS2 loci that flank the 5.8S region.

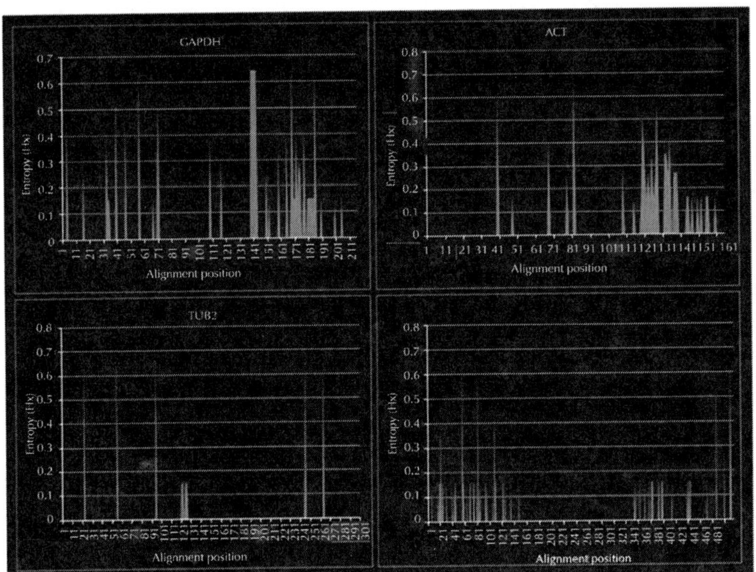

Figure 3: Entropy plots showing the relative entropy distributions within the GAPDH, ACT, TUB2 and ITS sequence alignment data sets.

Recombination Detection

Aligned sequences of each gene were analyzed with recombination detection software, using the three methods of analyses i.e. RDP, GENECONV and MaxChi, with the aim of identifying possible recombinant isolates and to locate the recombined regions in these recombinant sequences. Recombination was detected in the aligned ACT gene sequences. No significant recombination events above the *P*-value threshold of 0.05were detected for MAXCHI analyses. 3Seq detected 2 isolates with significant ($P < 0.05$) recombination events; PAW-Cg-107 and PAW-Cg-41; however, as none of the other methods detected these sequences, it is not certain whether these isolates were actual recombinants. There was no evidence of recombination in any of the other gene sequences.

Intra- and Interspecific Differentiation

The genetic distances for all sequences of the isolates in the data set for each marker were calculated under the Kimura 2-parameter (K2P) model using MEGA5 (Table 5). Frequency histograms were generated to examine the intra- and possible inter-specific variation based on pair-wise genetic distances. Summary statistics for genetic distance data are presented in Table 3. There is an apparent difference in the ability of a given markers to distinguish between intra-specific and interspecific sequences according to marker. The ACT and GAPDH markers were able to produce a discernible modal gap in the frequency histograms which indicate separation of intra- and inter-specific isolates (Figure 4). There is some uncertainty for TUB2 whether a true gap exists because the separation is not an easily discernible bimodal distribution as for the other markers. No discernible gap was observed for the ITS marker.

Table 5: Comparison of genetic distance statistics for each marker

Comparison of genetic distance statistics for each marker					
Marker	**Max GD1**	**Min GD**	**Med GD**	**Std Dev GD**	**≥5%2**
GAPDH	0.167	0.000	0.085	0.0495	97
TUB2	0.018	0.000	0.010	0.0093	0
ACT	0.098	0.000	0.052	0.0323	122

ITS	0.057	0.000	0.007	0.0263	0

[1]Summary statistics of genetic distances calculated using the Kimura 2-paramter (K2P) model and include: Maximum genetic distance (Max GD), Minimum genetic distance (Min GD), Median genetic distance (Med GD) with standard deviation (Std Dev GD) and thenumberofpair-wisecomparisonsofgeneticdistancesgreaterthanorequalto5%(≥5%).
[2]The number of pair-wise comparisons of genetic distances greater than or equal to 2% (≥2%) was also determined to be zero for ITS.

Rampersad *et al.*

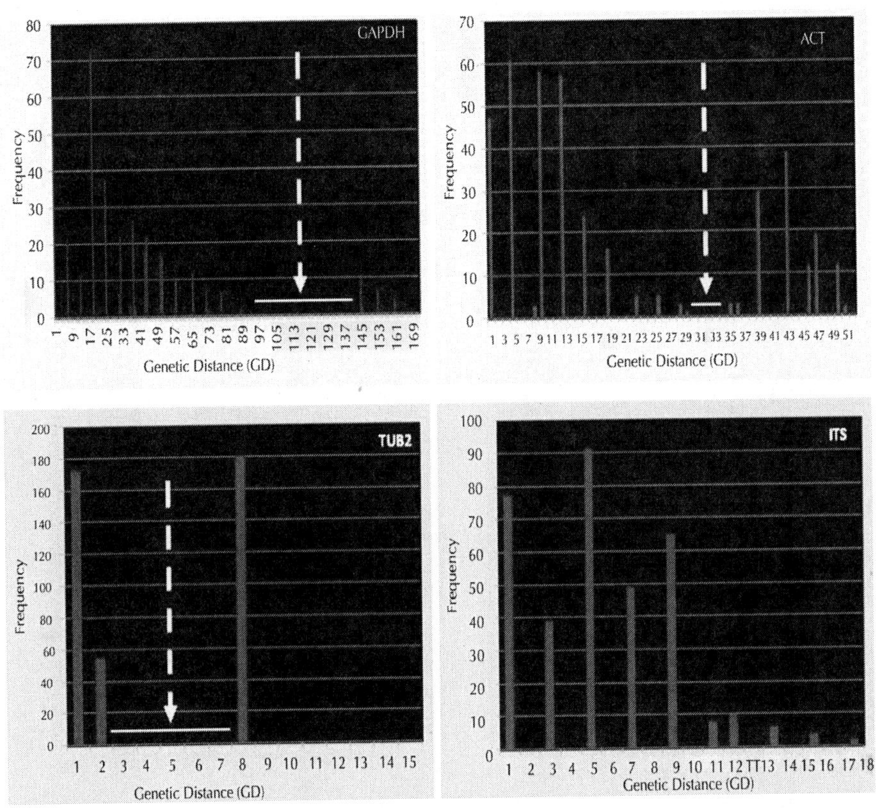

Figure 4: Frequency histograms to examine the intra- and inter-specific variation for each marker based on pair-wise genetic distances. The arrow indicates the frequency gap that separates intra-specific sequences.

Phylogenetic Analysis

The phylogenetic hypotheses for the Bayesian MCC trees are described here (Figures 5, 6, 7, 8, 9and 10). Individual loci were analyzed separately to assess the topological congruence among the different sequence data sets. The ability of each data set to efficiently resolve terminal lineages with robust branch support for relationships within the *C. gloeosporioides sensu lato* species complex was determined. The topologies for the Bayesian MCC, ML and NJ trees were compared for each marker and were found to be congruent.

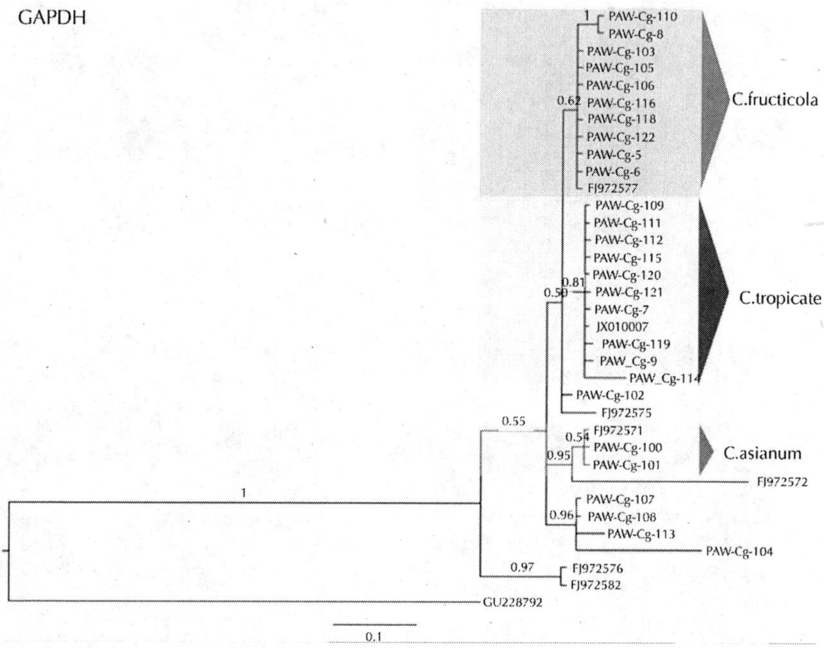

Figure 5: Bayesian MCC phylogenetic tree for the GAPDH marker.Numbers above branches are clade credibility scores.

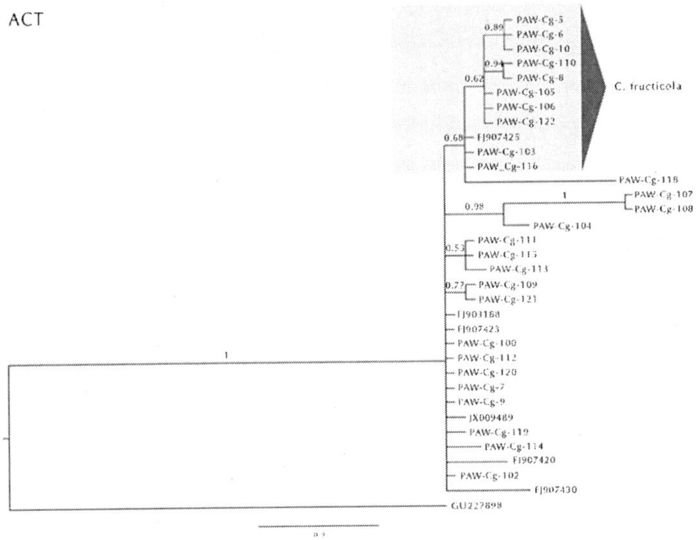

Figure 6: Bayesian MCC phylogenetic tree for the ACT marker.Numbers above branches are clade credibility scores.

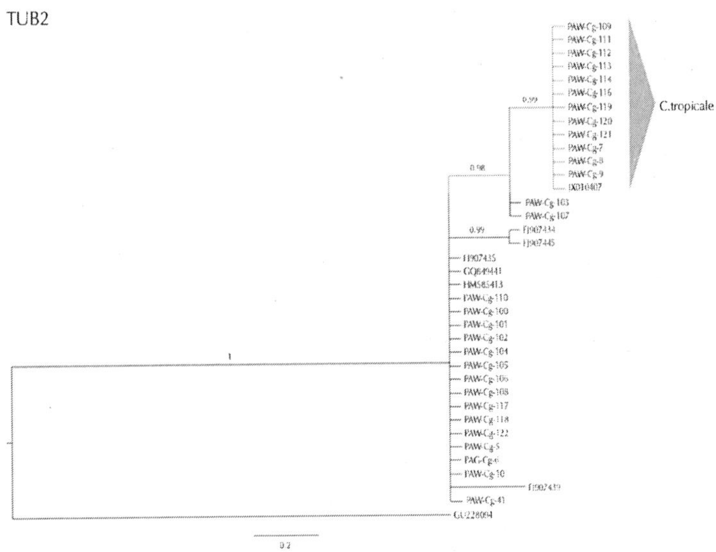

Figure 7: Bayesian MCC phylogenetic tree for the TUB2 marker.Numbers above branches are clade credibility scores.

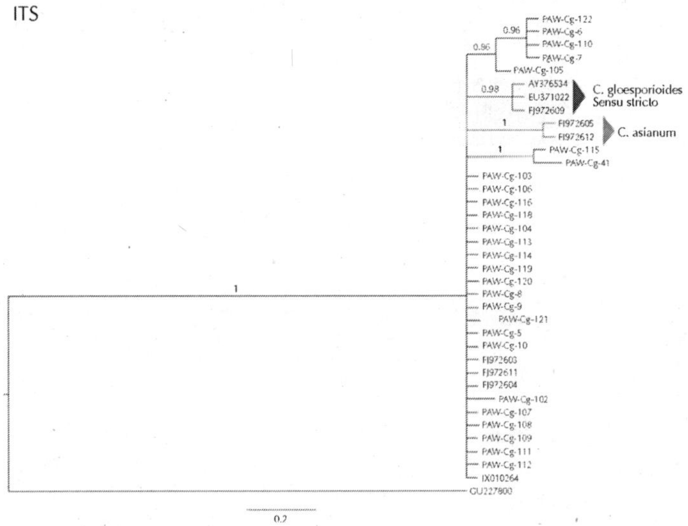

Figure 8: Bayesian MCC phylogenetic tree for the ITS marker (entire ITS1-5.8S-ITS2 region). Numbers above branches are clade credibility scores.

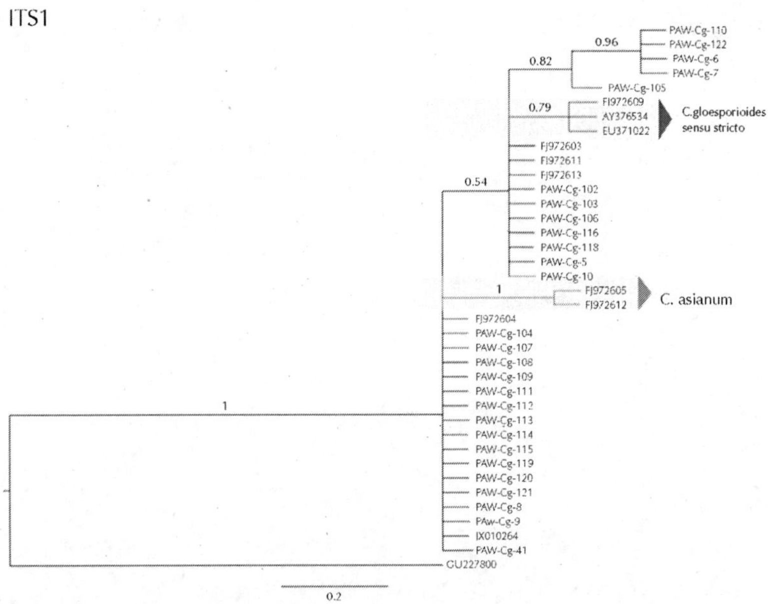

Figure 9: Bayesian MCC phylogenetic tree for the ITS1 marker.Numbers above branches are clade credibility scores.

ITS2

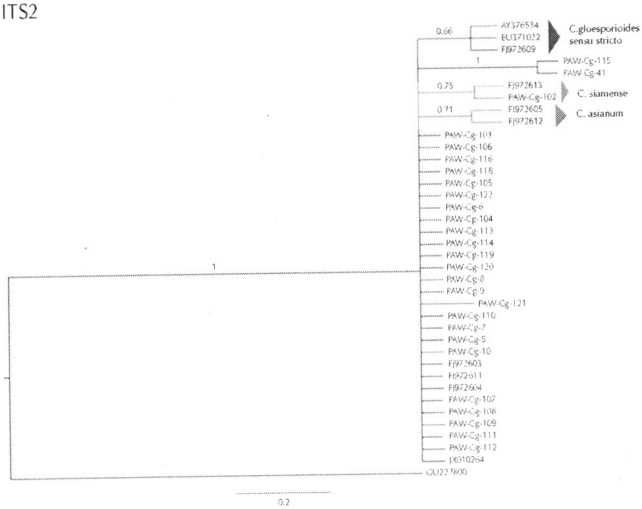

Figure 10: Bayesian MCC phylogenetic tree for the ITS2 marker.Numbers above branches are clade credibility scores.

The GAPDH Bayesian MCC tree (Figure 5) also revealed a distinct *C. fructicola* clade with one subclade consisting of sister taxa PAW-Cg-110 and PAW-Cg-8 similar to the ACT tree. *C. tropicale*formed a separate clade along with nine isolates but forming a polytomic phylogeny. The *C. asianum* clade consisted of PAW-Cg100 and PAW-Cg-101 with one epitype sequence. For the ACT and GAPDH analyses, four isolates (PAW-Cg-104, PAW-Cg-107, PAW-Cg-108, PAW-Cg-113) were commonly positioned in both trees.

For the ACT tree (Figure 6), there was a distinct *C. fructicola* clade but with only moderate clade support, and with two subclades. *C. siamense*, *C. gloeosporioides sensu stricto*, *C. asianum* and *C. tropicale* epitype sequences formed a polytomic phylogeny with seven isolates in the data set. PAW-Cg-107 and PAW-Cg-108 appeared to be sister taxa and formed a separate strongly-supported clade with PAW-Cg-104. Similarly PAW-Cg-109 and PAW-Cg-121 also appeared to be sister taxa forming a well-supported clade. PAW-Cg-118, PAW-Cg-107 and PAW-Cg-108 appear to be faster evolving or more genetically divergent isolates than the others given the comparatively longer branch lengths.

The TUB2 Bayesian MCC tree (Figure 7) indicated a strongly-supported separate *C. tropicale* clade, consisting of 12 isolates which was identical to the *C. tropicale* clade of the GAPDH tree. There may be an apparent *C. siamense* clade consisting of 15 isolates but these formed a basal polytomic phylogeny.

The ITS Bayesian MCC tree (Figure 8) displayed with a distinct *C. gloeosporioides sensu stricto*clade consisting only of the three epitype sequences used in the data set. Similarly, a distinct *C. asianum* clade was discernible with strong branch support. *C. siamense, C. fructicola* and *C. tropicale* epitype sequences formed a basal polytomic phylogeny with the majority of the isolates in the data set.

The separate ITS1 and ITS2 loci were considered. For the ITS1 region tree (Figures 9 and 10), *C. fructicola* and *C. tropicale* formed a polytomic phylogeny with the majority of the isolates. *C. siamense* formed a separate polytomic clade with seven isolates; the three *C. gloeosporioides sensu stricto* and the *C. asianum* epitypes formed two separate clades with higher branch support than for the ITS2 region tree. For the ITS2 region tree, *C. fructicola* and *C. tropicale* together with the majority of isolates formed a polytomic phylogeny; the three *C. gloeosporioides sensu stricto*epitypes, one *C. siamense* epitype and one query isolate and the two *C. asianum* epitypes formed separate moderately-supported clades.

Overall, analyses indicated that the ACT marker resolved only the *C. fructicola* clade; the GAPDH marker resolved the *C. fructicola, C. tropicale* and *C. asianum* clades; the TUB2 marker resolved the *C. tropicale* and *C. siamense* clades; and the ITS marker (entire) resolved the *C. gloeosporioides sensu stricto* and the *C. asianum* clades; the ITS1 region resolved the *C. gloeosporioides sensu stricto* and the *C. asianum* clades and the ITS2 region resolved the *C. gloeosporioides sensu stricto, C. siamense and* the *C. asianum* clades.

DISCUSSION

It was hypothesized that information bias among currently used molecular markers is one reason why it is difficult to generate phylogenetic trees that accurately reflect the genetic diversity and phylogenetic relatedness of individuals of the *C. gloeosporioides* species complex and as such intra-specific demarcation among such individuals remain a challenge. Tests that detect information bias were conducted.

The results indicated differences in saturation levels, entropy and heterogeneity among the genetic markers tested. Comparisons with phylogenetic analyses demonstrate an apparent correlation between the detected information bias and resolution of phylogeny. It is, therefore, important to conduct such tests in support of any proposed phylogenetic hypothesis.

Our data also showed a distinct difference in the number of isolates in the test population achieving phylogenetic placement which was dependent on the marker. None of the isolates in the test population were resolved phylogenetically based on the entire ITS, or the separate ITS1 or ITS2 region. However, differences in the ability of the separate ITS loci to separate epitype sequences were detected. One explanation for the branch support difference between the ITS1 and ITS2 region trees with respect to separating *C. gloeosporioides sensu stricto* from *C. siamense* can be found in the level of nucleotide differences (Pi) between the two loci.

GAPDH was able to position the greatest number of isolates in the test population, it had the highest entropy distribution, it was the second marker with the least saturation and it was characterized by a distinct frequency gap that correlated to the largest genetic distance compared to the other markers. Importantly, the data also suggested that ACT and TUB2 served to confirm the GAPDH phylogeny of the isolates in the test population for the *C. fructicola* and *C. tropicale*clades, respectively. Weir et al. (2012) and Rojas et al. (2010) also concluded that these markers were able to resolve some species but not others, and many species could not be discriminated using a single gene.

The phylogenetic relationships of closely related taxa that resulted from rapid radiation over a very short period time are difficult to define as is the case for *C. gloeosporioides* species complex. It is possible that some member species in the *C. gloeosporioides* species complex may have been pre-maturely given new species designations (i.e. before they developed independent evolutionary and phylogenetic routes) which would help to explain the low level of phylogenetic signal and poor or incomplete resolution. Hoelzer and Meinick (1994) purported that regardless of the model and concepts used to define and identify a species, geographically isolated and genetically divergent subspecies will have complicated patterns of gene flow with other subspecies over evolutionary time and will emerge prior to actual "speciation".

Examination of a greater number of loci in the descendants will still lead to an inaccurate estimation of the pattern of bifurcations in the phylogenetic trees even though the bifurcations may have robust statistical support (Rokas and Carroll 2006).

As with any approach that imposes structure on the data (Hoelzer and Meinick 1994; Whitfield and Lockhart 2007) bifurcations result from an imposition of the tree-building method and necessarily reality (Chan and Moore 2005). If there is a real dichotomous structure in the data, the bifurcations will be apparent and the unresolved nodes will occur usually at or near the terminal branches, as seen in the trees generated in this study. "Hard" polytomies in phylogenetic reconstructions, as seen in this study and others (Weir et al. 2012; Kliman et al. 2000; Takahashi et al. 2001; Coyne et al. 2004) should be viewed as real multifurcations or multiple, simultaneous divergence events, rather than incomplete or unsuccessful resolution of evolutionary history (Maddison 1989). Polytomies may represent the most accurate possible representation of historical relationships among the taxa under consideration (Rokas and Carroll 2006; Felsenstein 1985; Grafen 1989). The low level of genetic diversity within this species complex as reflected by the short branch lengths suggests that the species recognised within the *C. gloeosporioides* complex are very recently evolved (Cannon et al. 2012; Silva et al. 2012), therefore, it is likely that even with potentially more informative genes such as ApMAT and Apn25L (Silva et al. 2012) the low levels of genetic divergence across the *C. gloeosporioides* complex may continue to test the limits of phylogenetic resolution. If speciation events have only recently occurred among related taxa, the amount of informative phylogenetic data is low and the phylogenetic signal is small, leading to short internal tree branches or polytomies that are difficult to resolve (Saitou and Nei 1986; Philippe et al. 1994). Moreover, increasing the number of gene sequences in an attempt to solve a difficult phylogenetic question may not necessarily be the best approach if the data set has been generated from markers with low and high levels of phylogenetic information; the low phylogenetic signal may become dominant and yield inconsistent, yet highly supported, phylogenetic trees (Jeffroy et al. 2006).

CONCLUSIONS

Sequence data exploration of parameters known to affect phylogenetic signal and resolution is important to understanding the phylogenetic placement of individuals. The findings of this study demonstrated that these parameters explained the ability of different markers to adequately identify individuals at the intra-specific level. This data is important to verify whether nucleotide sequence alignments and phylogenetic trees accurately reflect the genetic diversity and phylogenetic relatedness of individuals and to establish if any of the markers and taxa included in the data set show a strong departure from basic assumptions in phylogenetic tree reconstruction. The findings of this work should be considered in the context of (i) the existence of information bias among the genetic markers used for generating the phylogenetic hypothesis and (ii) time of sequence divergence of isolates at a rate that appears to be faster than the rate of speciation of these isolates or conversely, divergence of certain gene sequences may occur at a rate that is slower than the time for speciation.

ACKNOWLEDGEMENTS

This work was supported by The University of the West Indies, St. Augustine, Campus Research and Publications Grant (Grant No. CRP.3.NOV11.8).

REFERENCES

1. Aguileta G, Marthey S, Chiapello H, Lebrun MH, Rodolphe F, Fournier E, Gendrault-Jacquemard A, Giraud T (2008) Assessing the performance of single-copy genes for recovering robust phylogenies. Syst Biol 57:613-627
2. Ayliffe MA, Dodds PN, Lawrence GJ (2001) Characterisation of a β-TUB2ulin gene from *Melampsora lini* and comparison of fungal β-TUB2ulin genes. Mycol Res 105:818-826
3. Boni MF, Posada D, Feldman MW (2007) An exact nonparametric method for inferring mosaic structure in sequence triplets. Genetics 176:1035-1047

4. Bremer B, Jansen R, Oxelman B, Backlund M, Lantz H, Kim KJ (1999) More characters or more taxa for a robust phylogeny-case study from the coffee family (*Rubiaceae*). Syst Biol 48:413-435
5. Bridge PD, Spooner BM, Roberts PJ (2005) The impact of molecular data in fungal systematics. Adv Bot Res 42:34-67
6. Brito PH, Edwards SV (2009) Multilocus phylogeography and phylogenetics using sequence-based markers. Genetica 135:439-455
7. Cai L, Hyde KD, Taylor PWJ, Weir BS, Waller J, Abang MM, Zhang JZ, Yang YL, Phoulivong S, Liu ZY, Prihastuti H, Shivas RG, Mckenzie EHC, Johnston PR (2009) A polyphasic approach for studying*Colletotrichum*. Fungal Divers 39:183-204
8. Cannon PF, Buddie AG, Bridge PD (2008) The typification of *Colletotrichum gloeosporioides*. Mycotaxon 104:189-204
9. Cannon PF, Damm U, Johnston PR, Weir BS (2012) *Colletotrichum*– current status and future directions. Stud Mycol 73:181-213 Carbone I, Kohn LM (1999) A method for designing primer sets for speciation studies in filamentous ascomycetes. Mycologia 91:553-556
10. Chan KM, Moore BR (2005) Symmetree: whole-tree analysis of differential diversification rates. Bioinformatics 21:1709-1710
11. Coyne JA, Elwyn S, Kim SY, Llopart A (2004) Genetic studies of two sister species in the *Drosophila melanogaster* subgroup, *D. yakuba* and *D. santomea*. Genet Res 84:11-26
12. Doyle VP, Oudemans PV, Rehner SA, Litt A (2013) Habitat and host indicate lineage identity in*Colletotrichum gloeosporioides s.l.* from wild and agricultural landscapes in North America. PLos One 8:e62394
13. Edgar RC, Haas BJ, Clemente JC, Quince C, Knight R (2011) UCHIME improves sensitivity and speed of chimera detection. Bioinformatics 27:2194-2200
14. Egger B, Koblmüller S, Sturmbauer C, Sefc K (2007) Nuclear and mitochondrial data reveal different evolutionary processes in the Lake Tanganyika cichlid genus *Tropheus*. BMC Evol Biol 7:137
15. Felsenstein J (1985) Phylogenies and the comparative method. Am Nat 125:1-15

16. Fitzpatrick D, Logue M, Stajich J, Butler G (2006) A fungal phylogeny based on 42 complete genomes derived from supertree and combined gene analysis. BMC Evol Biol 6:99
17. Glass NL, Donaldson GC (1995) Development of primer sets designed for use with the PCR to amplify conserved genes from filamentous ascomycetes. Appl Environ Microbiol 61:1323-1330
18. Grafen A (1989) The phylogenetic regression. Philos Trans R Soc Lond B Biol Sci 326:119-157
19. Hedges BS, Kumar S (2003) Genomic clocks and evolutionary timescales. Trends Genet 19:200-206
20. Hoelzer GA, Meinick DJ (1994) Patterns of speciation and limits to phylogenetic resolution. Trends Ecol Evol 9:104-107
21. Huelsenbeck JP, Ronquist F (2001) MRBAYES: Bayesian inference of phylogeny. Bioinformatics 17:754-755
22. Hugenholtz P, Huber T (2003) Chimeric 16S rDNA sequences of diverse origin are accumulating in the public databases. Int J Syst Evol Microbiol 53:289-293
23. Jeffroy O, Brinkmann H, Delsuc F, Philippe H (2006) Phylogenomics: the beginning of incongruence? Trends Genet 22:225-231
24. Jumpponen A (2011) Analysis of ribosomal RNA indicates seasonal fungal community dynamics in*Andropogon gerardii* roots. Mycorrhiza 21:453-464
25. Kliman RM, Andolfatto P, Coyne JA, Depaulis F, Kreitman M, Berry AJ, McCarter J, Wakeley J, Hey J (2000) The population genetics of the origin and divergence of the *Drosophila simulans* complex species. Genetics 156:1913-1931
26. Krüger D, Kapturska D, Fischer C, Daniel R, Wubet T (2012) Diversity measures in environmental sequences are highly dependent on alignment quality—data from its and new LSU primers targeting basidiomycetes. PLoS One 7:e32139
27. Landvik S, Eriksson OE, Berbee ML (2001) *Neolecta*: a fungal dinosaur? Evidence from -TUB2ulin amino acid sequences. Mycologia 93:1151-1163
28. Lockhart PJ, Steel MA, Hendy MD, Penny D (1994) Recovering evolutionary trees under a more realistic model of sequence evolution. Mol Biol Evol 11:605-612

29. Maddison DR (1989) Reconstructing character evolution on polytomous cladograms. Cladistics 5:365-377
30. Martin D, Rybicki E (2000) RDP: detection of recombination amongst aligned sequences. Bioinformatics 16:562-563 PubMed Abstract | Publisher Full Text
31. Maynard SJ (1992) Analyzing the mosaic structure of genes. J Mol Evol 34:126-129
32. O'Donnell K, Nirenberg HI, Aoki T, Cigelnik E (2000) A multigene phylogeny of the *Gibberella fujikuroi* species complex: detection of additional phylogenetically distinct species. Mycoscience 41:61-78
33. Padidam M, Sawyer S, Fauquet CM (1999) Possible emergence of new geminiviruses by frequent recombination. Virology 265:218-225
34. Philippe H, Chenuil A, Adoutte A (1994) Can the Cambrian explosion be inferred through molecular phylogeny? Development 120:S15-S25
35. Phillips MJ, Delsu F, Penny D (2004) Genome-scale phylogeny and the detection of systematic biases. Mol Biol Evol 21:1455-1458 Phoulivong S, Cai L, Chen H, McKenzie EHC, Abdelsalam K, Chukeatirote E, Hyde KD (2010) Colletotrichum gloeosporioides is not a common pathogen on tropical fruits. Fungal Divers 44:33-43
36. Posada D (2008) jModelTest: Phylogenetic Model Averaging. Mol Biol Evol 25:1253-1256
37. Posada D, Crandall KA (2002) The effect of recombination on the accuracy of phylogeny estimation. J Mol Evol 54:396-402
38. Rampersad SN, Perez-Brito D, Torres-Calzada C, Tapia-Tussell R, Carrington VF (2013) Genetic structure and demographic history of *Colletotrichum gloeosporioides sensu lato and C. truncatum*isolates from Trinidad and Mexico. BMC Evol Biol 13:130
39. Rojas EI, Rehner SA, Samuels GJ, Van Bael SA, Herre EA, Cannon P, Chen R, Pang J, Wang R, Zhang Y, Peng YQ, Sha T (2010) *Colletotrichum gloeosporioides* s.l. associated with *Theobroma cacao* and other plants in Panama: multilocus phylogenies

distinguish host-associated pathogens from asymptomatic endophytes. Mycologia 102:1318-1338

40. Rokas A, Carroll SB (2006) Bushes in the tree of life. PLoS Biol 4:e352
41. Saitou N, Nei M (1986) The number of nucleotides required to determine the branching order of three species, with special reference to the human-chimpanzee-gorilla divergence. J Mol Evol 24:189-204
42. Salichos L, Rokas A (2013) Inferring ancient divergences requires genes with strong phylogenetic signals. Nature 497:327-333 PubMed Abstract | Publisher Full Text
43. Shannon CE (1948) A mathematical theory of communication. Bell Syst Tech J 27:379-423
44. Silva DN, Talhinas P, Varzea V, Cai L, Paulo OS, Batista D (2012) Application of the Apn2/MAT locus to improve the systematics of the *Colletotrichum gloeosporioides* complex: an example from coffee (*Coffea* spp.). Mycologia 104:396-409
45. Slowinski J, Lawson R (2002) Snake phylogeny: evidence from nuclear and mitochondrial genes. Mol Biol Evol 24:194-202
46. Smith SA, Donoghue MJ (2008) Rates of molecular evolution are linked to life history in flowering plants. Science 322:86-89
47. Sutton BC (1992) The genus *Glomerella* and its anamorph *Colletotrichum*. In: Bailey JA, Jeger JJ (eds) Colletotrichum: Biology, Pathology and Control, Wallingford, UK: CAB International. pp 1-26
48. Swofford DL (2003) PAUP*. Phylogenetic Analysis Using Parsimony (*and Other Methods). Sinauer Associates, Sunderland, Massachusetts.
49. Takahashi K, Terai Y, Nishida M, Okada N (2001) Phylogenetic relationships and ancient incomplete lineage sorting among cichlid fishes in Lake Tanganyika as revealed by analysis of the insertion of retroposons. Mol Biol Evol 18:2057-2066
50. Tamura K, Peterson D, Peterson N, Stecher G, Nei M, Kumar S (2011) MEGA5: Molecular evolutionary genetics analysis using maximum likelihood, evolutionary distance, and maximum parsimony methods. Mol Biol Evol 28:2731-2739 Tanabe Y,

Watanabe MM, Sugiyama J (2002) Are *Microsporidia* really related to Fungi: a reappraisal based on additional gene sequences from basal fungi. Mycol Res 106:1380-1391

51. Tanabe Y, Saikawa M, Watanabe MM, Sugiyama J (2004) Molecular phylogeny of *Zygomycota*based on EF-1alpha and RPB1 sequences: limitations and utility of alternative markers to rDNA. Mol Phylogenet Evol 30:438-449 Taylor JW, Jacobson DJ, Kroken S, Kasuga T, Geiser DM, Hibbett DS, Fisher MC (2000) Phylogenetic species recognition and species concepts in Fungi. Fungal Genet Biol 31:21-32
52. Templeton MD, Rikkerink EHA, Solon SL, Crowhurst RN (1992) Cloning and molecular characterization of the glyceraldehyde-3-phosphate dehydrogenase-encoding gene and cDNA from the plant pathogenic fungus *Glomerella cingulata*. Gene 122:225-230
53. Thomas JA, Welch JJ, Woolfit M, Bromham L (2006) There is no universal molecular clock for invertebrates, but rate variation does not scale with body size. Proc Natl Acad Sci U S A 103:7366-7371
54. Townsend TM, Alegre RE, Kelley ST, Wiens JJ, Reeder TW (2008) Rapid development of multiple nuclear loci for phylogenetic analysis using genomic resources: An example from squamate reptiles. Mol Phylogenet Evol 47:129-142 van Oppen M, McDonald B, Willis B, Miller D (2001) The evolutionary history of the coral genus*Acropora* (*Scleractinia, Cnidaria*) based on a mitochondrial and a nuclear marker: reticulation, incomplete lineage sorting, or morphological convergence? Mol Biol Evol 18:1315-1329
55. Vialle A, Feau N, Allaire M, Diduk M, Martin F, Moncalvo J-M, Hamelin RC (2009) Evaluation of mitochondrial genes as DNA barcode for Basidiomycota. Mol Ecol Resour 9:99-113
56. Weir BS, Johnston PR, Damm U (2012) The *Colletotrichum gloeosporioides* species complex. Stud Mycol 73:115-180
57. Whelan S, Liò P, Goldman N (2001) Molecular phylogenetics: state-of-the-art methods for looking into the past. Trends Genet 17:262-272
58. White TJ, Bruns T, Lee S, Taylor JW (1990) PCR Protocols: A Guide to Methods and Applications. Academic Press Inc., NY.

59. Whitfield JB, Lockhart PJ (2007) Deciphering ancient rapid radiations. Trends Ecol Evol 22:258-265

60. Xia X, Xie Z (2001) DAMBE: software package for data analysis in molecular biology and evolution. J Hered 92:371-373

61. Xia X, Xie Z, Kjer K (2003) 18S ribosomal RNA and tetrapod phylogeny. Syst Biol 52:283-295

62. Xia X, Xie Z, Salemi M, Chen L, Wang Y (2003) An index of substitution saturation and its application. Mol Phylogenet Evol 26:1-7

Chapter 9

Effect of ATG Initiation Codon Context Motifs on the Efficiency of Translation of mRNA Derived from Exogenous Genes in the Transgenic Silkworm, *Bombyx mori*

Ken-ichiro Tatematsu, Keiro Uchino, Hideki Sezutsu, and Toshiki Tamura

Transgenic Silkworm Research Unit, National Institute of Agrobiological Sciences, Tsukuba, Ibaraki 305-8634, Japan

ABSTRACT

The context sequence motif surrounding the ATG initiation codon influences mRNA translation efficiency and affects protein production; however, the optimal sequence differs among species. To determine the optimal sequence for production of recombinant proteins in a transgenic silkworm, we compared 14-nucleotide context motifs around the ATG (ATG-context) in 50 silkworm genes and found the following consensus: (A/T)AN(A/T)ATCAAAatgN. We were also able to define the least-common motif: CCN(C/G) CGN(C/T/G)(G/C/T)(T/G) atgC, which served as a negative control. To examine the regulatory role of these motifs in protein expression, we constructed reporter plasmids containing different ATG-context motifs together with either the luciferase gene or an enhanced green fluorescent protein (EGFP) gene. These constructs were then used for comparison of luciferase reporter activity and EGFP production in BmN4 cells *in vitro* as well as in transgenic silkworms *in vivo*. We detected 10-fold higher luciferase activity in BmN4 cells transfected with the consensus ATG-context motif construct, compared to the negative control plasmid. ELISA measurements of EGFP translation products with the corresponding constructs in BmN4 cells showed consistently similar results. Interestingly, the translation efficiency of the novel consensus ATG-context motif did not show the highest activity in the transgenic silkworms *in vivo*, except for the fat body. The highest efficiency in the middle and posterior silk glands was produced by the sericin 1 context. Our results show that the ATG-context motifs differ among silkworm tissues. This result is important for the further improvement of the transgenic silkworm system for the production of recombinant proteins.

BACKGROUND

There is an increased demand to develop an efficient bioreactor for the production of recombinant proteins for pharmaceutical and/or diagnostic uses. Transgenic silkworms have several important properties, making them good candidates for such applications. Silk glands represent a highly efficient system for the production of large quantities of proteins, with a capacity of more than 500 mg of silk protein/larva. Also, the larval fat body is able to synthesize about 100 mg of hemolymph protein/larva. Other advantages include the low cost of silkworm rearing (less than 5 cents per larva) and the short time required for the generation of transgenic silkworms (60 days). Transgenic silkworms also allow the development of different protein production systems for various purposes (Tomita 2011;Tatemastu et al. 2012).

The current recombinant protein production systems using transgenic silkworms utilize mainly silk glands. The silk gland secretory products are categorized into two groups—sericins and fibroins. The sericins are glue proteins coating the surface of the silk thread and contribute as much as 25% of the cocoon silk. The fibroins form the silk thread and constitute the remaining 75% of the cocoon silk proteins. The sericins are produced in the middle silk gland (MSG) region, while the fibroins are secreted in the posterior part of the silk glands (PSG). The sericins and fibroins have different properties; the sericins are easily dissolved in ordinary buffer, but dissolving fibroins requires the use of strong protein-denaturing agents. The protein secretory mechanisms also differ in the MSG and PSG. The recombinant proteins produced in the MSG are secreted more easily compared to those in the PSG. For example, intact human collagen produced in the PSG is not transferred to the lumen, whereas there is no problem with its secretion in the MSG (Adachi et al.2010;Tomita 2011). Therefore, the MSG as a production system is more useful for proteins, which require easy purification without losing their biological activity, whereas the PSG seems to be more suitable for the production of specialized proteins, like modified silks (Tatemastu et al. 2012). Consistently, human serum albumin, mouse IgG antibody, and full-length collagen have been produced in MSGs (Ogawa et al. 2007;Iizuka et al. 2009;Adachi et al. 2010). Projects involving the production of fluorescent color proteins, spider silks, cytokine, human growth factor and mini-collagen as a fusion protein with fibroin H or L chains were performed in PSGs (Iizuka et al.2013;Teule et al. 2012;Kurihara et al. 2007;Tomita et al. 2003). The production of human μ-opioid receptor was also reported in the fat body (Tateno et al. 2009). Although the production of recombinant proteins in transgenic silkworms has already been established, problems regarding increasing productivity and posttranslational modifications need to be addressed (Tatemastu et al.2012).

The sequence context motif surrounding the ATG initiation codon (ATG-context) is an important factor that increases protein production (Kozak 1987a 1987b;Cavener 1987;Cavener and Ray1991;Ranjan and Hasnain 1995;Sugio et al. 2010;Agarwal et al. 2009;Cherbas and Cherbas 1993). However, there is no study on the effects of this sequence in the recombinant protein production systems of the transgenic silkworm. The optimal ATG-context motifs that show the highest efficiency in mRNA translation differ among protozoa, yeasts, vertebrates, invertebrates, and plants (Seeber 1997;Kozak 1984 1991;Joshi et al. 1997;Cavener and Ray 1991;Mankad et al.1998). In vertebrates, the 13-base nucleotide sequence surrounding the ATG is important, and has been determined to be the GCCGCC(A/G) CCatgG motif, also called the Kozak sequence (Kozak1987a 1987b). Positions -3 and +4 (the A of the ATG initiation codon is marked as +1) are the most important nucleotide positions of the Kozak sequence; mutation of the nucleotides at these positions significantly reduces the efficiency of translation (Kozak 1986). The optimal sequence context in insects is completely different from that in vertebrates. In *Drosophila melanogaster*, the consensus sequence surrounding ATG is (C/A) AA(A/C)atg (Cavener 1987) and reports of the function of the context sequence are rare, with the exception of one report regarding the investigation of the effects of point mutations in the ATG-context motif of the *Drosophila* alcohol dehydrogenase gene (Feng et al. 1991). In that report, it was shown that the mutant containing the A to T substitution at position -3 showed a 2.4-fold reduction in translation efficiency, whereas five other mutations in the ATG-context motif showed 5.9- to 12.5-fold reductions compared to the original ATG-context. In Lepidoptera, sequence contexts were examined only in baculoviral protein expression systems. It was reported that the consensus ATG-context motifs in *Spodoptera frugiperda* and *B. mori* are (A/G)NC(C/T) N(A/C)CA(A/C)(C/G)atg(G/A) and ANCAAAatg, respectively (Sano et al. 2002;Chang et al. 1999).

In the present study, we compared the 14-base sequence context motif at positions -10 to +4 of the ATG initiation codon of 50 *B. mori* genes and determined the consensus as well as the least-common motif. The observed ATG-context motifs differed from those of vertebrates, but were similar to those of *D. melanogaster* and Lepidoptera described earlier (Cavener 1987;Chang et al.1999;Sano et al. 2002). To evaluate their effects, we constructed a series of expression vectors containing luciferase or EGFP reporter genes together with various ATG-context motifs, including the novel *B. mori* consensus, variations thereof, a sericin 1 gene ATG-context, and the vertebrate Kozak sequence. Finally, we examined the effects of these ATG-context motifs on the production of the recombinant proteins in cultured cells and in the transgenic silkworms.

RESULTS

Consensus Sequence Surrounding the ATG Initiation Codon in 50 b. Mori Genes

To determine the consensus ATG-context motif we compared the 14-base initiation codon context motifs of genes encoding fibroin H chain, fibroin L chain, fibrohexamerin, sericin 1, and 46 other randomly selected *B. mori* genes. The consensus motif obtained from the alignment showed that 'A' was the most frequent nucleotide at positions -10, -9, -7, -6, -3, -2, and -1 (A of the ATG initiation codon corresponds to +1) within the context sequences of -10 to -1 using the binominal test (Figure 1). Specifically, the frequencies of 'A' at positions -3, -2 and -1 were higher than 50% and each was identified as a solo consensus nucleotide position by analysis of the 50/75 rule (Cavener 1987), while nucleotide 'A' at -7 was considered a co-consensus. From the analysis, the consensus motif (PC) was determined as (A/T)AN(A/T) ATCAAAatgN and the least-common motif, serving as a negative control (NC), was CCN(C/G)CGN(C/T/G)(G/C/T)(T/G)atgC.

	-10	-9	-8	-7	-6	-5	-4	-3	-2	-1	A	T	G	+4
A(%)	38	40	32	48	48	26	26	74	60	52	100	0	0	26
C(%)	8	16	28	12	14	26	34	0	14	24	0	0	0	16
G(%)	20	20	20	12	18	10	18	16	10	16	0	0	100	28
T(%)	34	24	32	28	20	38	22	10	16	8	0	100	0	30
lesst common sequence	C	C	N	C/G	C	G	N	C/T/G	G/C/T	T/G				C
consensus sequence	A/T	A	N	A/T	A	T	C	A	A	A				N

Figure 1: Compilation of the 14-base sequence context of the ATG initiation codon of *B. mori* genes. Sequences surrounding the ATG initiation codon were

compiled from 50 genes in GenBank. For reference, the ATG initiation codon corresponds to +1 through +3. The frequency of each nucleotide at each position is presented as a percentage. Solo consensus (italicized letter) and co-consensus nucleotides (underlined letter) were determined by the 50/75 rule (Cavener 1987). The consensus and least-common nucleotides determined by a binominal test are indicated by red and blue boxes, respectively. The consensus and least-common nucleotides were determined using Cavener's 50/75 rule and/or by a binominal test and are shown at the bottom. The accession numbers of genes used for the analysis are as follows: AY769299, BMU06073, D10953, D90454, DQ311154, DQ311189, DQ311242, DQ311250, DQ311264, DQ311306, DQ311321, DQ311322, DQ311328, DQ311332, DQ311333, DQ311340, DQ311341, DQ311350, DQ311356, DQ311358, DQ311360, DQ311363, DQ311365, DQ311378, DQ311384, DQ311386, DQ311388, DQ311397, DQ311402, DQ311407, DQ311412, DQ311418, DQ311430, DQ311436, DQ311438, DQ311439, DQ343760, DQ358079, DQ424947, M64336, NM_001044023, NM_001044041, NM_001113262, NM_001145941, S74376, S77508, U30289, U94993, X74320, and X95604.

Effects of ATG Initiation Codon Sequence Context on Reporter Genes in BmN4 Cells

To examine the effects of the ATG-context motif, we designed a number of variants of context sequences (shown in Figure 2a). The context motifs PC and NC were selected to represent the consensus *B. mori* context and the least-common motif, respectively. 7N3P and 7P3N were designed to compare the effects of nucleotides from -1 to -3 and -4 to -10 between the least-common and consensus motifs, respectively. AAT, AGT, CAA, and CGA were designed to identify the most important nucleotide at positions -1 to -3. Ser1 is the context sequence from the sericin 1 gene. Vert3 and Vert9 were generated to determine the effects of vertebrate Kozak sequences at positions -1 to -3 and -1 to -9. The context sequences of atgA, atgC, and atgT were designed to examine the effects of the nucleotides at position +4.

a

Name of context sequence (X)	Context sequence
NC	CCgGCCgCGTATGg
PC	AAaAATCAAAATGg
7N3P	CCgGCCgAAAATGg
7P3N	AAaAATCCGTATGg
AAT	CCgGCCgAATATGg
AGT	CCgGCCgAGTATGg
CAA	CCgGCCgCAAATGg
CGA	CCgGCCgCGAATGg

Name of context sequence (X)	Context sequence
SerI	gACCgcCAAcATGg
Vert3	ccgGCCgACcATGg
Vert9	CgcCgcCACcATGg
atgA	CCgGCCgAATATGa
atgC	CCgGCCgAATATGc
atgT	CCgGCCgAATATGt

b

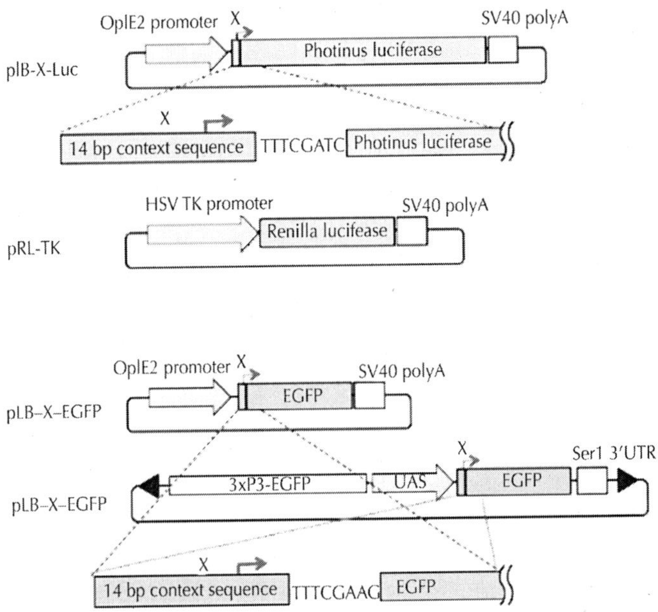

Figure 2: Construction of expression vectors.a. 14-base context motifs used in the experiment. Red and blue letters indicate the most- and least-frequent nucleotides at each position, respectively. Capital letters indicate the nucleotide of the consensus or the least-common nucleotide at each position. b. Physical maps of plasmids including a sequence around the translation initiation site of luciferase and EGFP constructs. X represents the newly designed context motif and the arrow points to translation initiation.

We constructed vectors containing the *Photinus* luciferase gene with various context motifs (Figure 2b). When the *Photinus* luciferase vector was transfected together with the *Renilla*luciferase vector (internal control) into BmN4 cells and the activity was measured

(Table 1), the*Renilla* luciferase activity appeared as constant values in each experiment, whereas the *Photinus*luciferase activity varied in the experiments, indicating that the transfection efficiency was similar in each experiment and that the difference in *Photinus* luciferase activity was caused by the variations in the context motifs. When we compared the normalized activity of the constructs (by the relative value of *Photinus* luciferase activity/*Renill*a luciferase activity), the highest activity was found in the PC-Luc construct, while NC-Luc had the lowest activity (Table 1). The activity of PC-Luc was approximately 10-fold higher than that of NC-Luc, indicating that the context motif is a critical factor affecting the activity of *Photinus* luciferase in BmN4 cells. 7N3P-Luc and 7P3N-Luc constructs demonstrated similar activities, which were ~65% of the PC-Luc activities, suggesting that the sequences from position -1 to -3 and -4 to -10 have similar effects, and that both sequences are important for luciferase expression. To examine the effects of 'A' at positions -1, -2, and -3 of the sequence context, five context motifs (7N3P, AAT, AGT, CAA, and CGA) were compared. Mutations of nucleotide 'A' at any of these positions reduced the activity, and a prominent reduction occurred when 'A' at position -3 was mutated, indicating that this position was the most important for expression. The effect of the nucleotide at position +4 was examined by comparing the effects among four sequence contexts (AAT, atgA, atgC, and atgT), but no significant differences were found among the four. The two sequence contexts derived from the Kozak sequence and the Ser1 context motif showed intermediate levels of activity, indicating that in BmN4 cells their effects on expression are weaker than that of the *B. mori* consensus motif.

Table 1: Effects of the ATG initiation codon sequence context of *Photinus* luciferase constructs on *Photinus* luciferase expression in BmN4 cells

Context name	***Photinus* luciferase activity (×10^4) (A)**	***Renilla* luciferase activity (×10^2) (B)**	***Photinus*/ *Renilla* (A/B)**	**Fold activity (%)**
NC-Luc	8.3±1.1	4.26±0.7	196.5±15.0	100
PC-Luc	88.3±26.6	4.34±0.5	2,008.7±393.1	1,022
7N3P-Luc	55.9±20.4	4.23±0.5	1,307.5±393.6	665
7P3N-Luc	56.8±10.2	4.49±0.8	1,272.5±184.9	648

AAT-Luc	38.7±27.4	4.44±0.6	834.6±539.9	425
AGT-Luc	39.4±6.7	4.28±0.6	941.6±252.9	479
CAA-Luc	15.0±2.7	4.80±0.4	316.8±84.3	161
CGA-Luc	11.1±2.4	4.82±0.4	228.2±30.8	116
Ser1-Luc	45.0±24.1	4.68±0.6	928.2±373.2	472
Vert3-Luc	25.5±3.1	4.66±0.3	547.9±61.0	279
Vert9-Luc	36.9±6.6	4.81±0.3	775.4±173.8	395
atgA-Luc	24.3±16.9	4.47±0.5	550.6±413.8	280
atgC-Luc	34.3±2.1	4.56±0.5	759.1±102.1	386
atgT-Luc	27.9±12.5	4.58±0.8	594.0±171.2	302

Values represent the mean ± standard deviation (SD) obtained from three individual experiments.

The fold efficiency is indicated as the activity of NC equal to 100. Measurement of each luciferase activity was performed at least three times. The value of *Photinus*luciferase activity/*Renilla* luciferase activity was calculated in each measurement.

Tatematsu *et al.*

To investigate the effect of the context motif on translational efficiency, we transfected EGFP constructs bearing 11 different sequence contexts (Figure 2a; Table 2) into BmN4 cells. Transfected cells were divided into two aliquots: one to measure the amount of EGFP mRNA and the other to measure the amount of EGFP protein (Table 2). In the experiment, the amounts of EGFP mRNA and ribosomal protein 49 (rp49) mRNA (internal control) were not significantly different. This indicates that the efficiency of transcription in each construct was constant, and that the difference in the amount of EGFP protein was caused by the translational efficiency of the sequence context of EGFP mRNA. When we compared the translational efficiency of various constructs, the highest amount of EGFP protein was observed in the PC-EGFP transgene, which was ~4-fold higher than that of NC-EGFP. This indicates that the sequence context of the ATG initiation codon significantly affects translational efficiency. The translational efficiencies of 7N3P-EGFP and 7P3N-EGFP showed intermediate values between PC-EGFP and NC-EGFP. In a comparison of 7N3P-EGFP, AAT-EGFP, AGT-EGFP, CAA-EGFP, and CGA-EGFP constructs, the CGA construct demonstrated lower activity compared to the other four, indicating that the mutation of 'A' at positions -3 and -2 resulted in a significant reduction in translational efficiency. The Ser1-EGFP, Ver3-EGFP, and Ver9-EGFP constructs showed intermediate efficiencies, indicating that these context motifs were suboptimal for the initiation of translation in BmN4 cells. These results were similar to those obtained with the *Photinus* luciferase construct; the optimal context for luciferase and EGFP construct expression in BmN4 cells was the *B. mori* consensus motif.

Table 2: Effects of the ATG initiation codon sequence context of EGFP constructs on translational efficiency of EGFP in cultured cells

Context name	EGFP protein (μg/well) (A)	EGFP mRNA ($\times 10^{-6}$) (pmol/ng) (B)	rp49 mRNA ($\times 10^{-7}$) (pmol/ng) (C)	EGFP/ rp49 mRNA (B/C)	Translational efficiency (A/{B/C})	Fold efficiency (%)
NC-EGFP	1.60±0.5	4.31±1.2	6.99±0.17	6.15±1.5	0.28±0.13	100
PC-EGFP	6.37±1.7	3.97±1.5	6.68±0.57	5.92±1.9	1.20±0.60	424
7N3P-EGFP	4.15±0.6	3.47±1.3	7.00±0.17	4.92±1.7	0.94±0.41	341
7P3N-EGFP	3.54±0.7	3.50±1.1	6.81±0.23	5.14±1.6	0.76±0.34	274
AAT-EGFP	4.58±1.2	4.32±1.2	6.89±0.47	6.22±1.5	0.79±0.34	286
AGT-EGFP	2.72±0.7	3.43±1.3	6.82±0.25	5.00±1.7	0.61±0.29	216
CAA-EGFP	4.04±1.0	4.26±1.2	6.63±0.14	6.43±1.9	0.68±0.30	248
CGA-EGFP	1.20±0.2	3.21±1.4	6.93±0.50	4.60±1.8	0.30±0.14	106
Ser1-EGFP	3.17±0.7	3.40±1.0	6.74±0.31	5.03±1.4	0.68±0.28	248
Vert3-EGFP	2.87±0.7	3.67±1.2	7.20±0.27	5.09±1.6	0.62±0.26	224
Vert9-EGFP	2.84±0.3	3.30±1.5	7.02±0.49	4.65±1.8	0.68±0.27	251

Values represent the mean±SD obtained from more than three individual experiments. The fold efficiency is indicated as the activity of NC equal to 100. Measurements of the amount of protein and mRNA were performed at least three times. Values of relative EGFP mRNA and translational efficiency were calculated in each measurement.

Tatematsu *et al.*

Effects of ATG-Context Motifs on the Efficiency of mRNA Translation in MSG, PSG, and Fat body of Transgenic Silkworm

To examine whether the results obtained in BmN4 cells can be applied to the production of recombinant protein in the transgenic silkworm, we generated transgenic silkworms containing the EGFP constructs with the NC, PC, AAT, and Ser1 context motifs using the *piggyBac* vector pBac-X-EGFP, as shown in Figure 2b. We chose the NC, PC, and AAT context motifs because they showed the lowest, highest, and intermediate efficiencies in BmN4 cells, respectively. We also examined the Ser1 context motif originating from the sericin 1 gene, because we expected it to be adapted for the MSG translation. We generated three lines of transgenic silkworms for NC-EGFP and two lines each for PC-EGFP, AAT-EGFP, and Ser1-EGFP by the ordinary methods using the transposon*piggyBac* as a vector. Because the reporter gene is under the control of the UAS sequence (Figure 2b), the EGFP gene is not expressed without the presence of GAL4 protein. To express these four constructs in different tissues, each transgenic strain was mated with the Ser1-GAL4 (Tatematsu et al. 2010), FibH-GAL4 (Sezutsu et al. 2009), or 30 k-GAL4 lines (H. Sezutsu, personal communication), which express the GAL4 gene at the fifth instar in the MSG, PSG, and the fat body, respectively. Silkworms that possessed both the UAS-X-EGFP (Figure 2b) and GAL4 constructs were easily selected by observing the larval stemmata because the UAS and GAL4 constructs were marked with 3 × P3-EGFP and 3 × P3-DsRed, respectively (Figure 3a). When comparing the effects of the four context sequences on EGFP expression, the transgenic strains with the NC-EGFP/Ser1-GAL4, PC-EGFP/Ser1-GAL4, and AAT-EGFP/Ser1-GAL4 constructs showed weak, strong, and intermediate levels of EGFP fluorescence in MSG, respectively (Figure 3b). The strongest fluorescence was observed in the Ser1-EGFP/Ser1-GAL4 silkworm in the MSG. Similar results were observed in PSG (Figure 3b); regarding the level of expression among the four constructs, the highest was Ser1-EGFP/FibH-GAL4, the lowest was NC-EGFP/FibH-GAL4, and the PC-and AAT-EGFP constructs had intermediate expression levels in the PSG. However, the effect of the sequence context on expression in the fat body was different (Figure 3b); the PC construct showed higher activity than that of Ser1-EGFP, and the expression levels of the NC and AAT constructs were much lower than those of PC and Ser1.

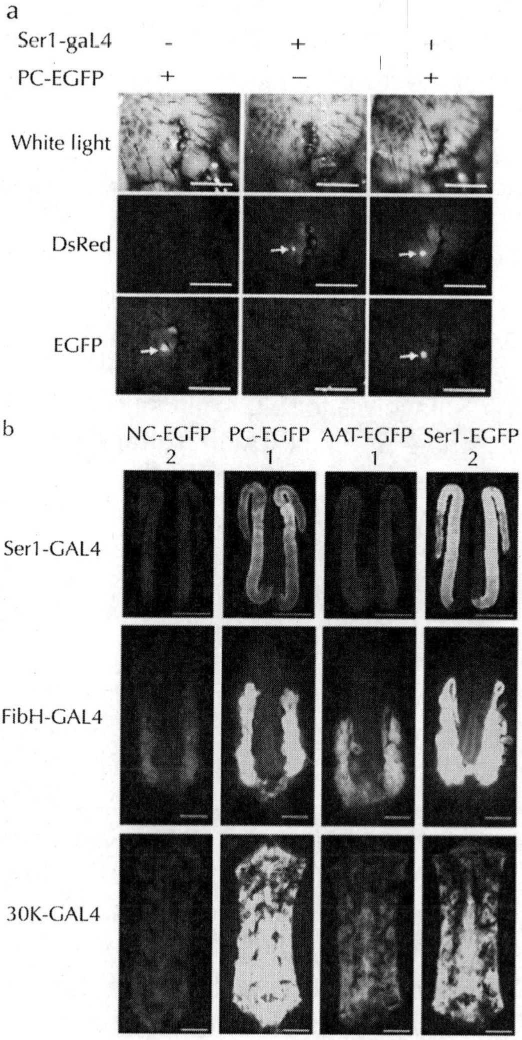

Figure 3: Expression of EGFP in the transgenic silkworm. a. Stemmata of the fifth instar larva. UAS-EGFP constructs with NC, PC, AAT, and Ser1 sequence contexts of the initiation codon were marked with a 3 × P3-EGFP marker construct allowing the expression of EGFP in the stemmata. The GAL4 driver construct under the control of the sericin 1, fibroin H chain, or 30 k protein gene promoter was linked with a marker construct, 3 × P3-DsRed, allow-

ing the expression of DsRed in the stemmata. The stemmata of a larva harboring PC-EGFP, Ser1-GAL4, or both PC-EGFP and Ser1-GAL4 are shown. The images were taken under white light or under a fluorescence microscope equipped with an EGFP or DsRed filter. Arrows indicate the stemmata with EGFP or DsRed fluorescence. Scale bar: 1 mm. b. EGFP expression in the MSG, PSG, and the fat body of the transgenic silkworm with different GAL4 driver constructs. MSG, PSG, and the fat body on the sixth day of the fifth instar are shown. Images were taken under a fluorescence microscope equipped with an EGFP filter. Expression of UAS-EGFP with different context motifs was only observed in silkworms with GAL4 drivers. The numbers above the tissue photographs indicate the transgenic silkworm lines. Scale bar: 10 mm.

To further analyze the effects of the context sequence, we measured the amounts of EGFP protein and mRNA in the MSG (Table 3), PSG (Table 4), and fat body (Table 5), and compared the translational efficiency among the four constructs. In the MSG, the highest translational efficiency was observed for the Ser1-EGFP construct and the lowest in NC-EGFP. PC-EGFP also showed very high translational efficiency, while AAT-EGFP was intermediate. The translational efficiency of Ser1-EGFP was ~20-fold higher than NC-EGFP, suggesting that the context motif of the ATG initiation codon is a critical factor for increasing the yield of recombinant proteins in transgenic silkworms. An amount of the EGFP protein proportional to the translational efficiency was produced from each construct in the MSG. The rank order of protein amounts was Ser1-, PC-, AAT- and NC-EGFP. Similar results were observed in the PSG with the same rank order of individual constructs: Ser1 > PC > AAT > NC-EGFP. The increase in the translational efficiency of Ser1-EGFP was 6-fold in the PSG when compared to the lowest-activity construct, NC-EGFP. In addition, the translational efficiency of each construct in the PSG was much higher than that in the MSG, and the protein yield was almost proportional to the translational efficiency. In the fat body, the rank order of the translational efficiency was PC > Ser1 > AAT > NC-EGFP. PC-EGFP was almost 50-fold higher in the translational efficiency compared to NC-EGFP. The EGFP protein yield was proportional to this order, suggesting that the context sequence was critical for recombinant protein production. NC-EGFP demonstrated the lowest EGFP protein production and translational efficiency in all tissues examined. Although PC-EGFP showed the highest EGFP protein production and translational efficiency in the fat body, the construct with the highest efficiency in the MSG and PSG was Ser1-EGFP. Thus, the effects of each sequence context on EGFP expression in the fat body were similar to those observed in BmN4 cells, but differed from the effects in MSG and PSG.

Table 3: Effects of sequence context of the ATG initiation codon on translational efficiency of EGFP expressed in MSG

	Line no.	EGFP protein (μg/larva) (A)	EGFP mRNA ($\times 10^{-7}$) (pmol/ng) (B)	rp49 mRNA ($\times 10^{-7}$) (pmol/ng) (C)	EGFP/rp49 mRNA (B/C)	Translational efficiency (A/{B/C})	Fold efficiency (%)
NC	1	31.4±5.8	70.8	7.83	9.05	3.47	
-EGFP	2	16.8±2.4	44.1	7.26	6.07	2.76	
	3	12.6±1.7	49.9	7.70	6.48	1.95	
	Average	20.3	54.9	7.60	7.23	2.73	100
PC	1	130.2±9.9	19.8	6.96	2.85	45.7	
-EGFP	2	158.9±12.1	29.6	6.71	4.40	36.1	
	Average	144.5	24.7	6.84	3.61	40.9	1,498
AAT	1	76.8±7.2	48.7	9.04	5.39	14.2	
-EGFP	2	89.0±12.9	34.4	8.14	4.23	21.0	
	Average	82.9	41.6	8.59	4.84	17.6	645
Ser1	1	276.3±36.4	40.8	8.14	5.01	55.1	
-EGFP	2	261.4±18.4	37.3	7.76	4.80	54.4	
	Average	268.9	39.0	7.95	4.91	54.8	2,007

Values obtained from each line and averages of lines are shown. The fold efficiency is indicated as the efficiency of the average of NC equal to 100.

Tatematsu *et al.*

Table 4: Effects of sequence context of the ATG initiation codon on translational efficiency of EGFP expressed in PSG

	Line no.	EGFP protein (μg/larva) (A)	EGFP mRNA ($\times 10^{-7}$) (pmol/ng) (B)	rp49 mRNA ($\times 10^{-7}$) (pmol/ng) (C)	EGFP/rp49 mRNA (B/C)	Translational efficiency (A/{B/C})	Fold efficiency (%)
NC	1	12.9±0.8	4.20	3.95	1.06	12.1	
-EGFP	2	9.9±0.3	1.73	3.54	0.49	20.2	
	3	12.6±0.7	0.48	3.60	0.14	93.0	
	Average	11.8	2.14	3.70	0.56	41.8	100
PC	1	80.6±15.3	1.35	2.93	0.46	175.5	
-EGFP	2	104.8±24.2	2.14	3.65	0.59	178.1	

	Average	92.7	1.75	3.29	0.52	176.8	423
AAT	1	25.5±3.6	1.64	3.23	0.51	50.3	
-EGFP-EGFP	2	53.5±15.8	0.84	3.50	0.24	220.5	
	Average	39.5	1.24	3.37	0.37	135.4	324
Ser1	1	94.4±10.4	1.53	3.52	0.44	216.9	
-EGFP	2	106.2±34.7	1.15	3.09	0.37	284.9	
	Average	100.3	1.34	3.31	0.40	250.9	600

Values obtained from each line and averages of lines are shown. The fold efficiency is indicated as the efficiency of the average of NC equal to 100.

Tatematsu *et al.*

Table 5: Effects of sequence context of the ATG initiation codon on translational efficiency of EGFP expressed in the fat body

	Line no.	EGFP protein (μg/larva) (A)	EGFP mRNA ($\times 10^{-7}$) (pmol/ng) (B)	rp49 mRNA ($\times 10^{-7}$) (pmol/ng) (C)	EGFP/rp49 mRNA (B/C)	Translational efficiency (A/{B/C})	Fold efficiency (%)
NC	1	4.9±0.6	25.8	5.15	5.01	0.98	
-EGFP	2	4.4±0.1	13.7	3.14	4.37	1.01	
	3	1.3±0.1	21.6	3.86	5.61	0.23	
	Average	3.5	20.4	4.05	5.04	0.70	100
PC	1	51.8±3.0	11.5	6.13	1.87	27.6	
-EGFP	2	70.2±17.3	12.4	6.89	1.80	39.0	
	Average	61.0	11.9	6.51	1.83	33.3	4,757
AAT	1	11.9±0.1	15.6	4.20	3.72	3.20	
-EGFP	2	4.2±0.2	11.8	4.83	2.44	1.71	
	Average	8.0	13.7	4.52	3.03	2.65	379
Ser1	1	44.6±3.8	35.3	6.20	5.70	7.82	
-EGFP	2	50.6±2.5	11.5	6.08	1.90	26.7	
	Average	47.6	23.4	6.14	3.82	12.5	1,785

Values obtained from each line and averages of lines are shown. The fold efficiency is indicated as the efficiency of the average of NC equal to 100.

Tatematsu *et al.*

DISCUSSION

We performed a compilation of the 14-nucleotide sequence motifs surrounding the ATG initiation codon of 50 *B. mori* genes and determined the consensus ATG context motif (A/T)AN(A/T)ATCAAAatgN, as well as the least-common ATG context CCN(C/G)CGN(C/T/G)(G/C/T)(T/G) atgC, which served as the control. The novel consensus motif was consistent with the previously reported consensus motifs of *B. mori* (Chang et al. 1999 (Cavener) and*Drosophila*1987), but differed from that of vertebrates (Kozak 1987a 1987b). When we compared the previously reported *B. mori* consensus motif (ANCAAAatgNNN) with our novel consensus sequence, a significant difference was found in the frequency of 'T' at position -5. The frequency of 'T' in our results was significantly higher than would be expected for a random occurrence (Figure 1). Our *B. mori* consensus motif was slightly more similar to the *D. melanogaster*consensus motif, (C/A)AA(A/C)atg, reported by Cavener (1987). When we reanalyzed Cavener's original data using our method, we received the consensus AANAAN(C/A)AA(A/C)atg. Differences between our *B. mori* consensus motif and the reanalyzed *D. melanogaster* consensus motif were at positions -1, -4, -5, -7, and -10.

We also determined which nucleotides of the consensus context motifs were important for efficient translation. Our results show that 'A' at position -3 was the most important nucleotide; substitution of this nucleotide caused a significant reduction in efficiency. The region of -10 to -4 also displayed a significant effect, while the nucleotide at position +4 did not have any detectable effect. Our results differ from those of Chang et al. (1999), who found that the nucleotides at positions -1 to -6 had no effect while those at +4 to +6 significantly increased the efficiency. Furthermore, we found that our new consensus motif of *B. mori* genes was more efficient than the Kozak sequence. However, Sano et al. (2002) reported that the use of the 5'-untranslated region including the Kozak sequence dramatically increased expression levels in baculovirus-infected cells. In both of these studies, baculovirus was used for transgene expression, whereas we used transient expression of plasmid. The differences between our results and previously reported data may be due to the different expression systems used in experiments; further studies are required to determine the exact reasons for the differences.

Three features differed between the vertebrate consensus Kozak motif and our *B. mori* consensus motif. First, the vertebrate ATG context shows a strong preference for nucleotide position +4 (Kozak 1987a 1987b 1997), but not in *B. mori*. Second, the vertebrate consensus motif is CG-rich throughout the sequence, but A-rich in *B. mori*. Third, three upstream triplet repeats, RCCRCCRCCatg, which are important for ribosomal recognition, are present in vertebrates (Kozak1987b), while no such repeats were found in our results, suggesting that mRNA recognition by the ribosome differs between insects and vertebrates.

The assays of context motif effects demonstrated that these sequences significantly affect the efficiency of translation initiation in BmN4 cultured cells, as well as the production of recombinant proteins in the MSG, PSG, and fat body of transgenic silkworms. This is the first report that optimization of context motifs is important for efficient recombinant protein production in transgenic silkworms. Such optimization is simple and does not affect the protein sequence. In addition, context motifs can be used for the suppression of translation initiation when a reduction in protein production is required. Furthermore, context motifs may facilitate precise translation initiation. Kanamori et al. (2010) reported that a leaky scanning mechanism for translation initiation sometimes causes the utilization of an internal ATG codon in *B. mori* cells.

In our experiment, the influence of context motifs on translational efficiency varied depending on the tissue. We designed NC and PC sequence contexts that represented the least-common and consensus sequences, respectively. The translational efficiency of the NC sequences was the lowest in all tissues, whereas the PC sequences demonstrated the highest efficiency in BmN4 cells and the fat body, but not in the MSG and PSG. The BmN4 cell line is derived from silkworm ovaries; however, it may retain similar characters to fat body cells because dexamethasone treatment induces accumulation of lipid in the cells (Akiduki and Imanishi 2007). Feng et al. (1991) reported that the effects of context sequences on translation in *D. melanogaster* were stage-dependent. These results suggest that stage- and tissue-specific adjustment of context motifs is needed for the maximum production of recombinant proteins. Silk glands represent an organ that is highly specialized for massive production of several secretory proteins within a short time interval (Julien et al. 2005). Our data show that this specialization also involves the adjustment of ATG initiation codon context motifs.

The optimization of different tissue-specific expression systems in transgenic silkworms is important from the viewpoint of posttranslational modifications, including glycosylations. The recombinant proteins produced by insect cells generally carry paucimannose-type n-glycans (Harrison and Jarvis 2006). The proteins produced by tissues other than the silk gland also have paucimanmose- or highmannose-type n-glycans (Tomita 2011). However, the recombinant proteins produced in the MSG contain N-acetylglucosaminylated complex N-glycans (Iizuka et al.2009). The optimization of the ATG context motifs is part of a large project developing recombinant protein production systems adapted for different tissues.

In conclusion, we identified the ATG context motif consensus in the silkworm *B. mori*, and showed that optimization of the sequence context is useful for increased production of recombinant protein in cultured cells and in transgenic silkworms. The context motifs best adapted to high production in cultured cells and the fat body differed from those in the MSG and PSG. Therefore, different optimized context sequences for different tissues may be required for the maximum expression of transgenes.

METHODS

Construction of Expression Vectors

To generate the plasmid pIB-X-EGFP (Figure 2b), the EGFP gene amplified from the plasmid pBac[UAS-ser_sig-EGFP/3 × P3-EGFP] (Tatematsu et al. 2010) using the primers EGFP-BstBI-U and EGFP Stop(+)L (Table 6) was inserted into the *Bam*HI-*Eco*RI site of pBluescript SK(-) (TOYOBO, Osaka, Japan). The constructed plasmid, pEGFP/pBS, contained the T*TTCGAA*G (italicized letter, BstBI site) sequence downstream of the context sequence instead of the original EGFP sequence TGAGCAAG (this sequence change did not alter the amino acid sequence of EGFP). The adapter, except for IB-BlnI in Table 6, was inserted into the *Bst*BI-*Bam*HI site of the plasmid pEGFP/pBS to generate the plasmid pBS-X-EGFP. Simultaneously, the IB-BlnI adapter

(Table 6) was inserted into the *Hind*III-*Kpn*I site of the plasmid pIB-V5/His (Life Technologies, Carlsbad, CA, USA) to introduce a *Bln*I site, and the resultant plasmid pIB-V5/His_BlnI was obtained. Then, the NheI fragment from pBS-X-EGFP was inserted into the *Bln*I site of the pIB-V5/His_BlnI. To generate the plasmid pBac-X-EGFP (Figure 2b), the NheI fragment from the plasmid pBS-X-EGFP was inserted into the *Bln*I site of the plasmid pBac[SerUAS/3 × P3-EGFP] (Tatematsu et al. 2010).

Table 6: List of primers and adapters used in the experiments

Name	Sequences
Primer	
EGFP-BstI-U	5′-GGGATCCGCTAGCACCATGGTTTCGAAGGGCGAG-3′
EGFP-Stop(+)-L	5′-GGAATTCGGACCGCTAGCTTACTTGTACAGCTCG-3′
pGL3-U	5′-GGATCGATCAAAAACATAAAGAAAGGCCCGGCG-3′
pGL3-L	5′-GGGCTAGCTTACACGGCGATCTTTCCGCCCTTC-3′
EGFPLCU2	5′-AACTTCAAGATCCGCCACAACATCGAGGAC-3′
EGFPLCL2	5′-AGGACCATGTGATCGCGCTTCTCGT-3′
rp49LCF2	5′-GGATCGCTATGACAAACTTAAGAGGA-3′
rp49LCR1	5′-TGCTGGGCTCTTTCCACGA-3′
Adapter	
NC	5′-GATCCGCTAGCCCGGCGGCGTATGGTTT-3′
	3′-GCGATCGGGCCGCCGCATACCAAAGC-5′
PC	5′-GATCCGCTAGCAAAAATCAAAATGGTTT-3′
	3′-GCGATCGTTTTTAGTTTTACCAAAGC-5′
7N3P	5′-GATCCGCTAGCCCGGCGGAAAATGGTTT-3′
	3′-GCGATCGGGCCGCCTTTTACCAAAGC-5′
7P3N	5′-GATCCGCTAGCAAAAATCCGTATGGTTT-3′
	3′-GCGATCGTTTTTAGGCATACCAAAGC-5′
AAT	5′-GATCCGCTAGCCCGGCGGAATATGGTTT-3′
	3′-GCGATCGGGCCGCCTTATACCAAAGC-5′
AGT	5′-GATCCGCTAGCCCGGCGGAGTATGGTTT-3′
	3′-GCGATCGGGCCGCCTCATACCAAAGC-5′
CAA	5′-GATCCGCTAGCCCGGCGGCAAATGGTTT-3′
	3′-GCGATCGGGCCGCCGTTTACCAAAGC-5′

CGA	5′-GATCCGCTAGCCCGGCGGCGAATGGTTT-3′
	3′-GCGATCGGGCCGCCGCTTACCAAAGC-5′
Ser1	5′-GATCCGCTAGCGACCGCCAACATGGTTT-3′
	3′-GCGATCGCTGGCGGTTGTACCAAAGC-5′
Vert3	5′-GATCCGCTAGCCCGGCGGACCATGGTTT-3′
	3′-GCGATCGGGCCGCCTGGTACCAAAGC-5′
Vert9	5′-GATCCGCTAGCCGCCGCCACCATGGTTT-3′
	3′-GCGATCGGCGGCGGTGGTACCAAAGC-5′
atgA	5′-GATCCGCTAGCCCGGCGGAATATGATTT-3′
	3′-GCGATCGGGCCGCCTTATACTAAAGC-5′
atgC	5′-GATCCGCTAGCCCGGCGGAATATGCTTT-3′
	3′-GCGATCGGGCCGCCTTATACGAAAGC-5′
atgT	5′-GATCCGCTAGCCCGGCGGAATATGTTTT-3′
	3′-GCGATCGGGCCGCCTTATACAAAAGC-5′
IB-BlnI	5′-AGCTTCCTAGGGTCGACGGTAC-3′
	3′-AGGATCCCAGCTGC-5′

Tatematsu *et al.*

To generate pIB-X-Luc constructs (Figure 2b), the *Photinus* (firefly) luciferase gene was amplified from the plasmid pGL3 (Promega, Madison, WI, USA) using primers pGL3 U and pGL3 L (Table 6). Then, the amplified fragment was inserted into the plasmid pZErO-2 (Life Technologies) in the same direction as the *lacZ* gene. A ClaI-EcoRI fragment from pLuc/pZero was inserted into the*Bst*BI-*Eco*RI site of pBS-X-EGFP to generate pBS-X-Luc constructs. The NheI fragment of pBS-X-Luc was inserted into the *Bln*I site of the plasmid pIB-V5/His, and pIB-X-Luc was constructed. Although the N-terminal amino acid sequence of the luciferase gene was altered in the process of plasmid construction, there was little or no effect on enzymatic activity. The sequence of the constructed plasmid was verified by DNA sequencing on an ABI3100 or ABI3130 DNA sequencer with a BigDye terminator DNA sequencing kit (Life Technologies).

Measurement of Luciferase Activity and EGFP Expression in BmN4 Cells

To measure the effect of the ATG-context motif sequence on luciferase activity, BmN4 cells were cultured in six-well plates until they were

~80% confluent. The transfection was performed with a mixture of 1-μg pIB-X-Luc vector, 0.1-μg pRL-TK vector (Promega) and 6-μL FuGENE HD reagent (Promega). Transfected cells were lysed in 200-μL 1 × passive lysis buffer (Promega) 3 days after transfection, and luciferase activity was measured using a dual luciferase assay kit (Promega). To measure EGFP protein and mRNA, cultured BmN4 cells were transfected with 1-μg pIB-X-EGFP vector and 6-μL FuGENE HD reagent (Promega). Cells were harvested 3 days after transfection and split into two aliquots to measure mRNA and protein by quantitative PCR and enzyme-linked immunosorbent assay (ELISA), respectively. Isogen (Nippon Gene, Tokyo, Japan) was used to extract total RNA from transfected BmN4 cells, and the cDNA was synthesized from isolated RNA using RevertraAce reverse transcriptase (TOYOBO). Quantitative PCR was performed as reported previously (Tatematsu et al. 2010). The primers EGFPLCU2 and EGFPLCL2 (Table 6) were used to amplify EGFP mRNA, and primers rp49LCF2 and rp49LCR1 (Table 6) were used to amplify rp49 (Table 2). To measure EGFP protein, transfected BmN4 cells were extracted with 300-μL 1 × passive lysis buffer (Promega), and an ELISA for EGFP protein was performed on Reacti-Bind anti-GFP-coated plates (Pierce, Rockford, IL, USA), as previously reported (Tatematsu et al. 2010). The measurements of luciferase activity and EGFP protein were repeated at least three times.

Generation of Transgenic Silkworms and Measurement of EGFP Expression in MSG, PSG, and Fat body

The silkworm strain w1-pnd, which is non-diapausing, with non-pigmented eggs and eyes, was used to generate transgenic silkworms. The diapausing strain w-1 was used to maintain each transgenic strain. These strains were maintained at the Transgenic Silkworm Research Unit, National Institute of Agrobiological Sciences. Silkworm larvae were reared on an artificial diet (Nosan, Yokohama, Japan) at 25°C.

Transgenic silkworms were generated as previously reported using transposon *piggyBac* as a vector (Tamura et al. 2007;Tatematsu et al. 2010). Transgenic silkworms harboring the EGFP construct with different sequence contexts were mated with adults from Ser1-GAL4 (Tatematsu et al. 2010), FibH-GAL4 (Sezutsu et al. 2009), or 30 K-GAL4 strains (H.

Sezutsu, personal communication), and their F1 larvae, harboring both EGFP and GAL4 constructs, were used for the experiment. The MSG, PSG, and fat body were dissected from 10 larvae on the sixth day of the fifth instar. The amounts of EGFP mRNA and protein in tissue were measured using quantitative PCR and ELISA, as reported previously (Tatematsu et al. 2010).

AUTHORS' CONTRIBUTIONS

KT: carried out the experiments, analyzed data and drafted the manuscript. KU: carried out the experiments. HS: carried out the experiments. TT: designed experiments, analyzed data and drafted the manuscript. All authors have read and approved the final manuscript.

ACKNOWLEDGMENTS

We thank Mr. Kaoru Nakamura, Mr. Koji Hashimoto, and Mr. Toshihiko Misawa for maintaining silkworms. We also thank Miss Mai Hayakawa for technical support and Dr. Michal Zurovec for critical reading of the manuscript. This work was supported by a grant from the Ministry of Agriculture, Forestry and Fisheries of Japan (Integrated Research Project for Plants, Insects, and Animals using Genome Technology, INSECT-2202).

REFERENCES

1. Adachi T, Wang X, Murata T, Obara M, Akutsu H, Machida M, Umezawa A, Tomita M (2010) Production of a nontriple helical collagen alpha chain in transgenic silkworms and its evaluation as a gelatin substitute for cell culture. Biotechnol Bioeng 106:860-870
2. Agarwal S, Jha S, Sanyal I, Amla DV (2009) Effect of point mutations in translation initiation context on the expression of recombinant human a1-proteinase inhibitor in transgenic tomato plants. Plant Cell Rep 28(12):1791-1798

3. Akiduki G, Imanishi S (2007) Establishment of a lipid accumulation model in an insect cell line. Archiv Insect Biochem Physiol 66:109-121
4. Cavener DR (1987) Comparison of the consensus sequence flanking translational start sites in Drosophila and vertebrates. Nucleic Acids Res 15(4):1353-1361
5. Cavener DR, Ray SC (1991) Eukaryotic start and stop translation sites. Nucleic Acids Res 19(12):3185-3192
6. Chang MJ, Kuzio J, Blissard GW (1999) Modulation of translational efficiency by contextual nucleotides flanking a Baculovirus initiator AUG codon. Virology 259(2):369-383
7. Cherbas L, Cherbas P (1993) The arthropod initiator: the capsite consensus plays an important role in transcription. Insect Biochem Mol Biol 23:81-90
8. Feng Y, Gunter LE, Organ EL, Cavener DR (1991) Translation initiation in Drosophila melanogaster is reduced by mutations upstream of the AUG initiator codon. Mol Cell Biol 11(4):2149-2153
9. Harrison RL, Jarvis DL (2006) Protein N-glycosylation in the baculovirus-insect cell expression system and engineering of insect cells to produce "mammalianized" recombinant glycoproteins. Adv Virus Res 68:159-191
10. Iizuka M, Ogawa S, Takeuchi A, Nakakita S, Kubo Y, Miyawaki Y, Hirabayashi J, Tomita M (2009) Production of a recombinant mouse monoclonal antibody in transgenic silkworm cocoons. FEBS J 276:5806-5820
11. Iizuka T, Sezutsu H, Tatematsu K, Kobayashi I, Yonemura N, Uchino K, Nakajima K, Kojima K, Takabayashi C, Machii H, Yamada K, Kurihara H, Asakura T, Nakazawa Y, Miyawaki A, Karasawa S, Kobayashi H, Yamaguchi J, Kuwabara N, Nakamura T, Yoshii K, Tamura T (2013) Colored fluorescent silk made by transgenic silkworms. Adv Funct Mater. doi:10.1002/adfm.201300365
12. Joshi CP, Zhou H, Huang Q, Chiang VL (1997) Context sequences of translation initiation codon in plants. Plant Mol Biol 35(6):993-1001
13. Julien E, Coulon-Bublex M, Garel A, Royer C, Chavancy C, Pudhomme JC, Couble P (2005) Silk gland development and

regulation of silk protein genes. In: Gilbert LI, Iatrou K, Gill SS (eds) Comprehensive molecular insect science, Amsterdam: Elsevier B.V. pp 369-384

14. Kanamori Y, Hayakawa Y, Matsumoto H, Yasukochi Y, Shimura S, Nakahara Y, Kiuchi M, Kamimura M (2010) A eukaryotic (insect) tricistronic mRNA encodes three proteins selected by context-dependent scanning. J Biol Chem 285(47):36933-36944
15. Kozak M (1984) Point mutations close to the aug initiator codon affect the efficiency of translation of rat preproinsulin in vivo. Nature 308(5956):241-246
16. Kozak M (1986) Point mutations define a sequence flanking the AUG initiator codon that modulates translation by eukaryotic ribosomes. Cell 44(2):283-292
17. Kozak M (1987) An analysis of 5′-noncoding sequences from 699 vertebrate messenger RNAs. Nucleic Acids Res 15(20):8125-8148
18. Kozak M (1987) At least six nucleotides preceding the AUG initiator codon enhance translation in mammalian cells. J Mol Biol 196(4):947-950
19. Kozak M (1991) Structural features in eukaryotic mRNAs that modulate the initiation of translation. J Biol Chem 266(30):19867-19870
20. Kozak M (1997) Recognition of AUG and alternative initiator codons is augmented by G in position +4 but is not generally affected by the nucleotides in positions +5 and +6. EMBO J 16(9):2482-2492
21. Kurihara H, Sezutsu H, Tamura T, Yamada K (2007) Production of an active feline interferon in the cocoon of transgenic silkworms using the fibroin H-chain expression system. Biochem Biophys Res Commun 20:976-980
22. Mankad RV, Gimelbrant AA, McClintock TS (1998) Consensus translational initiation sites of marine invertebrate phyla. Biol Bull 195(3):251-254
23. Ogawa S, Tomita M, Shimizu K, Yoshizato K (2007) Generation of a transgenic silkworm that secretes recombinant proteins in the sericin layer of cocoon: production of recombinant human serum albumin. J Biotechnol 128:531-544

24. Ranjan A, Hasnain SE (1995) Influence of codon usage and translation initiation codon context in theAcNPV-based expression system: computer analysis using homologous and heterologous genes. Virus Genes 9(2):149-153

25. Sano KI, Maeda K, Oki M, Maeda Y (2002) Enhancement of protein expression in insect cells by a lobster tropomyosin cDNA leader sequence. FEBS Lett 532(1–2):143-146

26. Seeber F (1997) Consensus sequence of translational initiation sites from Toxoplasma gondii genes. Parasitol Res 83(3):309-311

27. Sezutsu H, Uchino K, Kobayashi I, Tatematsu K, Iizuka T, Yonemura N, Tamura T (2009) Conservation of fibroin gene promoter function between the domesticated silkworm Bombyx mori and the wild silkmoth Antheraea yamamai. J Insect Biotechnol Sericol 78:1-10

28. Sugio T, Matsuura H, Matsui T, Matsunaga M (2010) Effect of the sequence context of the AUG initiation codon on the rate of translation in dicotyledonous and monocotyledonous plant cells. J Biosci Bioeng 109(2):70-173

29. Tamura T, Kuwabara N, Uchino K, Kobayashi I, Kanda T (2007) An improved DNA injection method for silkworm eggs drastically increases the efficiency of producing transgenic silkworms. J Insect Biotechnol Sericol 76(3):155-159

30. Tatemastu K, Sezutsu H, Tamura T (2012) Utilization of transgenic silkworms for recombinant protein production. J Biotechnol Biomaterial S9:004 doi:10.4172/2155-952X.S9-004

31. Tatematsu K, Kobayashi I, Uchino K, Sezutsu H, Iizuka T, Yonemura N, Tamura T (2010) Construction of a binary transgenic gene expression system for recombinant protein production in the middle silk gland of the silkworm Bombyx mori. Transgenic Res 19(3):473-487 doi:10.1007/S11248-009-9328-2

32. Tateno M, Toyooka M, Shikano Y, Takeda S, Kuwabara N, Sezutsu H, Tamura T (2009) Production and characterization of the recombinant human μ-opioid receptor from transgenic silkworms. J Biochem 145(1):37-42 doi:10.1093/jb/mvn147

33. Teule F, Miao YG, Sohn BH, Kim YS, Hull JJ, Fraser MJ, Lewis RV, Jarvis DL (2012) Silkworms transformed with chimeric silkworm/spider silk genes spin composite silk fibers with

improved mechanical properties. PNAS 109(3):923-928 doi: 2012PNAS..109..923T

34. Tomita M (2011) Transgenic silkworms that weave recombinant proteins into silk cocoons. Biotechnol Lett 33:645-654

35. Tomita M, Munetsuna H, Sato T, Adachi T, Hino R, Hayashi M, Shimizu K, Nakamura N, Tamura T, Yoshizato K (2003) Transgenic silkworms produce recombinant human type III procollagen in cocoons. Nat Biotechnol 21:52-56

Citations

CHAPTER 1

Almeida RPP, Daane KM, Bell VA, Blaisdell GK, Cooper ML, Herrbach E and Pietersen G (2013) Ecology and management of grapevine leafroll disease. Front. Microbiol.4:94. doi:10.3389/fmicb.2013.00094.

CHAPTER 2

Jing Chen and Ge-Xia Qiao, "Galling Aphids (Hemiptera: Aphidoidea) in China: Diversity and Host Specificity," Psyche, vol. 2012, Article ID 621934, 11 pages, 2012, doi:10.1155/2012/621934.

CHAPTER 3

Nicaise V (2014), Crop immunity against viruses: outcomes and future challenges, Front. Plant Sci. 5:660. doi: 10.3389/fpls.2014.00660.

CHAPTER 4

Fengping Chen, Ping Han, Pengfei Liu, Naiguo Si, Junli Liu and Xili Liu, Activity Of the Novel Fungicide SYP-Z048 against Plant Pathogens, doi:10.1038/srep06473.

CHAPTER 5

Yuri Drygin, Olga Kondakova, and Joseph Atabekov, "Production of Platinum Atom Nanoclusters at One End of Helical Plant Viruses," Advances in Virology, vol. 2013, Article ID 746796, 7 pages, 2013. doi:10.1155/2013/746796.

CHAPTER 6

Björn Krenz, Holger Jeske, and Tatjana Kleinow, The induction of stromule formation by a plant DNA-virus in epidermal leaf tissues suggests a novel intra- and intercellular macromolecular trafficking route, doi: 10.3389/fpls.2012.00291.

CHAPTER 7

Ian M. Mackay, Katherine E. Arden, and Andreas Nitsche, Real-time PCR in virology, doi: 10.1093/nar/30.6.1292.

CHAPTER 8

Sephra N Rampersad1*, Fazeeda N Hosein1, and Christine VF Carrington2 Sequence exploration reveals information bias among molecular markers used in phylogenetic reconstruction forColletotrichum speciesdoi:10.1186/2193-1801-3-614.

CHAPTER 9

Ken-ichiro Tatematsu, Keiro Uchino, Hideki Sezutsu, and Toshiki Tamura, Efficiency of Translation of Mrna Derived from Exogenous Genes in the Transgenic Silkworm, Bombyx Mori, doi:10.1186/2193-1801-3-136.

Index

A

Abundant nutrition 40
Actin (ACT 254, 255
Actual temperature 208
A cytoplasmic marker 189
Affect regional pattern 15
Agronomic importance 73
Alternanthera mosaic virus (AltMV 171
Amplified fragment length polymorphism (AFLP) 88
A natural lateral 58, 61
A novel member 137
Aphid classification system 41
Attacking different plant 62

B

Bean dwarf mosaic virus (BDMV 172
Bi-molecular fluorescence complementation (BiFC 174
Binominal test 291, 292

C

California Department of Food and Agriculture (CDFA 23
Calmodulin (CAL 254
Carbohydrate metabolism 177, 188
Central axial hole 150, 163
Central hole 154, 159, 162
Chemical Industry 129, 131, 138

Chemical reduction 153, 163
Citrus Tristeza closterovirus (CTV 76
Coat protein (CP 150
Coat protein (CP) 172
Coiled-coil (CC) 80
Colletotrichum gloeosporioide 251, 252, 257, 258, 259, 280, 283, 285
Control plate 139
Conventional assay 209
Crop genetic resistance 74
Current technologie 207
Cytomegalovirus (CMV) 223, 240

D

Detection processe 211
DICER-like (DCL) 84
Diseased fruit 142
Disease epidemiology 9
Dramatic morphological 177

E

Easily propagated plant 6
Economically important practice 4
Elliptical dilation 170, 178, 185, 187
Endogenous macromolecule 170
Endoplasmic reticulum (ER) 172
Enhanced green fluorescent protein (EGFP 288
Entropy variability 255
Epidemics often 78
Experimental condition 153
External source 18

F

Faced with viral attack 79
Fluorescence-generating 222
Fluorescence recovery after photobleaching (FRAP 179
Fluorescent chemistrie 222
Fluorescent protein 174, 181, 183, 202, 204
Fluorogenic hybridisation 213
Fold reduction 290
Foundation Plant Services (FPS 23
Fruit quality 2, 4, 15, 19, 31

G

Gall formation 40
Gall morphology 40, 41, 44, 62, 64, 66
Gall sink strength 60
Geminiviral cellular transfer 189
Genetically modified (GM) 90
Genetic variation 252
Geometric relation 189
Germ tube elongation 130, 133, 141
Global food problem 76
Global food security 73, 101
Glutamine synthetase (GS 254
Glyceraldehyde 3-phosphate dehydrogenase (GAPDH 254
Glycosylphosphatidylinositol (GPI) 183
Grape-growing region 2, 3, 6, 28
Grapevine leafroll disease (GLD 2
Great diversity 40, 62, 64
Great variety 62

H

Heat shock cognate 170, 173, 175, 190, 197
Herpes simplex viruses (HSV) 225
Highlighting scientific gap 2
Highly dynamic structure 178
Highly effective 130
Highly species specific 40, 60, 62, 64
High priority research 22
Host machinery 74
Host plant 40, 41, 42, 44, 53, 59, 60, 62, 63, 64, 65, 70, 71
Human life 76
Hypersensitive response (HR) 81

I

Important factor 60
Important nucleotide 290, 292, 302
Index of substitution saturation (ISS 263, 268
Index of substitution saturation (ISSc 263, 268
Infection process 170, 177
Initiation codon 287, 289, 290, 291, 294, 295, 296, 298, 299, 300, 301, 303, 309, 310, 311
Intercellular pathway 170
Internally transcribed spacer (ITS 254
International Committee on Taxonomy of Viruses (ICTV 5
Interplant study site 13
Intra-specific 251, 252, 254, 255, 264, 271, 272, 277, 279
Intra-specific level 252, 254, 255, 279

L

Larval stemmata 297
Leaf blade 59, 60
Lesion area 134
Life cycle 62, 70
Life Technologie 255
Lower epidermi 136, 143

M

Major culture 73
Management of plant 75
Marker-assisted selection (MAS) 89
Markov chain Monte Carlo (MCMC 265
Maximum Clade Credibility (MCC 265
Melting temperature 212, 213, 241
Membrane envelope 169
Middle silk gland (MSG 289
Minor groove binding (MGB 218
Molecular beacon 219, 235
Molecular biology technique 77
Molecular diagnostic 206
Motor protein 169
Multiple virus 3, 10, 14
Multiplexing 224
Musical organ 165

N

Nanoparticles formation 150
National Zoological Museum of China, Institute of Zoology,

Chinese Academy of Sciences 41
Natural biodiversity 89
Natural enemie 9, 26, 31
Natural Science 64
New Zealand Winegrowers (NZW 19
Non-fluorescent quencher (NFQ) 210
Novel structure 138
N receptor-interacting protein 1 (NRIP1 81
Nuclear-shuttle protein (NSP) 171
Nucleotide binding site-leucine-rich repeat (NBS-LRR) 80

O

Obligatory intracellular 74
Optical density at 650 nm (OD650 140
Optimal context 296

P

PAMP-triggered immunity (PTI 75, 94
Paper present 40
Part of the silk glands (PSG 289
Pathogenesis-related (PR) 87
Pattern recognition receptors (PRR) 94
Phylogenetic constraint 63
Phylogenetic study 40
Phylogenetic tree construction 254
Plant cell 74, 106
Plant hormone 86
Plant-pathogenic fungi 130, 131, 132, 139
Plant vaccination 93
Plant virus 171, 194, 197
Plant viruse 78
Plateau phase (PP) 208
Platinum atom cluster 150
Polymerase chain reaction (PCR) 206, 235
Potato potexvirus X (PVX) 80
Potato virus X (PVX 150
Potential virus 14
Potyvirus A (PVA) 83
Pox poty virus (PPV 76
Produces gall 59
Pt nanoparticles (Pt-NPs 154
Pt nanoparticles, Pt-NPs 150
Punctate structure 184
Putative chimera 264

Q

Quantitative, competitive PCR (qcPCR 221
Quarantine importance 251
Quinone outside inhibitor fungicides (QoIs 137

R

Random amplified polymorphic DNA (RAPD) 88
Reaction condition 163
Reactive oxygen species (ROS) 81
Real-time 206, 209, 222, 226, 234, 236, 241, 242, 243, 249, 250, 314
Recessive resistance 82
Resistance genes (R genes) 80
Restriction fragment length polymorphism (RFLP) 88

RNA interference (RNAi 80, 99

S

Salicylic acid (SA) 80, 81
Semi-persistent transmission 10
Several subgall 61
simple sequence repeats (SSR) 88
Single adjacent vine 17
Single nucleotide polymorphisms (SNPs 218
Smaller gall 60
Specific horticultural practice 15
Squash leaf curl virus (SLCV 172
Sterile water 141
Stromal component 185
Stromule formation 179, 183, 188, 189, 201, 314
Study block 20, 21, 22
Succinate dehydrogenase inhibitors (SDHIs 137
Systemic acquired resistance (SAR 80, 82

T

Tap water 141, 143
Terminal cytosine 216
Terminal protein 150, 157, 161, 164
Tobacco mosaic virus 150, 151, 157, 164, 165, 166, 167, 168
Tobacco mosaic virus (TMV 150
Tobacco mosaic virus (TMV) 171
Tobamovirus (TMV) 81
Tobamovirus [turnip vein-clearing virus (TVCV) 171
Toll–interleukin-1 receptor (TIR) 80
Tomato golden mosaic virus (TGMV 172
Transcriptional gene silencing (TGS) 85
Transgenic silkworm 288, 290, 297, 298, 299, 309, 310
Translational efficiency 295, 296, 299, 300, 301, 303, 309
Transmission efficiency 10, 11, 12, 15
Transmission electron microscopy (TEM 153
Treatment controlled 134
Tubulin (TUB2 254, 255

U

Upper epidermis 136, 143
Urea treatment 161

V

Varicella zoster virus (VZV) 225
Varied experimental method 10
Various component 3
Vector-borne virus 2
Vegetable crop 130
Vine Improvement Association (VIA) 16
Viral agent 77
Viral infection 177, 193
Viral particle 75
Viral protein 79, 172, 176, 183, 193
Viral transport 171, 199
Virus detection 206, 211, 226
Viruse 73, 75, 76, 86, 113
Virus-infected plant material 3, 6
Virus-like particles (VLPs) 151
Visual symptom 21
Vitro experiment 137